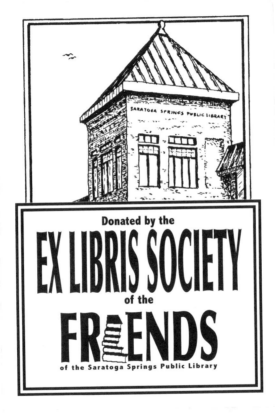

MAGNIFICENT
PRINCIPIA

COLIN PASK

MAGNIFICENT
PRINCIPIA

Exploring
ISAAC
NEWTON'S
Masterpiece

Prometheus Books

59 John Glenn Drive
Amherst, New York 14228–2119

Published 2013 by Prometheus Books

Cover image © Media Bakery/Martin Barraud
Cover design by Jacqueline Nasso Cooke

Inquiries should be addressed to
Prometheus Books
59 John Glenn Drive
Amherst, New York 14228–2119
VOICE: 716–691–0133 • FAX: 716–691–0137
WWW.PROMETHEUSBOOKS.COM

17 16 15 14 13   5 4 3 2 1

Library of Congress Cataloging-in-Publication Data

Pask, Colin, 1943-
    Magnificent Principia : exploring Isaac Newton's masterpiece / by Colin Pask.
        pages cm
    Includes bibliographical references and index.
    ISBN 978-1-61614-745-7 (hardcover)
    ISBN 978-1-61614-746-4 (ebook)
    1. Newton, Isaac, 1642-1727. Principia. 2. Mechanics—Early works to 1800.
3. Celestial mechanics—Early works to 1800. I. Title.

QA803.P37 2013
531—dc23

                                                          2013012120

Printed in the United States of America

To Johanna

CONTENTS

WHY YOU SHOULD READ THIS BOOK

In a 1987 book celebrating the three hundredth anniversary of the first publication of Isaac Newton's masterpiece, Steven Hawking writes:

> The *Philosophiae Naturalis Principia Mathematica* by Isaac Newton, first published in Latin in 1687, is probably the most important single work ever published in the physical sciences.[1]

The impact of the *Principia*, as it is usually called, can be summed up in Alexander Pope's immortal words marking Newton's death:

> Nature and Nature's laws lay hid in night:
> God said, Let Newton be! and all was light.

In case you think time has changed things, here, two hundred years after Pope, is Albert Einstein:

> The whole evolution of our ideas about the processes of nature, with which we have been concerned so far, might be regarded as an organic development of Newton's ideas.[2]

Yet very few people, including professional scientists and mathematicians who would readily agree with Pope and Einstein, have ever looked in the *Principia* or have a good idea of its actual contents. I suspect that most people have little idea beyond "Newton's Laws of Motion and Gravitational Theory of the Solar System." If so, that is a tragedy, not just for professionals, but for everyone who is curious about exactly how science has developed and the nature of Newton's seminal contributions.

The literature on Newton's life and work is vast. One reason is that he was such a wonderfully complex and intriguing character. Another is that, in addition to his major books *Principia* and *Opticks*, Newton churned out a very large number of notes, letters, papers, and draft documents, and he hoarded all of them. It is possible therefore to follow his thought processes and map out the various routes, stages, and controversies in his scientific (and other) endeavors. All of this makes fascinating reading, but it can mean that Newton's actual final products seem to get less attention than their history!

My aim here is simpler: I take the third and final edition of Newton's *Principia* and explain how it sets out his (and now our) approach to science; how the framework of classical mechanics is established; how terrestrial phenomena like the tides and projectile motion are explained; and how we can understand the dynamics of the solar system and the paths of comets. In this way I concentrate on the results themselves (with just a little of their historical context as appropriate). In each case, I explain the ideas behind and the conceptual importance of the result and how it fits into the scheme of mechanics that Newton is building. I also briefly elaborate on particular results that have interesting modern applications, or those through which Newton identified problems and puzzles that have shaped later developments of science.

Newton's mathematical methods are notoriously difficult to follow. A famous story tells of the comment made by a student as Newton himself passed by:

> There goes the man that writ the book that neither he nor anyone else understands.[3]

I will present details for a few examples in order to illustrate Newton's approach, but mostly I will avoid his more complicated mathematical arguments. That way I can concentrate on the concepts and ideas as Newton introduces them. It must be remembered that in science, it is the underlying concepts and ideas that are most important, rather than the type of mathematical formalism used to represent them. In addition, I will give the modern formulation in separate sections so a reader wishing to do so can appreciate in more detail that Newton's mechanics is essentially the same as that which we use today. For some readers, that may represent an introduction to mechanics or a review of the subject, probably emphasizing some ideas and methods that you had previously glossed over.

I have given references to the extensive literature so that interested readers may further explore particular topics, their history, and Newton's detailed working. It will be apparent that I rely on a range of wonderful Newtonian scholars, both past and present: Bruce Brackenbridge, S. Chandrasekhar, I. Bernard Cohen, Dana Densmore, Niccolo Guicciardini, A. R. Hall, Robert Iliffe, Michael Nauenberg, Bruce Pourciau, George E. Smith, Richard Westfall, Tom Whiteside, Curtis Wilson, and many others. In particular, if you are interested in how the *Principia* evolved over its earlier editions, or in relevant work that Newton completed and might have put into the *Principia*, then you should consult their impressive scholarly writings. Parts of the *Principia* are intricate and subtle, and for in-depth analysis you should consult their interpretations.

We shall also come to appreciate the originality of Newton's overall approach. He shows us how to develop a mathematical formalism, and how the theorist can explore the properties and predictions of that formalism using a selection of examples and mathematical manipulations. With that basis established, Newton shows us how experimental and observational data should be used to select the most appropriate theory. Few people realize that *Principia* itself contains details of experiments and Newton's attempts at data analysis. The situation is beautifully summarized by the great Newtonian scholar I. Bernard Cohen:

> Newton's *Principia* is a book of mathematical principles applied to nature insofar as nature is revealed by experiments and observation. As such, it is a treatise based on evidence. Never before had a treatise on natural philosophy so depended on an examination of numerical predictions and numerical evidence.[4]

Most people do not realize that Newton saw his approach of using a mathematical theory of dynamics, suitable forces, and experimental verifications as the one to use more generally in physics, and not as one restricted to the motion of projectiles, pendulums, planets, and comets. Indeed, in *Book II* of the *Principia* he attempts to apply his methods to liquids and gases, to understand their behavior, how they resist motion of bodies through them, and how they support sound and other waves. He set the challenges and gave us the directions. According to Nobel laureate Steven Weinberg:

> Newton's hope, "I wish we could derive the rest of the phenomena of nature . . ." has not been fulfilled, but we are working on it, very much in the Newtonian tradition: the formulation of increasingly comprehensive quantitative laws. From this high viewpoint, all that has happened since 1687 is a gloss on the *Principia*.[5]

We recognize that the *Principia* is a work of monumental importance, far too significant to allow us to be frightened off by stories of mathematical complexities.

Before embarking on my tour through the *Principia*, there are three introductory chapters. It is impossible to read the *Principia* without considering just who was (in Einstein's words) "this brilliant genius, who determined the course of western thought, research and practice like no one else before or since."[6] So the first chapter is a short description of Newton the man and his life. To fully appreciate the steps made by Newton, it is necessary to know the state of science as he began his work. The second chapter briefly reviews the situation as Newton found it. The third chapter gives a general introduction to the *Principia*, its publishing history, its form, and its major objectives and achievements. After the tour, I will present some conclusions and tell you how the *Principia* was received and how it influenced Enlightenment thinking.

One final point: it is true that in extreme situations, like the processes at atomic and subatomic levels or the cases where speeds approach the speed of light, Newton's mechanics must be modified. However, Newton's mechanics remains the theory we use in everyday life and beyond, as when we send people to the Moon or plan missions to Mars. It is worth noting Einstein's comment when writing about his theory of relativity:

> No one must think that Newton's creation can be overthrown in any real sense by this or any other theory. His clear and wide ideas will forever retain their significance as the foundation on which our modern conceptions of physics have been built.[7]

Each time I delve into the *Principia*, I find myself more and more amazed at the extent, level, and sophistication of the results that Newton produced. Over and over, I find myself asking how any one man could have taken such a leap forward and produced such innovative and groundbreaking work. I understand why the Marquis de l'Hôpital asked, "Does he eat and drink and sleep? Is he like other men?"[8] I invite you to share my sense of wonder and understand Newton's supreme achievement.

Finally, I recognize that this is an ambitious book and readers will find errors, significant omissions, or points that I should have developed more carefully. Let me end with Newton's concluding words from his preface to the first edition of the *Principia*: "*I heartily beg that what I have here done be read with candour; and that the defects in a subject so difficult be not so much reprehended as kindly supplied, and investigated by new endeavours of my readers.*"

ACKNOWLEDGMENTS

This book began life in material used for public lunchtime lectures and in undergraduate lecture courses. I thank everyone who discussed the subject with me and made useful suggestions. I am greatly indebted to Peter McIntyre and Ravi Sood for their careful reading of the manuscript and giving me many corrections and suggestions for improving the text. Of course, the remaining errors are entirely my own. Other colleagues supplied advice and ideas and, most importantly, encouragement to continue when I was struggling. Great thanks are due to the wonderful Annabelle Boag for turning my rough sketches into beautiful figures. (Unless otherwise indicated, the figures in this book have been created by her.) Much must be done to turn a manuscript into a polished book, and I wish to thank Mariel Bard for her meticulous editorial work and for her patience in dealing with an inexperienced author. Many scholars have studied Newton and his work, and I have made great use of their writings, as explained in my preface.

It is over fifty years since I first met Johanna, and she remains my greatest love and inspiration. Quite simply, without her, this book would not exist.

PLAN OF THE BOOK
AND READING PLANS

This is a large book, so to help you understand its structure, I have divided it into seven parts. Before the relevant chapters for each part, you will find a short introduction and some advice about which chapters to read to suit your particular interests. Below is a list of the parts. You may wish to read their brief introductions in order to formulate a reading plan if the idea of reading the whole book seems a little daunting! Also, remember that most chapters have introductory sections and concluding remarks that can be easily read to get an idea of the material being covered.

PART 1. INTRODUCTORY MATERIAL
  Newton the man, the scientific scene, a first look at the Principia
PART 2. HOW THE *PRINCIPIA* BEGINS
  approach to science, conceptual matters, definitions, the laws of motion
PART 3. DEVELOPING THE BASICS OF DYNAMICS
  describing how a force determines the motion of a body, the theory of orbits
PART 4. ON TO MORE COMPLEX SITUATIONS
  motion with constraints and two or more interacting bodies, large bodies
PART 5. ABOUT *PRINCIPIA BOOK II*
  theory, measurement and origin of resistance forces, motion and waves in fluids
PART 6. THE MAJESTIC *PRINCIPIA BOOK III*
  gravity and the solar system, terrestrial phenomena and comets, conclusions

PART 7. RECEPTION AND INFLUENCE OF *PRINCIPIA*
how the Principia *was received and used, how mechanics evolved*

A very basic plan would be to read chapter 1, chapters 3 through 8, and all of part 6; then return to other chapters later. Experts might appreciate the wonders in chapters 12 to 17. Part 5 may be a surprise and of interest to everyone, experts or not.

PART 1

INTRODUCTORY MATERIAL

Most people are curious to learn at least a little about Newton the man (chapter 1). The magnitude of Newton's achievement is clearer when we know something about the development of science leading up to the situation as it was when he began work at Cambridge University (chapter 2). The third chapter gives a few details about how Newton came to write the *Principia*, its basic structure, and its major themes. As such, it is a useful chapter to read before coming to grips with the book's details.

Chapter 1. INTRODUCING OUR HERO
Chapter 2. SETTING THE SCENE
Chapter 3. A FIRST LOOK AT THE *PRINCIPIA*

Reading plans: Though it is not essential to read the first two chapters, I think they help us to appreciate the achievement of Newton the man and the scientist. Chapter 2 is long and can be left out if you are keen to get directly to the Principia. *Chapter 3 is important for appreciating the* Principia *as a whole.*

INTRODUCING OUR HERO

t is virtually impossible to read the *Principia* without thinking about just what sort of man could have written it. Furthermore, there are places where knowing something of the nature of the author makes it easier to understand why certain topics and approaches are discussed. So a brief introduction is in order. It would be impossible to cover Newton and his life in detail here; therefore, this sketch is more in the nature of scene setting, with suggestions for further reading at the end.

In 1927, on the two hundredth anniversary of the death of Isaac Newton, Einstein wrote that "we feel impelled at such a moment to remember this brilliant genius, who determined the course of western thought, research, and practice like no one else before or since."[1]

What sort of a man could have gained such a reputation? Was he a boy genius, like Mozart, taught and encouraged by a talented father? No; there was nothing in Newton's origins to indicate that he would become the man who changed our view of the world and set us on the pathway to modern science.

1.1. ORIGINS

Hannah Newton gave birth to son Isaac on Christmas Day 1642. (A table of important dates in Newton's life is given at the end of the chapter.) The birthplace was a manor house at Woolsthorpe, about seven miles south of the midlands town of Grantham in Lincolnshire (see figure 1.1). Isaac was named after his father, an illiterate but prosperous yeomen farmer who had died three months earlier. The baby was premature, tiny and sickly. An unlikely start for the man who was to die eighty-four years later revered as the greatest man of his times.

Three years later, Hannah married the reverend Barnabas Smith and moved to the rectory in nearby North Witham. Isaac was left with his maternal grandmother, Margery Ayscough, until 1653, when Barnabas Smith died and Hannah moved back to Woolsthorpe with her three Smith children. The young Isaac did not have a fond relationship with his grandmother, and grandfather Ayscough left him out of his will. Between the Newtons, Ayscoughs, and Smiths, there was a rising level of prosperity, so Isaac grew up in relatively well-off circumstances.

It seems that Isaac was destined to take over the family farming business, but there were educated members on the Ayscough side of the family and, after a little time in a nearby village school, Isaac was sent to the Free Grammar School of King Edward VI (also called the King's School) in Grantham. Isaac was now twelve and lodged with Joseph Clark, an apothecary. Latin was a major subject, which was lucky for Isaac, as Latin was something of a universal language of science at the time. It is possible that the school's headmaster, Mr. Stokes, may have been interested in mathematics and helped Isaac along that road too. There is evidence of Isaac's skills at making models of all kinds and at drawing.

In 1659, as he was approaching seventeen, Isaac was taken home by his mother to learn how to run the estate and gain some farming experience. By all accounts, this was not a success and Isaac neglected his duties in favor of things like model building and reading. The situation must have been summed up by his uncle Reverend William Ayscough, who urged that Isaac be sent back to school to prepare for university studies. Headmaster Stokes also told Newton's mother that such talents should not be buried under various rural pursuits. It worked, and in 1660, Isaac returned to school in Grantham, although none could have guessed what a critical and momentous step that was.

Figure 1.1. The author seeks inspiration at Newton's birthplace. (Photograph by Johanna Pask.)

1.2. A LIFE AT CAMBRIDGE

Isaac Newton finally escaped rural life in June 1661 when he entered the University of Cambridge. He would remain at Cambridge first as a student, then as a fellow and as a professor for the next thirty-five years. Newton was exposed to a variety of subjects as a student, but most importantly he studied mathematics and science with great intensity and was largely self-taught in many important areas. He graduated with a bachelor of arts degree in early 1665.

An outbreak of the plague caused a closure of the university, and Newton spent most of 1665 to 1668 back in the quiet of farms in Lincolnshire. Here came the famous anni mirabiles when Newton developed so many of his early, brilliant ideas in mathematics, optics, and mechanics. As the plague subsided, Cambridge University reopened in spring 1667 and Newton returned to be made a fellow of Trinity College.

Two years later, at the age of just twenty-six, he was appointed as the Lucasian Professor of Mathematics. This gave Newton security in his position and the chance to devote his life to study and research. The chair came with an income of one hundred pounds per annum and, on top of that, he had income from his fellowship and from the Lincolnshire estate. His required duties were comparatively light, with one lecture per week to be given for three terms. Ten lectures had to be deposited in the university library each year. Newton does not seem to have been an inspiring lecturer, and it is said that sometimes he gave the lectures to an empty room. (What would you or I give now to have heard a lecture by Isaac Newton!)

Newton continued his mathematical studies and teaching, although little was published. He was also working on optics, and his invention of the reflecting telescope was an early triumph. It was presented to the Royal Society in late 1671, and Newton was made a fellow in 1672 with seven of his papers published by the society that year. In 1672 he published *Light and Colours*, which described his theories on those subjects, and his *Hypothesis Explaining the Properties of Light* was published in 1675. But it was not until 1704 that the comprehensive treatise *Opticks* appeared. Newton now had the opportunity to mix and interact with major figures like Robert Boyle, but around this period, he also spent much time in secret work in alchemy and religious studies, interpreting the various versions of the Bible and evaluating chronological studies of ancient kingdoms.

The year 1679 marked important changes in Newton's life. He returned to Woolsthorpe to nurse his dying mother and spent much of the year attending to family affairs after her funeral in June. Later in the year, correspondence began between Robert Hooke and Newton on questions about planetary motion. In the years 1681–1682 Newton observed comets (that in 1682 being the famous Halley's comet), and his interest in dynamics was increasing. All of this was to culminate in the publication of the first edition of the *Principia* in 1687 after urging on by Edmond Halley, who actually funded the publication. With this event, Newton's reputation was ensured.

Despite his retiring nature, Newton did play a part in university affairs. In 1688 he was elected Member of Parliament for Cambridge University (not that he seems to have made many great contributions in that line). He also made a notable friend in John Locke, the famous thinker and philosopher, and Newton is most kindly mentioned in Locke's major work *An Essay concerning Human Understanding.*

Newton was now deep into his chemical and religious studies, and 1692 saw important events in this area. Newton had interacted with Robert Boyle about chemical matters, and he attended Boyle's funeral in 1692. In his will, Boyle left money for annual lectures supporting the Christian religion and defending it against "infidels" and others. Richard Bentley was chosen as the first Boyle lecturer. Bentley had been in earlier contact with Newton, and now he sought Newton's help in the preparation of his lecture material. The result was a series of letters in which Newton discussed issues in science and religion. These private letters also reveal some of Newton's most important concerns about his system of dynamics.

In 1693 Newton had what today we would probably call a mental breakdown. The reasons are not completely clear, but someone with Newton's self-imposed, almost-manic workload, while at the same time playing with strange chemicals and dealing with numerous personal disputes and problems, would be a candidate for physical and mental problems. He recovered and continued his various activities, but clearly his greatest inventive days were over. His friends persuaded him to move on and, in 1696, he left Cambridge for London to assume the position of warden of the Royal Mint.

1.3. THE FINAL PHASE

Newton was now changing from retiring scholar to public figure, and he was exerting power and influence in a number of ways. He took his mint position very seriously, supervising the recoinage that was necessary to deal with a monetary crisis and taking steps to deal severely with "coiners" who debased the coins in circulation. In 1700 his title changed to master of the Royal Mint.

Newton continued with his scientific and mathematical work, and during this phase of his life, the second and third editions of the *Principia* were published, along with *Opticks* and *Arithmetica Universalis* (based on his earlier lectures on algebra) and other mathematical works. He also befriended and promoted a number of younger scholars who spread the Newtonian doctrine and supported Newton in his various intellectual struggles.

In 1703 Newton was elected as president of the Royal Society. The Royal Society (of London) is one of the oldest such organizations, having been founded in 1662. It had become a little run-down in the years before Newton took on the presidency, but his prestige and ever-present drive and organizational skills meant that it again began to flourish. Between 1703 and 1726 there were 175 society council meetings, of which Newton attended 161 (compared with his predecessor, who had attended none during the previous five years). He also attended most ordinary meetings ensuring that the society, and with it science, flourished. He presided over his last Royal Society meeting in 1727, just weeks before he died.

Newton was now at the height of his public esteem, and Queen Anne made him Sir Isaac Newton in 1705. Newton died in 1727 on the twentieth of March. He was buried in Westminster Abbey on the fourth of April. On his tomb is written:

Let Mortals rejoice That there has existed
such and so great an Ornament to the Human Race

1.4. NEWTON'S INTERESTS

Isaac Newton was the Lucasian Professor of Mathematics, and his work in many areas of mathematics has had a profound effect on the development of the subject. However, Newton's interests ranged far beyond mathematics. His works on dynamics and astronomy (culminating in the *Principia*) and in optics (expounded in *Opticks*, first in English rather than scientific Latin) were also fundamental and set the course of science for future generations. To have written even just a part of any of those works would have been a claim to fame and immortality, so it will come as a surprise to many people to learn that Newton was equally, if not more so, involved in even more areas.

During his adult life, Newton had a deep interest in religion and theological questions. He made incredibly detailed and thorough studies of the Bible and the writings of the early founders of the Christian church. His writings were voluminous but remained secret during his lifetime, and only now is the full extent of his work in this area becoming evident as more manuscripts are edited and published. Newton was particularly interested in chronological matters and prophesies that had been set out in the Bible and ancient writings. The *Chronology of Ancient Kingdoms Amended* was published in 1728. However, Newton's 323-page book, *Observations upon the Prophecies of Daniel and the Apocalypse of St. John*, was not published until 1733. It was not until 1756 that the *Four Letters from Sir Isaac Newton to Dr. Bentley* was published. (I return to these religious matters shortly.)

When Newton left Cambridge for London, his papers were packed into a large box and were untouched until Bishop Horsley in 1779 and the physicist Sir David Brewster in 1855 took a look at them. It is generally said that they soon closed the box in horror, and it was many years later that the documents were finally revealed. It was bad enough that there were extreme religious writings, but it was also apparent that Newton had been an alchemist. In fact, he had explored chemical and alchemical matters over a long period with the same energy, thoroughness, and zeal (and secrecy) that characterized his more conventional scientific and mathematical studies. Newton had his own chemical laboratory and spent large amounts of time there. He interacted with Robert Boyle and owned twenty-three of his works.

Some idea of the spread of Newton's interests can be gained by looking at the books in his personal library. (He also had access to the Trinity College library of course.) For example, Newton owned thirty Bibles, almost as many as books on astronomy. Table 1.1 gives some details.

What should we make of all of this? One striking opinion was offered by John Maynard Keynes, the great economist. Mathematical works from the box of Newton's papers went to the Cambridge University library in 1888, and the remainder finally came up for auction in 1936. Lord Keynes was responsible for buying and preserving as many of those papers as he could, and he became involved in their analysis. (A Jewish scholar bought many other documents, which are now in Israel—they too are now available on a National Library of Israel website. For details about these manuscripts, see the section in this chapter titled "Further Reading.") Keynes wrote an article called "Newton, the Man" for a meeting planned to celebrate the three hundredth anniversary of Newton's birth, but World War II intervened. The paper was finally presented at the postponed meeting in 1946 by Geoffrey Keynes, brother of the then deceased Lord Keynes. After poring over the contents of the box, Keynes was moved to write:

> Newton was not the first of the age of reason. He was the last of the magicians, the last of the Babylonians and Sumerians, the last great mind that looked out on the visible and intellectual world with the same eyes as those who began to build our intellectual inheritance rather less than 10,0000 years ago.[2]

Later in the article, he explains:

> Why do I call him a magician? Because he looked on the whole universe and all that is in it as a riddle, as a secret which could be read by applying pure thought to certain evidence, certain mystic clues which God had laid about the world to allow a sort of philosopher's treasure hunt to the esoteric brotherhood. . . . By pure thought, by concentration of the mind, the riddle, he believed, would be revealed to the initiate.[3]

Whether or not we see this as an extreme viewpoint, clearly the genius of Isaac Newton was amazingly diverse and complex. It would be unfair to classify all of Newton's work in alchemy and chemistry as strange and magical, since in that work would also be his thinking about matter and interactions, thinking that led to some of his greatest ideas.

Topic	Number of Books	Proportion (%)
Scientific works	538	30
Nonscientific works	1,214	70
Alchemy	138	8
Chemistry	31	1.5
Physics/optics	52	3
Astronomy	33	1.5
Medicine/physiology	57	3
Geography/travel	76	4.5
Theology/Bible	477	27.5
Mathematics	126	7
Classics	149	9
History	143	8
Reference	90	5

Figure 1.2. Books in Newton's personal library

1.5 THE ROLE OF RELIGION IN NEWTON'S LIFE

We must remember that religion played a large part in most people's lives in Newton's time, and he was certainly a very religious person. Early in his time at Cambridge, on Whit Sunday 1662, he recorded forty-nine sins that he had committed, ranging from actions ("*eating an apple at Thy house, making pies on Sunday night*") to thoughts ("*having uncleane thoughts words actions and dreamese*"). However, he soon began his theological studies, and they took him down his own individual religious road.

Newton's studies led him to a Unitarian religious viewpoint, so that he recognized one God the Father rather than the Trinity of conventional religion as set out in the Nicene Creed. In particular, Newton did not accept Jesus as God but as an agent and creation of God. Among his documents is *Twelve Articles on Religion* in which he makes his position very clear:

> 1. *There is one God the Father everliving, omnipresent, omniscient, almighty, the maker of heaven and earth, and one Mediator between God and Man the Man Christ Jesus.*
>
> . . .
>
> 12. *To us there is but one God the Father . . . we are to worship the father alone as God Almighty and Jesus alone as the Lord the Messiah.*[4]

Newton decided from his biblical studies that such things formed the old position until the Council of Nicaea in 325, after which the doctrine of the Trinity became official Church teaching. One of the great figures in establishing this position was Athanasius, a deacon at the Council of Nicaea and later the bishop of Alexandria. Newton researched these events with great intensity and wrote large numbers of documents about this change, conspiracies, and the evil ways of Athanasius. Today I suppose we might label it as an obsession.

If such religious questions were purely personal matters, Newton would have had no difficulties. However, the 1688 Act of Toleration specifically excluded from protection those who were against "the Blessed Trinity." A following act in 1689 barred such people from holding public office. Newton kept his views secret to avoid such problems. At Cambridge, it was a requirement for continuation in a fellowship that the holder become ordained, something that Newton's personal religious beliefs could never allow. It appeared that he would leave Cambridge, but powerful supporters organized a royal mandate that excused the Lucasian Professor from this requirement. (In fact, Newton's successor in the Lucasian Professorship, William Whiston, came to views much like Newton's, which he made public with the result that in 1710 he lost his position and was banished from the university.) Newton maintained his religious stance to the very end and even on his deathbed refused the sacrament of the Anglican Church.

Newton appears to have kept religion and science separate to a large degree in his official writings, as we will learn when reading the *Principia*. In fact, he begins his *Seven Statements on Religion* as follows:

1. *That religion and Philosophy are to be preserved distinct. We are not to introduce divine revelations into Philosophy, nor philosophical opinions into religion.*[5]

There are some references to God in Newton's scientific work, and no doubt his deep religious views did both motivate and influence his scientific thinking. The idea that studying the natural world was like reading "God's other book" (other than the Bible, that is) was a widespread one, and Newton began his first letter to Richard Bentley with:

When I wrote my treatise [meaning the *Principia*] *about our system, I had an eye upon such principles as might work with considering men, for the belief of a deity, and nothing can rejoice me more than to find it useful for that purpose.*[6]

These matters are further discussed in chapters 26 and 27.

1.6. NEWTON'S CHARACTER

In recent years much has been discovered and written about Newton the man and his character. The picture that has emerged is not a particularly attractive one. Today some would say that his disrupted and emotionally disturbing early childhood set the tone for his development and character formation.

There was conflict in Newton's early life, with the young Isaac's separation from his mother and her wishing for a farmer and manager while he had interest only in mechanisms, models, and studies. His absentminded and shoddy approach to his duties would not have endeared him to the servants who had to cover for him. In his 1662 list of sins, we find:

12. *Refusing to go to the close at my mother's command*
13. *Threatening my father and mother Smith to burne them and the house over them*

. . .

24. *Punching my sister*

. . .

30. *Falling out with the servants.*

At school and in his free time, Isaac was happy making models and dolls' furniture for girls, but he seems to have spent little time with other boys.

The move to Cambridge did not change Newton's way of life. He met a similarly solitary and dejected figure in Nicholas Wickins, and they shared rooms for twenty years. Newton did have a few select friends, but it says much about him that even after twenty years there appears to have been no lasting, deep friendship between Newton and Wickins. Newton appears to have been the fabled intense, driven, eccentric, and absentminded scholar and professor, as these irresistible recollections of his assistant Humphrey Newton (no relation) indicate:

I never knew him to take any recreation or Pastime, either in riding out to take the air, Walking, bowling, or any other exercise whatever,

Thinking all Hours lost, that was not spent in his Studyes, to which he kept so close, that he seldom left his Chamber. . . .

So intent, so serious upon his Studies, that he eat very sparingly, nay, oftimes he has forget to eat at all, so that going into his Chamber, I have found his Mess untouch'd, of which when I reminded him, would reply, have I; and then making to the Table would eat a bit or two standing. . . .

He very rarely went to Bed, till 2 or 3 of the clock, sometimes not till 5 or 6. . . .

He very rarely went to dine in the Hall unless upon some Publick Dayes, and then if he has not been minded, would go very carelessly, with shoes down at the Heels, stockins unty'd, surplice on, & his head scarcely comb'd. . . .

At some seldom Times when he design'd to dine in the Hall, would turn to the left hand, & go out into the street, where making a stop, when he found his mistake, would hastily turn back, & then sometimes instead of going into the Hall, would return to his chamber again.[7]

However, adjectives such as *kindly*, *jolly*, or *entertaining* that so often go along with the absentminded-professor image do not seem to be applicable in Newton's case. Over five years, Humphrey Newton saw Isaac Newton laugh but once. There was a darker side to Newton, and the picture of a secretive, fearful, lone scholar with his hidden interests in alchemy and religious issues is not an attractive one.

Newton was reluctant to publish his findings, although he seems to have craved recognition and respect. His infamous disputes with many people are revealing. Robert Hooke managed to upset Newton in several ways, and there were bitter disputes about Hooke's contributions to optics and mechanics (and one cannot help believing that Hooke was a little hard-done-by). There are suggestions that Newton's famous conciliatory statement in a letter to Hooke ("*if I have seen further it is by standing on ye shoulders of Giants*") contained a veiled insult, because Hooke was short and somewhat deformed. Newton was certainly dissuaded from publishing by disputes, and perhaps that is why *Opticks* did not appear until 1704, the year after Hooke's death.

Newton had great arguments over data availability with the Astronomer Royal, John Flamsteed, a similarly strong-willed and bad-tempered

man. Flamsteed once described Newton as "insidious, ambitious, excessively covetous of praise and impatient of contradiction."[8] After Newton's
knighthood, Flamsteed referred to Sir Isaac Newton by his initials, SIN.
Newton's best-known disputes were with Leibnitz, particularly over the
invention of calculus. The Royal Society set up a committee to investigate the various claims, and it appears that Newton strongly influenced
both the choosing of a subservient group of people and the writing of their
report. The extent of Newton's deviousness can be judged by the fact that
he then wrote an anonymous review of that report for publication in the
Royal Society's journal *Philosophical Transactions*.

In later life, there is evidence of Newton's kindness and generosity to his
friends, but clearly he was selective and careful. Newton recommended that
William Whiston be his successor as the Lucasian Professor, but then Newton
did nothing to come to Whiston's aid when religious matters led to his dis-
missal. Whiston described Newton as having a "prodigiously fearful, cautious,
and suspicious temper."[9] It was Whiston who finally publicly revealed Newton's
secret religious attitudes, but not until 1727–1728, after Newton's death.

Newton makes a good hero for the mathematician or scientist, but
as a man there is a mass of contradictions and flaws. Perhaps William
Whiston accurately summarizes the matter in his 1749 *Memoirs*:

> So did I then enjoy a large portion of his favour for twenty years together.
> But, he then perceiving that I could not do as his other darling friends
> did, that is, learn of him without contradicting him, when I differed in
> opinion from him, he could not in old age bear such contradiction, and
> so he was afraid of me the last thirteen years of his Life.[10]

Two hundred years later, the playwright George Bernard Shaw gave
this pithy summary: "Newton was able to combine prodigious mental
faculty with credulities and delusions that would disgrace a rabbit."[11]

In terms of his achievements and influence on the future course of
science, Newton certainly warrants the label of "hero." However, when
we consider Newton the man as well as Newton the mathematician and
scientist, we are left with the impression of hero, yes, but a flawed hero.

Figure 1.3. Portrait of Newton from the studio of Enoch Seeman, circa 1726. He holds a copy of the *Principia*, the third edition of which was published that year. (Image reproduced by permission from the National Portrait Gallery, London.)

1.7. CHRONOLOGY

Date	Age	Event
1642		Born at Woolsthorpe, near Grantham, England
1655	12	Attends the King's School, a grammar school in Grantham
1661	18	Goes up to Trinity College, Cambridge
1665	22	January: Graduates with a bachelor of arts, Cambridge
		August: Outbreak of the plague causes move back to Lincolnshire
1667	24	Returns to Cambridge; elected as a fellow at Trinity College
1668	25	Constructs first reflecting telescopes
1669	26	Is elected Lucasian Professor of Mathematics
1672	29	Is elected as a fellow of the Royal Society;
		Light and Colors read to Royal Society; disputes with Hooke begin
1673	30	Begins lectures on mathematics
1682	39	Observes Halley's comet
1684	41	Visit by Halley leads to writing of the *Principia*
1687	44	Publication of the first edition of the *Principia*
1689	46	Becomes member of parliament for Cambridge University
1693	50	Suffers a mental breakdown
1696	53	Moves to London as warden of the Royal Mint
1700	57	Appointed as master of the Royal Mint
1703	60	Elected president of the Royal Society
1704	61	Publication of *Opticks*
1705	62	Is knighted by Queen Anne
1712	69	Royal Society report favors Newton in calculus-priority dispute
1726	83	Third and final edition of *Principia* published
1727	84	Dies; buried in Westminster Abbey, London

Figure 1.4. Some important dates in Newton's life.

1.8. FURTHER READING

The literature on Newton and his work is vast, so the following represents a very limited personal choice. The definitive biography comes from Richard S. Westfall and is now in three forms as follows:

Isaac Newton. Oxford: Oxford University Press, in the VIP series, 2007. [small-format book, 113 pages, a good and quick read]

The Life of Isaac Newton. Cambridge: Cambridge University Press, 1993. [the abridged version, 328 pages]

Never at Rest: A Biography of Isaac Newton. Cambridge: Cambridge University Press, 1980. [magnificent, 908 pages]

Other works include:

Fauvel, J., R. Flood, M. Shortland, and R. Wilson, eds. *Let Newton Be!* Oxford: Oxford University Press, 1988. [excellent collection of chapters by experts in fields related to this and later chapters of this book]

Gjertsen, Derek. *The Newton Handbook.* London: Routledge and Kegan Paul, 1986. [an invaluable reference for all things Newtonian]

Iliffe, Rob. *Newton: A Very Short Introduction.* Oxford: Oxford University Press, in the VSI series, 2007. [small-format book, 141 pages, but still manages to be quite comprehensive, ch. 6 is a good introduction to Newton's work on religion and prophecies]

Janiak, Andrew, ed. *Isaac Newton: Philosophical Writings.* Cambridge: Cambridge University Press, 2004. [gives a selection of Newton's writings, including the four letters to Bentley and his review of the report on the conflict with Leibnitz]

Keynes, John Maynard. *Newton, the Man.* Available (in part) in *The World of Mathematics.* Vol. 1. Edited by James R. Newman. London: Allen and Unwin, 1956.

Manuel, Frank E. *A Portrait of Isaac Newton.* Cambridge, MA: Harvard University Press, 1968. [concentrates on Newton's personality and sometimes referred to as psychological study]

Westfall, Richard. *Science and Religion in Seventeenth-Century England.* Ann Arbor: University of Michigan Press, 1973. [includes a chapter called "Isaac Newton: A Summation"]

White, Michael. *The Last Sorcerer.* New York: Basic Books, 1997. [a "popular" biography of Newton with emphasis on his personality and his alchemical studies and how they influenced his other work; refers to the relevant "scholarly works" of Betty Jo Dobbs]

The Newton Project (http://www.newtonproject.sussex.ac.uk) is a website directed by Professor Rob Iliffe with the goal of compiling Newton's known works into one electronic edition. [this is a wonderful website where you can read original writings by Newton (and others, like William Whiston and Humphrey Newton) including many of the "Keynes documents" that came out of the famous leaving-Cambridge box]

Many of the other documents from the famous box are now available on a Newton-manuscripts website at the National Library of Israel: web .nli.org.il/sites/NLI/English/collections/Humanities/pages/newton.aspx.

The story of Newton's papers and their sale is given in the following source:

Spargo, P. E. "Sotheby's, Keynes and Yahuda—the 1936 Sale of Newton's Manuscripts." In *The Investigation of Difficult Things: Essays on Newton and the History of the Exact Sciences, in Honour of D. T. Whiteside.* Edited by P. M. Harman and Alan E. Shapiro. Cambridge: Cambridge University Press, 1992.

Finally, the seventeenth century was a period of enormous change and significance for the development of England; to gain an impression of Newton's times, one much-acclaimed reference is the following:

Hill, Christopher. *The Century of Revolution.* London: Routledge, 2006.

SETTING THE SCENE

This is a long chapter and is not essential for reading the rest of the book. However, it is also not very demanding and will help to give added meaning to many things that follow. Understanding what the scientific scene was like when Newton began work also lets us appreciate further the immensity of the advances he made.

Newton's *Principia* was first published toward the end of the seventeenth century, and many people view it as the culmination of the Scientific Revolution. This book discusses the revolutionary ideas that we find in the *Principia*, but before doing that we should try to understand what was being overthrown in the Scientific Revolution. Only then will we be able to appreciate fully the steps made in the *Principia*.

The great questions driving the *Principia* are:

How do we understand and describe motion?
How do we use those ideas to discuss motion on Earth, motion in the heavens, and the connections between them?

When tackling those questions, it is necessary to involve a third question:

How do we do science?

Those questions did not originate with Newton, as they are at the heart of science, stretching back to the very earliest attempts of mankind to come to terms with the world and how we live in it. Thus, Newton had to contend with many centuries of work by earlier mathematicians and scientists; some of that work was extremely helpful, and some led to confusion and errors that hindered progress. This is the context in which Newton began his education and research, and it is necessary to understand that context in order to appreciate the greatness of Newton's achievements and the importance of the *Principia*. The aim of this chapter is to briefly introduce that context. A long time period is involved, so only a few of the most salient points can be mentioned here. The interested reader is referred to the literature cited in the "Further Reading" section at the end of the chapter.

2.1. ANCIENT BEGINNINGS

Civilizations like those in Babylon and Egypt were skilled in technology, building, surveying, practical astronomy and calendar making, using mathematics, and developing writing systems for keeping records. For example, the Egyptians used information about the rising of the star Sirius to design an accurate calendar. However, we have no evidence of a widespread search to understand the natural world and to explain why things happen. Those initial stirrings in science appear to belong to the ancient Greeks, who absorbed and built on knowledge from those earlier civilizations. The Greeks also had an alphabetical system of writing, so that even today we can read the books they created. A key step involved looking beyond myths, gods, and supernatural beliefs to consider the cosmos as a natural place and, thereby, as a comprehensible one. This applied equally in medicine, as the Hippocratics wrote around 400 BCE: "Every disease has a natural cause, and no event occurs without a natural cause."[1]

An indication of the timing of these developments is given by the (often approximate) lifetime dates for some of the major contributors: Thales (624–546 BCE), Pythagoras (582–497 BCE), Empedocles (492–432 BCE), Democritus (470–390 BCE), Hippocrates (460–370 BCE), Socrates (469–399 BCE), Plato (427–348 BCE), Aristotle (384–322 BCE), Eudoxus (408–355 BCE), Euclid (325 BCE–unknown), Archimedes (287–212 BCE), Eratosthenes (276–195 BCE), Apollonius

(262–190 BCE), and Ptolemy (100–170 CE). These men (for they were all men) and many others contributed to the beginnings of the science, mathematics, and philosophy that has come down to us today. Five of them are particularly important for this book.

2.1.1. Euclid

Euclid's *Elements* is the first great mathematics book. It is an enduring masterpiece, going through more editions than any book except the *Bible*. The development of geometry that it presents remains valid today. There are three reasons for the enormous influence of the *Elements*. First, it shows us how a subject may be developed on the basis of definitions, axioms, and logical rules for working. Euclid provided a model for use in other fields. Second, it gave us the first systematic approach to a branch of mathematics—geometry in a plane and in three dimensions—as well as certain aspects of number theory.

The third thing Euclid gave us is an accurate and consistent piece of theoretical physics, the first such work, according to Einstein. There was debate about the nature of Euclid's diagrams, but clearly they can be (and were) interpreted in a practical sense. They can also be used to describe the physical world either as direct properties of objects or as paths traced out by moving particles.

2.1.2. Plato and Aristotle

The philosopher-mathematician A. N. Whitehead famously wrote in 1929 that the European philosophical tradition "consists of a series of footnotes to Plato." His influence on science also lingers today. Plato championed the importance of mathematics in science (with "let no one ignorant of geometry enter here" allegedly written over the entrance to his academy), suggesting a path for scientists and philosophers ever since. Plato held that the most perfect shapes, like the circle and the five regular or perfect solids (the "Platonic solids"), should be at the heart of the forms underlying the most basic elements of the world. In his *Timaeus*, he identified the four basic constituents—earth, water, air, and fire—with perfect solids, and, hence, to two basic triangles used in their formation. The idea of elementary particles fitting into a mathematical scheme related to symmetry remains with us today.

Aristotle was a pupil of Plato, but he set out to create his own universal picture and procedures. Aristotle was a towering figure whose ideas and writings dominated intellectual life for many centuries. His output was truly amazing, with works ranging from *Physics* to *On the Soul*, from *Prior Analytics* to *Nicomachean Ethics*, from *On the Heavens* to *On Length and Shortness of Life*, from *Meteorology* to *Poetics*. Aristotle's biological books (such as *History of Animals*, *Parts of Animals*, *On Respiration*, *Movement of Animals* and *Generation of Animals*, with his emphasis on observing, recording, and interpreting) has led many to see him as the founder of biology. His *Analytics* books gave us the basics of logic and the exploration of the form of arguments such as syllogisms.

In contrast to his "modern" approach to biology, Aristotle approached physics in a more philosophical manner, perhaps believing that observation and experimentation could not help with some of his key philosophical questions such as: What is change? What is place? What is time? He was less mathematically inclined than Plato, although he did retain Plato's ideas about things like the perfection of the circle. The result is that much of Aristotle's work in physics is now seen as wrong (and maybe even irrelevant and bizarre). But that does not mean we can ignore it. **Such was Aristotle's reputation that his work was accepted as the final word for many people over hundreds of years and right up to Newton's time. Four facets of Aristotle's most influential thinking are particularly important for our story.**

(i) Causes

In his *Physics*, Aristotle sets out a principle that dominates much of his work: "A prerequisite for knowing anything is understanding why it is as it is—in other words, grasping its primary cause."[2] He then goes on to define four kinds of cause that he will use to explain events in the physical world.

One way in which the word "cause" is used is for that from which a thing is made . . . for example, the bronze of a statue.

A second way in which the word is used is for the form or pattern (i.e., the formula for what a thing is, . . . for example, the ratio 2:1 and number in general, cause the octave).

A third way in which the word is used is for the original source of change or rest. For example, a deviser of a plan is a cause, a father causes a child. . . .

A fourth way in which the word is used is for the end.[3]

In a remarkably modern-sounding example, Aristotle writes "if asked, 'why is he walking?' we reply, 'to get healthy', and in saying this we mean to explain the cause of his walking."[4]

These causes embody Aristotle's idea of teleology; and, as it had to be used in all cases, there came a problem that was answered by his famous first-agent proposal:

There cannot be an infinite regress of movements caused by other movements . . . since everything that changes must be changed by something.

There is a first agent of change which is eternal, and is not changed even coincidentally.

The eternal first agent of change has no magnitude, and is located at the outer edge of the universe.[5]

(ii) Motion

Aristotle had to reconcile his need for causes with the idea of motion, and this led to: "Proof that there is no void separate from bodies," which involves the argument that "if each of the simple bodies naturally has its own proper motion (as fire moves upwards and earth moves downwards and towards the centre of the world), it is easy to see that void is not responsible for their motion."[6] For Aristotle, there is no vacuum possible.

A little later he makes the points: "What about the fact that when things are thrown they continue to move when the thrower is no longer touching them? . . . It would be impossible to explain why something which has been set in motion should stop anywhere: why should it stop here rather than there?"[7]

Here we have Aristotle wrestling with the problem of how motion actually "works," and, using his causes approach, he concludes that "the agent of change and the object changed must be in contact."[8] **This conceptual difficulty remained up to and beyond Newton's time.**

Notice that Aristotle is also introducing the concept of natural motion, and it is a distinguishing motion, "for if the natural motion is upward, it will be fire or air, and if downward, water or earth." There is also the link to weight and which bodies fall fastest: "A given weight moves a given distance in a given time; a weight which is as great and more moves the same distance in a less time, the times being in inverse proportion to the weights. For instance, if one weight is twice another, it will take half as long over a given movement."[9]

(iii) Circular Motion

Aristotle followed Plato and thus imposed ideas about perfect circles on future generations trying to understand motion. There can be no doubting his viewpoint:

> Circular movement is the primary kind of movement.
>
> As for the fact that circular movement is the primary kind of movement, this is obvious. Every movement is either circular or rectilinear or a combination of the two. . . . Also, the higher degree of simplicity and completeness possessed by circular movement means that it has priority over rectilinear movement.[10]

Aristotle is led directly to the most vital property of circular motion:

> Only circular motion can be continuous and eternal. . . . We now go on to claim that it is possible for there to be an infinite change, which is single and continuous, and that this is circular movement.[11]

(iv) Earth and the Heavens

Aristotle believed there was a sublunar region that was changing and made up of earth, water, fire, and air; he also believed there was a heavenly region that was perfect, unchanging, and materially different.

> On all these grounds, therefore, we may infer with confidence that there is something beyond the bodies that are about us on this earth, different and separate from them; and that the superior glory of its nature is proportionate to its distance from this world of ours.
>
> These premises clearly give the conclusion that there is in nature some bodily substance other than the formations we know, prior to all of them and more divine than they.
>
> The shape of the heaven is of necessity spherical; for that is the shape most appropriate to its substance and also by nature primary.[12]

2.1.3. Archimedes

Archimedes was the great mathematician and applied mathematician of the ancient Greek era. His reputation is undiminished today. He built on Euclid's *Elements*, and Archimedes's *On the Sphere and the Cylinder*, *Measurement of a Circle*, *On Conoids and Spheroids*, *On Spirals*, and *Quadrature of the Parabola* are masterpieces of geometry and contain early versions of aspects of calculus. His books *On the Equilibrium of Planes or the Centre of Gravity of Planes* and *On Floating Bodies* are the founding works in statics. They established what we might call a scientific approach by following Euclid in setting out definitions, postulates, and propositions to be established using logical development of the subject. This was the example for Galileo and others to follow.

2.1.4. Ptolemy

There were various views about how the Earth, Sun, planets, and stars were arranged (including some with the Sun at the center), but the authoritative description was produced by Ptolemy working in Alexandria toward the end of the ancient Greek era. Ptolemy maintained the ideas of Plato and Aristotle, but in his *Almagest*, he showed how detailed mathematical models could be constructed to describe the motion of the planets and to fit and predict a large variety of astronomical results. Even today, the *Almagest* impresses by its size and comprehensive approach.

Ptolemy followed the rule that heavenly motions are perfect and so must be circular. However, he knew that the simple model of Sun and planets moving on circular paths centered on the Earth could never give accurate results. The solution was a combination of circles. A planet moves around a circle, the epicycle, whose center moves around another circle, the deferent, whose center is a little displaced from the Earth. Adjusting these parameters gave Ptolemy an accurate astronomical model. The time taken for motion around the deferent gave the zodiacal period and the epicycle fine-tuned, orbital-shape details and could account for retrograde motion of the planets. Motion around the deferent was uniform with respect to a point called the equant, which did not coincide with the deferent center, and this allowed for speeds to differ for different parts of an orbit. The diagram in figure 2.1 shows the Ptolemaic model. This brilliant piece of what today we might call "mathematical modeling" was to be the dominant picture for over a thousand years.

Those various works contained big ideas that the Greeks passed on to future scientists and philosophers. It should not be assumed that they represent the sum of Greek science, for there were many other contributors over those early centuries and not all of them agreed with the views expressed by Aristotle and Ptolemy, for example. There were those who believed in an atomic theory and others who suggested different astronomical models, including those with a moving Earth in them. Nevertheless, those works by the likes of Plato, Aristotle, Archimedes, and Ptolemy were the ones that had the greatest influence for future generations of thinkers.

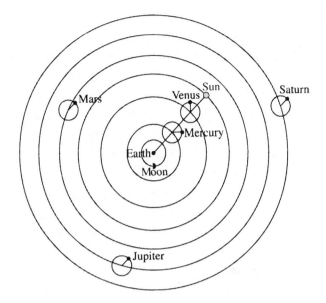

Figure 2.1. Ptolemy's model of the solar system. (From C. M. Linton, *From Eudoxus to Einstein: A History of Mathematical Astronomy* [Cambridge: Cambridge University Press, 2004], reprinted with the permission of Cambridge University Press.)

2.2. BEYOND THE GREEKS

The decline of the Greek world and the fall of the Roman Empire saw Europe move into a low period sometimes called the Dark Ages, although that term is not favored today. For our story, the important trend is the shift of activity to the Arab world and the development of Islamic science. The years from 400 CE onward saw Greek works translated into Arabic, and the influence of Indian and Chinese science and mathematics also absorbed into a new intellectual tradition. Places like Baghdad and Cordoba became the new intellectual centers.

Greek works were not only preserved but they were also given commentaries, corrected, and extended. Islamic science could build on the works of Euclid and Ptolemy, and it is only in recent times that this role of Islamic science has come to be appreciated. For example, the Arab mathematician al-Khwarizmi (circa 780–850) lived in Baghdad and produced work that brought the Hindu decimal system into use and gave us basic ideas in algebra. There was also progress in astronomy and optics.

Around the twelfth century, there was a major development: the works preserved in the Arab world were translated into Latin and introduced into Europe. For example, Gerard of Cremona (1114–1187) translated about seventy books of all kinds (including works by Aristotle, Ptolemy, Galen, and many Islamic writers), and William of Moerbeke (1215–1286) did similarly for about fifty such books.

There were challenges to Aristotle, and critical commentaries were written by people like John Philoponus, who was living in Alexandria around 550 CE. Of great interest is his statement about falling bodies:

> Our view may be corroborated by actual observations more effectively than by any sort of verbal argument. For if you let fall from the same height two weights, one many times as heavy as the other, you will see that the ratio of the times required for the motion does not depend on the ratio of the weights, but that the difference in time is very small. And so, if the difference in the weights is not considerable, that is, if one is, let us say, double the other, there will be no difference, or else an imperceptible difference, in time.[13]

These views were both attacked and defended by Islamic scholars, and the Aristotelian tradition had its strong supporters, such as Averroes. Similarly, when the Greek and Islamic works reached Europe in translations, there were commentaries written about them and debates about their validity.

Two aspects of motion were contentious over the centuries: (1) how to understand the causal processes such as the way in which an arrow is maintained in its flight, and (2) how to describe the details of motion. The early challenges of people like Philoponus and Avicenna raised questions about motion in air or in voids and the role of a medium in maintaining or resisting motion. The Parisian professor Jean Buridan (1295–1358) introduced the term *impetus* to stand for an internal impressed force that maintained motion. Some see this as the first groping toward an idea of inertia and its role in motion. Others, such as Thomas Bradwardine (1290–1349), devised the Merton rule (uniformly accelerated motion can be related to the mean speed over the time of motion), and Nicole Oresme (1320–1382) gave it a geometrical picture and proof (which surfaced in much the same form in Galileo's writings). Contrary to the impression often created, there was scientific activity over this period, but what did not emerge was a comprehensive and consistent treatment of dynamics.

The debates about the validity of Aristotle's ideas were also tangled up with religious ideas and the power of the Church. As one of the Fathers of the Church, Saint Augustine (354–430) set the trend when he said that in matters where reason and faith collided, faith must win, "for if reason be found contradicting the authority of the Divine Scriptures, it only deceives by a semblance of truth, however acute it be, for its deduction cannot in that case be true."[14] When the translations of Aristotle reached the West, there were new challenges for the Church. But eventually Saint Thomas Aquinas (1225–1274) wrote a commentary that served to combine Church, Augustinian, and Aristotelian ideas in a satisfactory manner so that it became something of an official Church doctrine. (As late as 1879, Pope Leo XIII pronounced "Thomism" to be the official doctrine of the Roman Catholic Church.)

At various times, there were heated debates and bans on the discussion of Aristotle's works. However, their study in universities continued, and the "schoolmen" maintained a dominant influence in education right up to Newton's time.

2.3. TWO MAJOR STEPS

After many centuries of reliance on Greek texts and Islamic modifications, it was time for some new works that set out to be major and comprehensive and to suggest a new way forward. These came from the Pole Nicolaus Copernicus (1473–1543) and the German Johannes Kepler (1571–1630).

2.3.1. Copernicus

Although there were Greek and Islamic scientists who suggested models with the Earth at the center of the universe, and Nicole Oresme had given a careful analysis of the possibility of a rotating earth, it was Copernicus who finally gave the complete challenge to Ptolemy in the former's *On the Revolutions of the Celestial Spheres*. Copernicus put the Sun in the center and the Earth and other planets in orbits around the Sun. The Earth was also given a daily rotation. Although this was a major step, Copernicus still retained elements from Plato and Aristotle, as is evident when we read section headings such as "The World Is Spherical" and "The Movement of the Celestial Bodies Is Regular, Circular, and Everlasting— or else Compounded of Circular Movements."

In fact, Copernicus used a scheme involving circles and epicycles much like Ptolemy, so his model was not much (if any) simpler to use. However, it had made that fundamental change in astronomical thinking, with the Earth now one of the planets orbiting the Sun. For some people, the year 1543—when Copernicus' book was published—marks the start of the Scientific Revolution.

Copernicus had long delayed publication of his book. He might be seen as a perfectionist and he worried over data to support his heliocentric model. In his dedication to Pope Paul III, he expected some readers of his book to "immediately shout to have me and my opinions hooted off the stage."[15] There also had to be religious concerns. Martin Luther had described Copernicus as the "fool who wanted to turn the art of astronomy on its head."[16] Although the ideas in *On the Revolutions* were counter to Church and Bible teachings, there was no immediate adverse reaction and, in part, this is due to a foreword or introduction inserted by the Lutheran theologian Andreas Osiander, who was responsible for seeing the book finally printed and who gave Copernicus, laying on his deathbed, a first copy. In *To the Reader concerning the Hypotheses of the Work*, Osiander suggested that the book should be viewed as a calculational scheme, and he wrote, "for it is not necessary that these hypotheses should be true, or even probable; but it is enough if they provide a calculus which fits the observations."[17] It was over seventy years later, in 1616, that the Catholic Church formally suspended its approval of On the Revolutions until it should be corrected. Among Copernicus's "dangerous ideas" was work on the scale of planetary orbits and statements about the vast distances to the stars. It might be recalled that Giordano Bruno used those ideas to suggest that there could be other suns with orbiting planets such as ours; he was burned at the stake in 1600!

2.3.2. Kepler

Kepler was something of a mystic and originally planned to be a Lutheran clergyman. His long, rambling books show that he still maintained his religious bent as he sought the glory and methods in God's ways in his astronomical discoveries. For Kepler, there was to be no religious conflict. A letter reveals that "for a long time I wanted to become a theologian, for a long time I was restless. Now, however, behold how through my effort God is being celebrated through astronomy."[18]

Kepler had available the extensive data of the great Danish astronomer Tycho Brahe, and he worked with staggering determination to fit planetary orbits to mathematical models. Despite his heroic efforts, he was defeated by Mars until he took the radical step of challenging the venerable domination of the circle. We now summarize Kepler's greatest achievements in his three laws:

Law 1: The orbit of each planet is in the shape of an ellipse with the Sun at one focus.

Law 2: In any equal time intervals, a line from the planet to the Sun will sweep out equal areas.

Law 3: The squares of times of revolution of any two planets around the Sun are proportional to the cubes of their mean distances from the Sun. (We now tend to say that, for a planet, the square of its period is proportional to the cube of its mean distance.)

By these discoveries, Kepler removed the requirement for orbits to be described in terms of circles. Furthermore, he removed the idea of uniform motion around circles since his second law tells us how the planet's speed varies around the orbit.

With these discoveries, Copernicus and Kepler made the great break from Greek ideas that had dominated thinking for almost two thousand years. What did remain, however, was the use of mathematics in science, as advocated by the Pythagoreans and Plato. The books of Copernicus and Kepler contain intensely mathematical chapters.

2.4. THE CHALLENGES

Even today we all naturally think of the Sun as rising and traveling across the sky rather than a fixed Sun only apparently moving as our Earth rotates. Similarly, it is not obvious that we are moving in an orbit around the Sun. The first great challenge of the new astronomy was to find evidence for the new scheme, something Copernicus was acutely aware of.

There was a second so-called common-sense argument against the new proposals. The description of motion and the underlying mechanisms needed for understanding the flight of projectiles and other such problems now became of even more importance as the work of Copernicus and Kepler became accepted. If the Earth really was rotating on its axis and also traveling around the Sun, then simple calculations showed that extremely large speeds are involved. This immediately raises a new problem: How is it that objects do not fly off the Earth, and why are cannonballs, clouds, and flying birds not left way behind? Copernicus had ideas about the air and objects somehow rotating with the Earth, but there was no coherent scheme of dynamics to understand how everything worked. Even comparatively slow motions, like galloping horses, could show how objects flew off and were left behind; to many people, such a rapidly moving Earth was incomprehensible.

There was also a new emphasis on what it was that controlled the motion of the Moon and the planets. Old Greek ideas of an ordered, perfect, unchanging heavens, crystalline spheres, and Aristotle's "causes" were now inadequate notions. Although Copernicus clearly wondered

about the role of the Sun in the planetary orbits, it was Kepler who saw
that astronomy had reached a turning point:

> Astronomers should not be granted excessive licence to conceive any-
> thing they please without reason: on the contrary, it is also necessary
> for you to establish the probable causes of your Hypotheses which you
> recommend as the true causes of Appearances. Hence, you must first
> establish the principles of your Astronomy in a higher science, namely
> Physics or Metaphysics.[19]

He tried to discover how the Sun might have an extending influence
that somehow swept the planets around in their orbits, and he considered
the possibility that magnetic effects might be involved. Kepler had found
an overall pattern into which all planetary motion fit, and he recognized
that the next step was to find the corresponding unifying underlying
mechanism, one that clearly involved the Sun. This was the great
challenge to which Newton would find a response, and so it is interesting
to read Kepler's wonderful statement:

> My goal is to show that the heavenly machine is not a kind of divine
> living being but similar to a clockwork insofar as almost all the mani-
> fold motions are taken care of by one single absolutely simple magnetic
> bodily force, as in a clockwork all motion is taken care of by a simple
> weight. And indeed I also show how this physical representation can be
> presented by calculation and geometrically.[20]

This was a great step forward, and here was the move from what we
might call "mathematical modeling" to astronomy as a branch of physics.
Kepler failed to find the appropriate physics, but he was the one who
pointed the way for those following him.

2.5. HOW TO RESPOND: GUIDING PRINCIPLES

There are three important methodological principles that guided scientists as the Scientific Revolution gradually unfolded (and it is that gradualness that makes some question the use of the word *revolution* with its implications of dramatic, rapid changes).

2.5.1. The Mathematization of Science

This takes us back to Pythagoras and Plato and is already apparent in the works of Archimedes and Ptolemy. Over the centuries, many people supported the views expressed, for example, by Roger Bacon (1219–1292) in his *On the Importance of Studying Mathematics*, which has chapter 3 headed "In which It Is Proved by Reason That Every Science Requires Mathematics." Copernicus clearly used a mathematical approach, and in his book, we can see a good example of two different interpretations: Instrumentalists believed that mathematical models are used to facilitate calculations and to make predictions, whereas Realists believed that a successful mathematical treatment then actually reveals how things must be. The foreword slipped in by Osiander suggested an Instrumentalist approach, but Copernicus himself was a Realist. He believed that the accuracy of his mathematical description did reveal things about how the planets were arranged and why physically we observe retrograde motion for some of them. Kepler too believed that his work revealed physical facts about the solar system.

Perhaps the most famous—and certainly one of the most beautifully expressed—statements of this guiding idea came from Galileo:

> Philosophy is written in that great book which ever lies before our eyes—I mean the Universe—but we cannot understand it if we do not first learn the language and grasp the symbols in which it is written. This book is written in the mathematical language, and the symbols are triangles, circles and other geometrical figures, without whose help it is humanly impossible to comprehend a single word of it, and without which one wanders in vain through a dark labyrinth.[21]

Mathematics provides the language or framework in which science should operate. On the religious side, nature obeys mathematical laws because God used those laws in creating Nature.

2.5.2. Experiment

Gradually, people began to challenge the authority of Aristotle and the idea that if something required identification or explanation, then the solution was to examine his writings on the subject. The philosopher Francis Bacon (1561–1626) wrote extensively on this, stating that "books must follow sciences and not sciences books."[22] The performing of experiments became a standard event at the early meetings of the Royal Society, and Albert Einstein wrote that one of the great developments of this period was "the discovery of the possibility to find out causal relationships by systematic experiment."[23]

2.5.3. The Mechanical Philosophy

There was a clear move away from the approach of Aristotle and his ideas about causes and teleology. The new approach saw the universe in terms of matter and motion. This did not remove a role for God, but now man could understand God's works with a new philosophy. One of the great proponents of this new approach was Robert Boyle (1627–1691), and under the title *About the Excellency and Grounds of the Mechanical Hypothesis*, he wrote:

> Thus the universe being once framed by God and the laws of motion settled and upheld by his perpetual concourse and general providence; the same philosophy teaches that the phenomena of the world are physically produced by the mechanical properties of the parts of matter, and, that they operate upon one another according to mechanical laws.[24]

Boyle supported the corpuscular view of matter, and so the ancient atomists were again of interest. Descartes, one of the most influential founders of the mechanical philosophy, did not see the need for the atomic hypothesis; however, he did set out clearly how a new analogy with machines might be invoked and used to see how we can delve deeper into the physical world:

> Men who are experienced in dealing with machinery can take a particular machine whose function they know and, but looking at some of its parts, easily form a conjecture about the design of the other parts, which they cannot see. In the same way I have attempted to consider the observable effects and natural parts of bodies and track down the imperceptible causes and particles which produce them.[25]

2.6. TWO GREAT PIONEERS

There was great and widespread scientific activity in the sixteenth and seventeenth centuries, but for our story, two people—Galileo and Descartes—are worthy of particular attention.

2.6.1. Galileo Galilei (1564–1642)

The person who best marks the transition between the scientific era begun by Aristotle and the ancient Greeks and the start of the modern era is surely Galileo. Galileo was a follower of Archimedes in his mathematical approach. He presented some of his work in *Dialogues* in which the views of Aristotle were exposed, discussed, and discredited. Galileo believed in the experimental approach to science and also refused to be cowed by Church doctrine, boldly stating that "I think that in discussions of physical problems we ought to begin not from the authority of scriptural passages, but from sense-experiences and necessary demonstrations."[26]

Galileo supplied observations that supported the astronomical revolution begun by Copernicus. In 1604, he observed a nova (as had Tycho Brahe in 1572) and that, together with records of comets, challenged the idea of a perfect and unchanging heavens beyond the Moon. Galileo provided more evidence when the telescope became available, and he observed features on the Moon, which he interpreted as mountains and craters, and spots on the Sun. He observed the phases of Venus, something that clearly demonstrated the superiority of Copernicus's heliocentric model. The astronomical realm was greatly extended as Galileo saw an enormous number of stars and observed the satellites around Jupiter.

In his *Dialogue on the Two Great World Systems, the Ptolemaic and Copernican*, Galileo set out the arguments that made the transition to the Copernican viewpoint inevitable (and caused his problems with the Catholic Church!). However, his book makes no mention whatsoever of Kepler's laws, and the planetary orbits are still taken as circles. (Drake has argued that this is all about ease of presentation; see the "Further Reading" section at the end of this chapter.) Galileo's diagram in the *Dialogue* is much like Copernicus's, except that four moons are now shown orbiting Jupiter. (Galileo knew nothing about the planets beyond Saturn.)

In his *Two New Sciences*, Galileo set out experimental and mathematical results that are often taken as the first coherent starting point for the science of mechanics. His experiments on motion down inclined planes and with pendulums, as well as his considerations of falling bodies, led him to two major results. First, in uniform motion, a body moves equal distances in equal times. Second, in uniformly accelerated motion, the distance moved depends on the square of the elapsed time. (How original some of Galileo's conclusions are is debatable, but he set out the results in a comprehensive and easily understandable form.) Galileo made a major step when he combined a uniform horizontal motion with a vertical motion under gravity to prove that the path of a projectile is a parabola (if, as he pointed out, we ignore air resistance).

In an important methodological step, Galileo used his parabolic-projectile theory to prove that maximum range occurs for a launch angle of forty-five degrees. He was obviously proud of this step and sets out the important underlying principle:

> The knowledge of a single fact acquired through a knowledge of its causes prepares the mind to ascertain and understand other facts without need of recourse to experiments, precisely as in the present case, where by argument alone the Author proves with certainty that the maximum range occurs when the elevation is 45 degrees.[27]

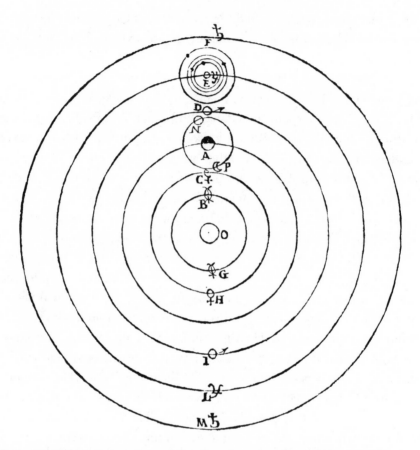

Figure 2.2. Galileo's diagram for modeling the solar system. (From R. Taton and C. Wilson, eds., *The General History of Astronomy: Volume 2, Planetary Astronomy from the Renaissance to the Rise of Astrophysics* [Cambridge: Cambridge University Press, 1989], reprinted with the permission of Cambridge University Press.)

Galileo was close to stating what today is often called the principle of inertia when he wrote:

Now it is evident from what has been said elsewhere at greater length that equitable [i.e., uniform] motion on this plane would be perpetual if the plane were of infinite extent.[28]

However, in practice Galileo saw such motion as occurring along a circular path:

> Hence, along the horizontal by which we understand a surface, every point of which is equidistant from this common centre, the body will have no momentum [i.e., impetus to change] whatever.[29]

Galileo made enormous steps and set the scene for those like Newton who followed. We should note, however, that Galileo was giving results and a mathematical formulation of **kinematics** rather than **dynamics**. He left behind the causal notions of Aristotle, but he did not introduce the idea of force and dynamical equations. His work contains the words *natural tendency* in his discussions of motions of falling bodies and projectiles. He did not supply the Sun-centered mechanism that Kepler sought for control of the planets. However, in terms of the experimental and mathematical approaches, his work was of the greatest importance.

2.6.2. René Descartes (1596–1650)

Descartes, as a founder of the mechanical philosophy, was critical of Galileo's work, despite praising his break with the ancient doctrines and his use of mathematical methods. He wrote:

> Without having considered the first causes of nature, he has merely sought reasons for certain particular effects; and thus he has built without a foundation. . . . As to what Galileo has written about balance and the lever, he explains very well what happens, but not why it happens, as I have done in my Principles.[30]

By "Principles," Descartes is referring to his masterpiece *Principia Philosophiae*, or *Principles of Philosophy*, which was first published in Latin in 1644. If Galileo's work is a prime example of the experimental and mathematical approach, then Descartes represents the mechanical philosophy in action. *Principia Philosophiae* is an impressive attempt to build a new universal intellectual structure for understanding everything. Descartes was the first person since Aristotle to attempt such an ambitious project. The work of Descartes was highly regarded and, since we shall see how it influenced Newton and more so the reception of his work, it is useful to focus on some key points.

Principia Philosophiae is arranged in four parts. Part I, *On the Principles of Human Knowledge*, covers the way we think and gain knowledge and how God is involved. It includes the following sections (among others):

1. *That whoever is searching after truth must, once in his life, doubt all things; insofar as this is possible.*

. . .

9. *What thought is*

. . .

14. *That from the fact that necessary existence is contained in our conception of God, it is properly concluded that God exists*

. . .

20. *That we were not created by ourselves, but by God, and that consequently He exists.*

All of this leading to the final section of part I:

76. *That divine authority is to be preferred to our perception: but that apart from divine authority, it does not become a philosopher to assent to things other than those which have been perceived.*

Part II, *On the Principles of Material Objects*, sets out the fundamentals, with sections such as:

1. *The reasons why we know with certainty that material objects exist.*

. . .

4. *That the nature of body does not consist in weight, hardness or colour, or similar properties; but in extension alone.*

. . .

11. *That space does not in fact differ from material substance.*

. . .

16. *That it is contradictory for a vacuum, or a space in which there is absolutely nothing, to exist.*

. . .

20. *That this also shows that no atoms can exist.*

Then come important sections in which Descartes sets out some significant steps in dynamics:

27. *That movement and rest are merely diverse modes of the body in which they are found.*

. . .

36. *God is the primary cause of motion; and that He always maintains an equal quantity of it in the universe.*

37. *The first law of nature: that each thing as far as is in its power, always remains in the same state; and that consequently, when it is once moved, it always continues to move.*

. . .

39. *The second law of nature: that all movement is, of itself, along straight lines; and consequently, bodies which are moving in a circle always tend to move away from the centre of the circle which they are describing.*

There is very little mathematics in the *Principia Philosophiae*, but part II ends with the following:

64. But I do not accept or desire in Physics any other principles than in Geometry or abstract Mathematics; because all the phenomena of nature are explained thereby, and certain demonstrations concerning them can be given.

Part III, *On the Visible Universe*, contains Descartes's mechanism for the solar system:

30. *That all the Planets are carried around the Sun by the heaven.*

Descartes postulates "that the matter of the heaven, in which the Planets are situated, unceasingly revolves, like a vortex having the Sun as its centre,"[31] and it is by this motion that the planets are moved around the Sun. This was a universal mechanism that achieved great prominence, and Descartes gave diagrams showing how it is constructed (see figure 2.3).

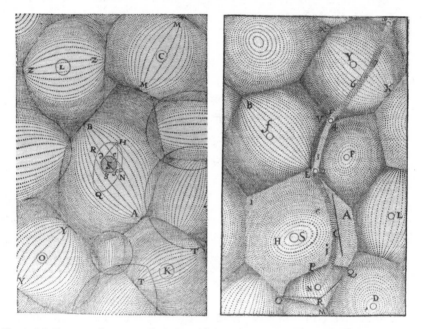

Figure 2.3. Descartes's system of vortices. The Earth moves in the Sun's vortex (*left*), and a comet's path is controlled by the vortices it travels through (*right*). (From "Tycho Brahe to Newton," in *The General History of Astronomy: Volume 2, Planetary Astronomy from the Renaissance to the Rise of Astrophysics*, ed. R. Taton and C. Wilson [Cambridge: Cambridge University Press, 1989], reprinted with the permission of Cambridge University Press.)

In part IV, *Of the Earth*, Descartes deals with a vast range of subjects including heat and fire, earthquakes, glass and steel, fluids and tides, and magnetism. In fact, he concludes: "199. That no phenomena of nature have been omitted in this treatise." His very last section reveals again the power of religion in that time: "207. But that I submit all my opinions to the authority of the Church."

It is easy to see why *Principia Philosophiae* was such a dominant force in the intellectual life of the times of Descartes and Newton, particularly in continental Europe. It provided what seemed like plausible explanations for the way the world works and how we can understand everything around us without offending those who revered the ancients and their ideas. (In part IV we find this section: "200. That I have used no principles in this treatise which are not accepted by all; and that this Philosophy is not new but extremely ancient and commonplace." And there he states that he has used "no principle that was not accepted by Aristotle.") There is virtually no mathematical modeling or analytical development of the ideas (for example, just how the vortex model of planetary motion works and leads to elliptical orbits). That certainly made it accessible to most readers, but it limited the rigor and technical scope of its achievements.

2.7. OTHER GIANTS AND THEIR SHOULDERS

Newton stated that one reason for his success was that he had "stood on the shoulders of giants," and we have already met some of those people. It would be unjust not to at least mention others who were known to Newton and who made significant contributions to the development of seventeenth-century science.

Robert Boyle (1627–1691) was the son of the Earl of Cork, and he had the time and money to be a scientific virtuoso. Newton was much influenced by Boyle and had twenty-three of his works in his personal library. The two men interacted, and Newton presented Boyle with a first edition of the *Principia* and even attended his funeral. Boyle's support for the mechanical philosophy (a term he coined) and the experimental method were important in forming Newton's own approach to science.

Pierre Gassendi (1592–1655) was a continental philosopher, a scientist, and a religious figure who championed the old Greek atomic hypothesis of Epicurus. This set him apart from Descartes. Gassendi also espoused the experimental approach and influenced both Boyle and Newton. He had clear ideas about inertia, and it was Gassendi who had performed the experiment of dropping a stone from the mast of a moving ship to show that the stone fell at the foot of the mast and was not left behind. Gassendi believed that the composition of motions was a universal effect. In astronomy, he observed—and named—the aurora borealis. In 1631 he recorded the first transit of Venus within five hours of the time predicted by Kepler and in so doing added strong support to the theories of Copernicus and Kepler.

Edmond Halley (1656–1742) was a talented astronomer (at age sixteen he showed how three observations could be used to fix the main features of a planet's orbit) and mathematician (he published the first printed version of Appolonius's *Conics* in 1710). Halley was appointed Savilian Professor of Geometry at Oxford in 1704. It was Halley who ensured that Newton's *Principia* was published (see chapter 3) and who supplied the editorial skills. In some ways, it is surprising that Newton and Halley maintained their relationship throughout Newton's life since Halley had a reputation as a libertine and an atheist. Of course, Halley is now best known for his work on comets and his great prediction for the return of the 1682 comet (Halley's comet) in 1758. He became Astronomer Royal in 1720.

Robert Hooke (1635–1703) was a gifted man across a whole range of activities. He excelled as an experimentalist and invented or improved numerous instruments and mechanical devices. He began his career helping Boyle to develop the air pump that was so important in the study of gases and their properties. He was long associated with the Royal Society, serving as secretary from 1677 to 1683, and, from 1662 until his death, he was Curator of Experiments. Hooke is known now for his work on springs and elastic effects and for his beautiful book *Micrographia* in which his excellent draftsmanship allowed him to display the microscopic structure of such things as insects, feathers, and fish scales. Hooke had many ideas in optics and mechanics, and he clashed with many people (especially Newton) about the validity and precedence of his ideas. In particular, Hooke had key ideas about orbital motion and the roles played by inertia and a central force. His weakness was his lack of the mathematical background needed to show where these ideas led.

Christiaan Huygens (1629–1695) was one of the most brilliant men of the period, working in dynamics and optics with emphasis on mathematical methods. His work on centrifugal forces, gravity, regular and cycloidal pendulums, and the theory of collisions was groundbreaking. His ideas are set out in a clear, logical, and mathematical style in his very impressive 1673 work titled *Horologium Oscillatorium*. Therein he advances new ideas for accurate clock design. As an astronomer, Huygens described the rings around Saturn and discovered Titan, one of Saturn's satellites. Newton had great respect for the work of Huygens.

John Wallis (1616–1703) was Oxford's Savilian Professor of Geometry from 1649 until 1703. His most important work was published in1655 as *Arithmetica infinitorum*, the first British work on calculus. That and other works by Wallis had a major impact on Newton. Wallis tried hard to get Newton to publish more (a failure) and incorporated some of Newton's discoveries into his own books, such as the 1685 *Algebra*. Wallis is also famous for his work in cryptography and for his bitter clashes with the philosopher Thomas Hobbes over Hobbes's claim to have squared the circle.

Christopher Wren (1632–1723) is best known as the architect who rebuilt London after the great fire of 1666. (Robert Hooke was also involved.) However, Wren was a multitalented man who held professorships in astronomy and geometry at Oxford. Newton classed Wren with Wallis and Huygens as the leading geometers of the time. Wren helped to found the Royal Society, probably preparing its preamble, and became president in 1681. He did important work on collision theory and was also a meteorologist and a naturalist, using the new microscope to study the structure of insects.

2.8. NEWTON'S STARTING POINT

The influence of Aristotle remained in universities when Newton moved to Cambridge in 1661. People like Boyle and Galileo worked outside the university system and so were less hampered by constraints, often legally imposed, on their activities. Newton had to study the old curriculum with its emphasis on logic, rhetoric, and ethics going back to the ancient

Greeks. Newton appreciated the value of some of this material but, by his famous notebook entry "*Amicus Plato amicus Aristoteles magis amica veritas* [*Plato is my friend, Aristotle is my friend, truth is a greater friend*]," he showed that he understood the limitations, and soon he had gone beyond the official curriculum.

Newton began to read very widely and tackled the works of Copernicus, Galileo, Kepler, Gassendi, Descartes, and others either directly or through translations and commentaries. It was then that he saw the breadth of Descartes's approach, but he also began to question its validity and thus began his long battle to establish a better world picture. Newton also read widely in mathematics, and his interest in history also began around this time.

2.9. FURTHER READING

There is an extensive literature covering the topic of this chapter; the following is a highly selective list of personal recommendations at different levels. To begin, there are three very readable books by leading experts in Newtonian studies:

Bertoloni Meli, Domenico. *Thinking with Objects: The Transformation of Mechanics in the Seventeenth Century.* Baltimore: Johns Hopkins University Press, 2006. [comprehensive work by a Newton expert]

Cohen, I. Bernard. *The Birth of a New Physics.* New York: Norton, 1985. [particularly good on evaluating Galileo's contributions]

Westfall, Richard. *The Construction of Modern Science.* Cambridge: Cambridge University Press, 1977.

A wonderful collection of relevant articles can be found in the following:

Park, Katherine, and Lorraine Daston, eds. *The Cambridge History of Science, Volume 3: Early Modern Science.* Cambridge: Cambridge University Press, 2006. [the article "Physics and Foundations" by Daniel Garber is particularly useful]

For study of the earlier periods, see:

Crombie, A. C. *Augustine to Galileo*. 2 vols. London: Penguin 1969. [first published in this revised version in 1959, so it's a little dated but still a great source; something of a classic]

Gregory, Andrew. *Eureka! The Birth of Science*. London: Icon Books, 2001. [a light once-over of ancient Greek science; easy and entertaining to read]

Hannam, James. *God's Philosophers: How the Medieval World Laid the Foundations of Modern Science*. London: Icon Books, 2009. [very readable "popular account"]

Lindberg, David C. *The Beginnings of Western Science*. 2nd ed. Chicago: University of Chicago Press, 2007. [a comprehensive account up to 1450; for the second edition, the chapter on the Islamic influence has been rewritten to better reflect recent findings]

Here are some examples of recent popular books dealing with the Arab influence:

Lyons, Jonathan. *The House of Wisdom*. London: Bloomsbury Publishing, 2009.

Masood, Ehsan. *Science and Islam: A History*. London: Icon Books, 2009.

Concentrating on astronomy:

Hoskin, Michael. *The History of Astronomy: A Very Short Introduction*. Oxford: Oxford University Press, 2003. [an expert's concise tour of the subject in one of the highly successful Oxford VSI series]

Linton, C. M. *From Eudoxus to Einstein: A History of Mathematical Astronomy*. Cambridge: Cambridge University Press, 2004. [gives a comprehensive coverage of the subject; manages to be readable and not too daunting as far as the mathematics is concerned]

There are many books and papers on the Scientific Revolution, and debates about it continue. Some recent books include:

Boas, M., and R. Hall. "Newton's 'Mechanical Principles.'" *Journal of the History of Ideas* 20 (1959): 167–78. [discusses how Newton uses the mechanical hypothesis and microscopic particle theory]

Dear, Peter. *Revolutionizing the Sciences: European Knowledge and Its Ambitions, 1500–1700.* Houndmills, UK: Palgrave, 2001.

Henry, John. *The Scientific Revolution and the Origins of Modern Science.* 3rd ed. Houndmills, UK: Palgrave Macmillan, 2008. [a short book but with a good coverage of the topics in this chapter and a very useful bibliography listing 332 items]

Jardine, Lisa. *Ingenious Pursuits: Building the Scientific Revolution.* London: Abacus, 2000. [readable, light introduction; particularly good on experimental work]

Osler, Margaret J. *Reconfiguring the World: Nature, God and Human Understanding from the Middle Ages to Early Modern Europe.* Baltimore: Johns Hopkins University Press, 2010.

Principe, L. M. *The Scientific Revolution.* Oxford: Oxford University Press in the VSI series, 2011.

Shapin, Steven. *The Scientific Revolution.* Chicago: University of Chicago Press, 1996. [includes an extended, comprehensive bibliographic essay]

The original works of authors such as Archimedes, Aristotle, Plato, Ptolemy, and Galileo are readily available. Major relevant works by Copernicus, Galileo, Kepler, and Newton (and Einstein) are conveniently gathered together in:

Hawking, Steven, ed. *On the Shoulders of Giants.* London: Penguin, 2002.

There are several collections of selections of significant scientific writings, and two good examples are:

Grant, Edward, ed. *A Source Book in Medieval Science*. Cambridge, MA: Harvard University Press, 1974. [a massive collection of original writings with expert introductions]

Sambursky, Shmuel, ed. *Physical Thought from the Presocratics to the Quantum Physicists: An Anthology*. New York: Pica Press, 1975.

Stillman Drake has written extensively about Galileo and translated his work into English. His recent book in the Oxford VSI series offers a new interpretation of Galileo's work and the religious connotations:

Drake, Stillman. *Galileo: A Very Short Introduction*. Oxford: Oxford University Press, 2001.

An entertaining picture of people and events involved in the founding of the Royal Society is given in:

Gribbin, John. *The Fellowship*. London: Allen Lane, 2005.

A FIRST LOOK AT THE *PRINCIPIA*

Before plunging into the Principia, *it may be useful to present an overview so that you have some idea of what is in store and how it will all fit together.*

3.1. PUBLISHING HISTORY

Newton's *Philosophiae Naturalis Principia Mathematica* was first published in 1687. The title page (figure 3.1) shows that the imprimatur, or license to print, was given for the Royal Society by its president, Samuel Pepys, on July 5, 1686, and the printer's date is given as 1687. Up to four hundred copies were printed, and a bound version cost nine shillings. Forty copies were sent to Newton on July 5, 1687. The second edition was published in July 1713, and the third edition in March 1726, a year before Newton died. It is the third edition that I discuss in this book.

The *Principia* (as it is usually called) was written in Latin. A translation into English by Andrew Motte was published in 1729, and that became the definitive version. A revision of Motte's translation was made by Florian Cajori and published in 1934. In 1999, a new translation was published by a great Newtonian scholar, the late I. Bernard Cohen, and Anne Whitman. That book gives a slightly more modern approach to the language, discusses translation matters, and contains a superb "A Guide to Newton's *Principia*" by I. Bernard Cohen. A detailed publishing history is given by Derek Gjertsen. (See this chapter's "Further Reading" section for more details.)

Newton used the second and third editions of the *Principia* to make corrections, improve certain sections, and insert new results. There is evidence that he planned a thorough revision, but he failed to carry out the sweeping changes that some people believe were needed to improve the flow of the book and make it easier to follow. My own view, as I explain later, is that, remarkably, the underlying structure of large parts of the *Principia* already has a quite modern feel about it; it was probably beyond the ageing Newton, with his other duties to consider, to carry through a whole rewrite.

Four editions of the Motte translation were published in New York by Daniel Adee in 1848, 1849, 1874, and 1885. In this book, I use the 1848 version, as published again in 1995 by Prometheus Books (see this chapter's "Further Reading" section). (There are copyright problems involved in using the Cajori or Cohen versions.) I have changed a few of the more archaic expressions and occasionally inserted a modern version or "translation" shown in brackets.

I will quote extensively from the *Principia* so that you get a flavor of Newton's own style. Although you can read this book without looking at the *Principia* itself, it would be useful if you have a copy of the *Principia* to actually look at and perhaps read a little more for yourself. There are many versions available, including that by Prometheus Books, costing very little. It is often possible to find copies in secondhand bookshops. An Internet search may turn up large parts of the book that can be downloaded for free.

3.2. THE STORY OF THE *PRINCIPIA*

My aim in this book is to look at the actual contents of the *Principia* rather than its history, but since the path that led to it is one of science's great stories, I will outline a few of the steps. (It is explored in detail by many historians of science—for example, see the 1980 book by Westfall.)

As a student, Newton naturally became interested in the great problems of the time, of which how to describe motion and how to account for the orbits of the planets were the most outstanding. We know Newton claimed that, in the plague year of 1666, he pondered motion and gravitation while at his home in Lincolnshire. It is likely that he continued on and off in those investigations, although optics and other interests took up much of his time. We know that Newton's correspondence with Robert Hooke in 1679 spurred his interest in dynamics, and he almost certainly gained more from Hooke's suggestions and probing than he would ever admit.

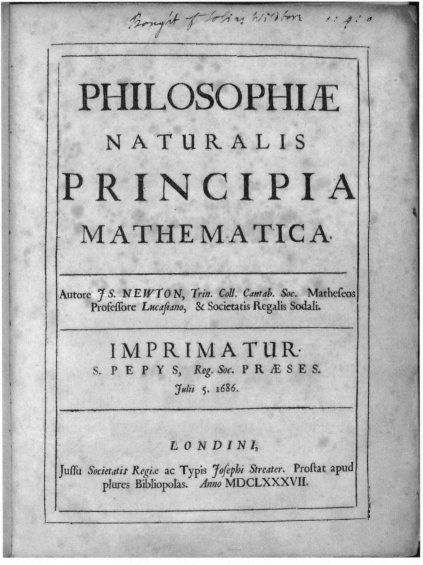

Figure 3.1. The title page of the first edition of the *Principia*. (Image reproduced by permission of the State Library of Victoria, Australia.)

Several people were coming to the conclusion that the force moving the planets around the Sun fell away inversely as the square of the distance of the planet from the Sun. Edmond Halley, Christopher Wren, and Robert Hooke discussed the matter in 1684 and agreed on a prize of a book worth forty shillings for the one who could demonstrate the link between the inverse-square law and the form of the planetary orbits. Hooke claimed to be able to do it but never produced his working. Thus it was that Halley had his momentous meeting with Newton at Cambridge in August 1684. According to one account:

> After they had been some time together, the Dr asked him what he thought the curve would be that would be described by the planets supposing the force of attraction towards the sun to be reciprocal to the square of their distance from it. Sr Isaac replied immediately that it would be an Ellipsis, the Dr struck with joy & amazement asked him how he knew it, Why saith he I have calculated it. Whereupon Dr Halley asked him for his calculations without any further delay. Sr Isaac looked among his papers but could not find it, but he promised him to renew it, & then send it to him.[1]

Given Newton's propensity for secrecy and his various squabbles, it is unlikely that he would immediately hand over any document to Halley. However, in November 1684, Newton did send Halley a document, *De motu corporum in gyrum (On the motion of bodies in an orbit)*. This paper was only nine pages long but showed how the inverse-square law and an elliptical path were linked. Nothing more was heard for over a year, but then, in April 1686, Halley was able to inform the Royal Society that *Book I* of the *Principia* had been received. The Society agreed on May 19 that it should be printed and that Halley should oversee the operation. (The *Principia* comprises three "*Books*"—see section 3.3, "Organization of the *Principia*.") There were various difficulties and upsets (not least of which was the claim by Hooke that it was really his work), but Halley managed to secure from Newton *Book II* on March 7, 1687, and *Book III* on April 4. Halley wrote to Newton on July 5 that he had "at last brought your book to an end."[2]

The role played by Halley was major and vital. He somehow managed to convince the ever-reticent Newton to complete his work and hand it over for publication. That must have been a most testing and exasperating task! However, Halley had to go even further, for the Royal Society did not have the funds at the time to publish the *Principia*; Halley himself covered the financial costs involved. Most people have heard of Halley's comet, but probably few realize his pivotal role in getting science's greatest book actually published.

Little is known about Newton's life during the period from 1684 to 1687, but during that time, the nine-page *De motu* was converted into the weighty *Philosophiae Naturalis Principia Mathematica* (460 pages in manuscript form). The foundations of dynamics and its applications were set out basically in the form that we use today. It is a staggering achievement, and it is hard to grasp that one man could have produced so much in sustained effort over just two or three years. Of course, Newton had been thinking about mechanics and planetary motion for many years; his route to the *Principia* is fascinating and covered in detail in references given at the end of this chapter.

3.3. ORGANIZATION OF THE *PRINCIPIA*

The contents page for the third edition of the *Principia* is shown in figure 3.2. The basic structure is:

Prefaces   Definitions   Axioms, or Laws of Motion   Books I, II, and III.

The *Preface* Newton wrote for the first edition was important, and, because it sets out his general approach and philosophy, I will discuss it in detail in the next chapter. He wrote additional brief prefaces for the second and third editions in which he simply mentioned some of the changes and those who had assisted him. Roger Cotes helped with the revision and publication of the second edition and also wrote a long preface for it. There is also a poem, "Ode to Isaac Newton," supplied by Edmond Halley.

The *Principia* begins with *Definitions*, and Newton deals with the basic concepts needed when discussing motion and dynamics. In *Axioms, or Laws of Motion*, Newton states his celebrated laws of motion and gives a general discussion of them and some immediate consequences. Now he has the basics in place for the study of dynamics.

It is in *Book I*, called *"Of the Motion of Bodies,"* that Newton explores the theory of dynamics. He begins by describing the mathematical methods to be used. Then he builds up from problems involving a single particle acted on by a force to successively more-involved situations, such as those with two or more interacting particles and the motion of bodies formed from many particles. Newton stresses that this is a theoretical development, and he expands the formalism and provides examples without worrying about physical applicability. To me, in places in *Book I*, Newton is the modern "theorist at play," but, as we shall see later, his deliberate delineation into theory and practice was an important methodological approach.

Book II is called *"Of the Motion of Bodies (continued)."* Newton now shows how to build in such things as air resistance and its effect on projectile and pendulum motion. He thereby starts to describe effects of continua, so that he can consider how his dynamics can apply to waves, fluids, and the propagation of sound. In this way, he also has the background for analyzing Descartes's vortex theory for planetary motion. In *Book II*, Newton describes more experimental work, so it is more of a mixture of theory and application.

It is in *Book III* that Newton moves from the mathematical formalism of the first book to astronomical science. He first sets out the *Rules for Reasoning*, which embody a framework for doing science—a framework that still makes sense today. He begins by presenting data about the solar system and then blends theory and phenomena to develop the theory of gravity. This is Newton's great triumph, as he explains how the solar system works, discusses the motion of the Moon and the paths of comets, derives the shape of the Earth, and gives the origin of terrestrial phenomena like tides. The book concludes with a *General Scholium*, in which Newton reviews the achievements of *Book III*, contrasts his approach with that of Descartes, and comments on the strengths and problems in his work. It is only in this *General Scholium* that Newton allows himself to delineate the role of God in his description of the universe.

THE AUTHOR'S PREFACE
DEFINITIONS
AXIOMS, OR LAWS OF MOTION

BOOK I. OF THE MOTION OF BODIES
SECTION
    I. *Of the method of first and last ratios, by the help whereof we demonstrate the propositions that follow*
    II. *Of the invention of centripetal forces*
    III. *Of the motion of bodies in eccentric conic sections*
    IV. *Of the finding of elliptic, parabolic and hyperbolic orbits, from the focus given*
    V. *How the orbits are to be found when neither focus is given*
    VI. *How the motions are to be found in given orbits*
    VII. *Concerning the rectilinear ascent and descent of bodies*
    VIII. *Of the invention of orbits wherin bodies will revolve, being acted upon by any sort of centripetal force*
    IX. *Of the motion of bodies in moveable orbits; and of the motion of the apsides*
    X. *Of the motion of bodies in given superficies, and of the reciprocal motion of funependulous bodies*
    XI. *Of the motion of bodies to each other with centripetal forces*
    XII. *Of the attractive forces of spherical bodies*
    XIII. *Of the attractive forces of bodies which are not of a spherical figure*
    XIV. *Of the motion of very small bodies when agitated by centripetal forces tending to the several parts of any very great body*

BOOK II. OF THE MOTION OF BODIES (continued)
SECTION
    I. *Of the motion of bodies that are resisted in the ratio of velocity*
    II. *Of the motion of bodies that are resisted in the duplicate ratio of their velocities*
    III. *Of the motion of bodies that are resisted partly in the ratio of the velocities, and partly in the duplicate of the same ratio*
    IV. *Of the circular motion of bodies in resisting mediums*
    V. *Of the density and compression of fluids; and of hydrostatics*
    VI. *Of the motion and resistance of funependulous bodies*
    VII. *Of the motion of fluids and the resistance made to projected bodies*
    VIII. *Of the motion propagated through fluids*
    IX. *Of the circular motion of fluids*

BOOK III. THE SYSTEM OF THE WORLD

RULES OF REASONING IN PHILOSOPHY
PHENOMENA, OR APPEARANCES
PROPOSITIONS
OF THE MOTION OF THE MOON'S NODES
GENERAL SCHOLIUM

Figure 3.2. Contents of the *Principia* (3rd edition).

3.3.1. DETAILED STRUCTURE

As figure 3.2 shows, the *Books* are divided up into *Sections*, each with a descriptive title. The basic structure of each section is a series of main results called *Propositions*. Often a *Proposition* is followed by one or more *Corollaries*, which contain results following from and complementing it. The central results in *Propositions* are also called *Theorems*, and the applications or special cases are called *Problems*. So, for example, we have things like "*Proposition III. Theorem III.*" and "*Proposition VIII. Problem III.*" Mixed in with the *Propositions* are *Lemmas*, which usually present additional mathematical results that are needed as the theory is developed.

Groups of *Propositions* are followed by a *Scholium*, or discussion, and it is in these that Newton expands on the more formal presentation with a variety of ideas and examples that help the reader to understand and appreciate what has just been done. Some of the scholiums are vital for the conceptual development of the work and have become famous for highlighting difficulties and identifying areas of concern that have had a profound effect on the growth of physics.

3.4. STYLE OF THE *PRINCIPIA*

The full title of the book, *Philosophiae Naturalis Principia Mathematica*, extends Descartes's *Philosophiae Principia* to emphasize that Newton is giving a far more mathematical exposition than Descartes had attempted.

It comes as no surprise to find that Newton, as a follower of Galileo, wrote a mathematical book. However, the modern reader will not find the extensive use of symbols, algebra, and equations that characterize modern books on dynamics. Rather, we find something more like Euclid's writing from over two thousand years ago. There are many diagrams with labeled points and lines, and arguments related to those figures are presented in prose form. Nor is the *Principia* written in the colloquial style employed by Galileo, rather it is more in keeping with the works of Euclid, Archimedes, and Huygens.

If the modern reader struggles with Newton's somewhat archaic mathematical approach, his contemporaries were, for the most part, totally bewildered, and very few of them could fight their way through the whole book. One approach was that followed by Newton's friend, the philosopher John Locke, who absorbed the thrust of the arguments and then checked with an expert, Christiaan Huygens, that the mathematical details were correct. Newton obviously recognized that dynamics could properly proceed only in a mathematical form, but his uncompromising approach was also linked to his dislike of criticism and controversy. As William Derham reported after talking with Newton:

> And for this reason, mainly to avoid being baited by little Smatteres in Mathematics, he told me, he designedly made his *Principia* abstruse; but yet so as to be able to be understood by Mathematicians, who he imagined, by comprehending his Demonstrations would concur with him in his Theory.[3]

I will say more about Newton's sensitivities and mathematical disputes in chapter 7.

The legendary mathematical difficulty of the *Principia* can overshadow the totality of the work (something I hope to combat for readers of this book). The *Principia* sets out ideas, concepts, methods, and a formalism for dynamics. It creates a path through the subject of dynamics that is in large part exactly the one we follow today. To a certain extent, there must be a mathematical setting, but the mathematics provides a language and a tool, so that it plays an essential but ancillary part in the whole exercise.

Much of the *Principia*, then, is presented in prose form. Unfortunately some parts are still demanding, and we can feel Newton struggling to explain his thinking and to present key conceptual points. The eminent Newtonian scholar D. T. Whiteside gave a blunt assessment, but it was also an excuse: "The logical structure is slipshod, its level of verbal fluency none too high, its arguments unnecessarily diffuse and repetitive and its very content on occasion markedly irrelevant to its professed theme. . . . But these faults can largely be excused by the very rapidity with which the *Principia* was written."[4]

Even Newton recognized some of this when he wrote in his *Preface* to the first edition of the *Principia*:

Some things, found out after the rest, I chose to insert in places less suitable, rather than change the number of the propositions and the citations. I heartily beg that what I have done may be read with forebearance.

We should remember that the *Principia* was written in a short period over three hundred years ago and it contains material that was largely completely new and being presented for the first time. It seems unfair to expect brilliant writing as well as brilliant science, especially in a book produced long before our era of word processors and expert supporting editorial processes. Nevertheless, it is hard to imagine Sir Isaac readily submitting to the stringent editorial review process used today!

3.5. MAJOR ACHIEVEMENTS

As we saw in the previous chapter, until Newton's time, most results were kinematic in form. Kepler gave his three laws to define and characterize the motion of the planets. Galileo showed how projectile motion on Earth corresponded to parabolic paths. There was no underlying causal theory; Kepler gave three independent laws, and certainly there was no connection between Galileo's parabolas and Kepler's ellipses. As Einstein pointed out, "We have to realize that before Newton there existed no self-contained system of physical causality which was somehow capable of representing any of the deeper features of the empirical world."[5]

Newton introduced the conceptual framework for dynamics; in particular, he introduced forces and a theory to tell how bodies in motion respond to the forces acting on them. There is also an important shift in viewpoint resulting from Newton's approach. Results such as Kepler's elliptical orbits and Galileo's parabolic projectile trajectories concern whole-of-path properties. Newton's laws tell us what happens at a given point, how the current motion and the impressed forces dictate the immediately following motion. In mathematical language, Newton gave us differential laws that we integrate to find the whole motion. And Newton showed the way this theory works for a whole array of forces and conditions.

Newton introduced a force to describe the effects of gravity and could then show the common origin of Kepler's three laws. He could also take the monumental step of showing how the motion of bodies on Earth, like pendulums and Galileo's projectiles, are linked to the motion of bodies like planets and comets in the heavens. The enormous step was made from specific examples of paths of motion to a universal and unifying theory capable of dealing with all cases. Newton recognized the significance of that step, and we shall see that he carefully set down the assumptions and tests needed to make it.

To make those advances, Newton also devised a program of working that many have followed ever since. The great Newtonian scholar I. Bernard Cohen called it the "Newtonian Style." Newton began by setting out a general mathematical theory of dynamics, and that allowed him to do two things: First, he could obtain results like the conservation of momentum Law, which are general and independent of specific force details. Second, he could assume a variety of mathematical forms for the force and theoretically explore the resulting motions. With that mathematical formalism in place, he could turn to physical phenomena and the results of observations and experiments in order to find the appropriate forces for describing actual physical motions. This split allowed Newton to match theory and experiment in a major way unseen before his time. In Cohen's words, "Newton's *Principia* is a book of mathematical principles applied to nature insofar as nature is revealed by experiment and observation. As such, it is a treatise based on evidence. Never before had a treatise on natural philosophy so depended on an examination of numerical predictions and numerical evidence."[6]

It is interesting to note that Einstein begins a famous essay, "The Theory of Relativity," with a statement echoing Newton's position:

> Mathematics deals exclusively with the relations of concepts to each other without consideration of their relation to experience. Physics too deals with mathematical concepts; however, these concepts attain physical content only by the clear determination of their relation to the objects of experience.[7]

Newton's split of his program into a theoretical, mathematical part followed by an application to natural phenomena also allowed him to deal with a philosophical problem that caused him great discomfort, as we shall see right from his initial *Preface* to the first edition to the concluding *General Scholium*. In the previous chapter, we saw that the Scientific Revolution was underpinned by three guiding principles: (1) mathematics must be used to describe the physical world, (2) experiments should be used to explore the physical world, (3) and the **mechanical philosophy** suggests that all is to be explained in terms of matter and motion. The latter principle was designed to remove mysterious causes and "occult effects" from consideration. Newton embraced these principles and introduced forces into the mechanical philosophy.

But now came the problem; a force is not a simple or physically obvious entity causing interactions. Contrast that with Descartes's use of continuous matter and effects transmitted through contacts and collisions. The nature of force, the question of physical mechanisms, was a whole new physical and philosophical problem. Newton planned to describe forces using a mathematical formula that can be inserted into the laws of motion so that the paths of bodies in motion under the influence of those forces can be determined. This he insists is a valid mathematical exercise. Beyond that, he also insists that the formalism so developed provides a useful way to describe and study natural phenomena. Dealing with those who found it hard to go beyond the origin-of-forces philosophical problem was a major task for Newton. The fact that he succeeded, and that the *Principia* came to be such a major step forward for science, is one of his greatest achievements.

3.6. SUMMARIZING *PRINCIPIA*

Newton introduced **the concepts and mathematical formalism** for a **general theory of dynamics**. One of his most influential critics, the mathematician and philosopher Ernst Mach (1838–1916) was quite clear about Newton's role:

He completed the formal enunciation of the mechanical principles now generally accepted. Since his time no essentially new principle has been stated. All that has been accomplished in mechanics since his day has been a deductive, formal, and mathematical development of mechanics on the basis of Newton's laws.[8]

That is a rather extreme point of view, and later (chapter 28) I will discuss the ways in which Newton's work had to be extended for a complete system of dynamics. However, it is true that Newton gave us the basis for dynamics. Arnold Sommerfeld puts it nicely in the introduction to his classic text *Mechanics*:

We commence with Newton's fundamental analysis in his "Philosophiae Naturalis Principia Mathematica" (London, 1687); not that Newton lacked important predecessors, such as Archimedes, Galileo, Kepler, and Huygens, to mention only a few. It was, nevertheless, Newton who first created a firm foundation for general mechanics. Even today, apart from some changes and reinforcements, the foundation laid down by him provides us with the most natural and didactically simplest approach to general mechanics.[9]

Newton's theory is **general**. It applies to discrete bodies such as the weight on the end of a pendulum, a projectile like a cannonball, and a planet orbiting the Sun. Beyond that, Newton indicated that it could extend to liquids and gases, and to phenomena like water waves and sound waves.

Newton introduced the idea of a **universal** theory; his **theory of gravity** applies to all bodies whether they are on the Earth or in the heavens. The success of this approach changed our thinking about the nature of the universe as a whole; we now accept that the physics we discover on Earth can be applied across the whole universe.

Finally, Newton insists that **the theory must be matched to experimental data and observations** in order for it to become part of physics rather than mathematics.

3.7. FURTHER READING

Cajori, Florian. *Newton's* Principia: *Motte's Translation Revised.* Berkeley: University of California Press, 1934.

Cohen, I. Bernard. *The Birth of a New Physics.* New York: Norton, 1985.

Einstein, Albert. "The Mechanics of Newton and Their Influence on the Development of Theoretical Physics." In *Ideas and Opinions.* New York: Crown Publishers, 1954.

Fauvel, J., R. Flood, M. Shortland, and R. Wilson, eds. *Let Newton Be!* Oxford: Oxford University Press, 1988. [ch. 2 by John Roche is an excellent short introduction to the *Principia*]

Gjertsen, Derek. *The Newton Handbook.* London: Routledge and Kegan Paul, 1986. [an invaluable reference for all things Newtonian— including good sections on the *Principia*]

Newton, Isaac. *Isaac Newton: The* Principia. Translated by I. Bernard Cohen and Anne Whitman. Berkeley: University of California Press, 1999. [this is a completely new translation; also contains the comprehensive and invaluable "A Guide to Newton's *Principia*" by I. Bernard Cohen]

———. *The* Principia. Amherst, NY: Prometheus Books, 1995.

Smith, George E. "Newton's *Philosophiae Naturalis Principia Mathematica.*" In *Stanford Encyclopedia of Philosophy.* Stanford University Metaphysics Research Lab. Article first published December 20, 2007. http://plato.stanford.edu (accessed May 2, 2013).

Sommerfeld, Arnold. *Mechanics: Lectures on Theoretical Physics.* Vol. 1. New York: Academic Press, 1964. First published 1942.

Westfall, Richard. *The Construction of Modern Science.* Cambridge: Cambridge University Press, 1977.

———. *Never at Rest: A Biography of Isaac Newton.* Cambridge: Cambridge University Press, 1980. [ch. 10 is a wonderful introduction to Newton's book]

PART 2

HOW THE
PRINCIPIA BEGINS

Newton begins with the *Preface* in which he sets out the various philosophical and methodological points that guide his work. Next come definitions of mass and other things needed in mechanics, plus a discussion of space and time. Now Newton is ready to give his laws of motion and to set out their immediate consequences, such as the principle of conservation of momentum. Finally, he sets out the mathematical methods he will use to fully develop the theory of mechanics and its applications.

Chapter 4. NEWTON'S *PREFACE*: SETTING
OUT METHODS AND AIMS
Chapter 5. FUNDAMENTALS
Chapter 6. NEWTON'S LAWS OF MOTION AND THEIR
IMMEDIATE CONSEQUENCES
Chapter 7. MATHEMATICAL METHODS

Reading plans: This is where Newton explains the basis for his whole work, so it is really essential reading. If your mathematical background is not too strong, you might lightly read over section 6.6, where I introduce the modern formulation, but I do suggest you try just to see how the symbolic form flows. Chapter 7 gets technical only in sections 7.3 and 7.4, but they can be lightly scanned without too much loss.

NEWTON'S *PREFACE*
SETTING OUT METHODS AND AIMS

Newton wrote The Author's Preface *for the first edition of
the* Principia, *and it remained in later editions. The writing
was completed at Cambridge, Trinity College, May 8, 1686.
Newton was apparently satisfied with his introduction because
his supplemental prefaces to the second and third editions are
merely short comments on minor changes and additions to the
book. (Roger Cotes wrote a separate fourteen-page preface for
the second edition.)*

*In around two pages, Newton sets out his methods, phil-
osophical approach, and overall aims. It is remarkable that
many of the ideas and concepts that Newton introduces in this
preface are those that have guided the work in science since
then and remain of importance today.*

4.1. NEWTON'S SCIENTIFIC MANIFESTO

Mathematical Framework

The *Preface*, and hence the *Principia*, opens with the following
sentence:

*Since the ancients (as we are told by Pappus) made great account of the
science of mechanics in the investigation of natural things; and the moderns,
laying aside [rejecting] substantial forms and occult qualities, have endeav-
ored to subject the phenomena of nature to the laws of mathematics, I have in
this treatise cultivated mathematics as far as it regards to philosophy.*

Newton is paying his respects to the ancient founding fathers of mathematics and science (called "natural philosophy" or "philosophy" in Newton's time), but he is telling us that he will follow those modern ideas underpinning the Scientific Revolution, as discussed in chapter 2, section 2.5. In particular, he is stressing the importance of mathematics.

Newton specifically notes the rejection of "occult qualities," and this takes us back to Aristotle and the insistence on causes for everything. Newton introduced forces, and challenging questions about the possible underlying physical processes and mechanisms were raised by his critics. This was a great worry for Newton; he deals with the lack of such things at various places in the book and finally in the great *General Scholium* with which he concludes the *Principia*. For some people, forces that were not related to contact between bodies and collisions but which operated over a distance still seemed to involve the forbidden "occult qualities."

The Two Aspects of Mechanics

Newton next considers the nature of his subject:

> The ancients considered mechanics in a twofold respect: as rational, which proceeds accurately by demonstration; and practical. To practical mechanics all the manual arts belong, from which mechanics took its name.

Today we still make such distinctions and talk about motor mechanics, for example.

Mechanics and Geometry

Newton sees a strong, natural link between geometry and mechanics, which I detail in section 4.5. For us, it is important to recognize that Newton saw geometry as the branch of mathematics to use when formulating a theory of dynamics. More details of this will come in chapter 7.

Newton slips in a statement showing the deep impression Euclid's geometry has made on him. Today most of us who see the edifice that

Euclid erected on just five postulates using simple rules of logic find ourselves similarly amazed and would agree with Newton when he writes that *"it is the glory of geometry that from those few principles, brought from without, it is able to produce so many things."*

The style of the *Principia* is clearly modeled on that of Euclid's *Elements*, and now we are in the position of admiring the great development that Newton too constructs on the basis of a relatively few stated definitions and axioms. We might say that Newton's *Principia* is to mechanics what Euclid's *Elements* is to geometry.

Newton sums up and says how geometry relates to mechanics:

> *Therefore geometry is founded in mechanical practice, and is nothing but that part of universal mechanics which accurately proposes and demonstrates the art of measuring. But since the manual arts are chiefly conversant [used] in the moving of bodies, it comes to pass that geometry is commonly referred to their magnitudes, and mechanics to their motion.*

Defining Rational Mechanics

After that somewhat difficult introduction to mechanics and geometry, Newton is ready to define his subject:

> *In this sense rational mechanics will be the science of motions resulting from any forces whatsoever, and of the forces required to produce any motions, accurately proposed and demonstrated.*

Notice that Newton says *"any forces whatsoever"*; he is aiming for complete generality. He makes it clear that he is going beyond the ancient world of levers and pulley systems:

> *This part of mechanics was cultivated by the ancients in the five powers which relate to manual arts, who considered gravity (it not being a manual power) no otherwise than as it moved weights by those powers. Our design, not respecting arts, but philosophy, and our subject not manual but natural powers, we consider chiefly those things which relate to gravity, levity, elastic force, the resistance of fluids, and the like forces, whether attractive or impulsive.*

Newton is to deal with motion as it occurs in all ways, as produced by man and as it is found in the natural world, and as it is generated or resisted by a whole range of forces. This is not a practical manual, but "*I offer this work as the mathematical principles of philosophy.*"

Newton's Methodology

Newton now gives a concise and remarkably modern view of the way to proceed:

> For all the difficulty of philosophy [what we today might call "natural philosophy" or "science"] *seems to consist in this—from the phenomena of motions to investigate the forces of nature, and then from these forces to demonstrate the other phenomena.*

That is a basic approach in science; that is, study certain phenomena and produce theories to explain them, then use those theories to predict and understand a wider range of phenomena. Recognizing this approach is vitally important for Newton, as we will see. In *Book III*, Newton returns to general questions of methodology and sets out rules for carrying out his program.

Applied in the *Principia*

The *Principia* is organized so that theories are first developed and then they are applied to a dazzling range of phenomena. Thus Newton continues the above methodological quote:

> And to this end the general propositions in the first and second books are directed. In the third book we give an example of this in the explication of the System of the World; for by the propositions mathematically demonstrated in the former books, we in the third derive from the celestial phenomena the forces of gravity with which bodies tend to the sun and the several planets. Then from these forces, by other propositions which are also mathematical, we deduce the motion of the planets, the comets, the moon, and the sea.

The Method in General

Newton knew that his approach was wonderfully successful for gravitational problems, and he believed that it is the one to be used in all other cases. He goes on:

> *I wish we could derive the rest of the phenomena of Nature by the same kind of reasoning from mechanical principles; for I am induced by many reasons to suspect that they may all depend upon certain forces by which the particles of bodies, by some causes hitherto unknown, are either mutually impelled towards each other, and cohere in regular figures, or are repelled and recede from each other; which forces being unknown, philosophers have hitherto attempted the search of Nature in vain; but I hope the principles here laid down will afford some light either to this or some truer method of philosophy.*

It is this suggested plan of action to which Nobel laureate Steven Weinberg is referring in the quote from this book's preface ("all that has happened since 1687 is a gloss on the *Principia*").

Newton is again confirming his support for the mechanical philosophy or hypothesis, but now he couches it in terms of forces yet to be discovered. Of course, his mathematical approach and theory-fitting ideas are the basis for applying that hypothesis. We will see examples in the *Principia* where Newton does tackle other than gravitational problems; the last three hundred years have shown that he was correct in his assumption that he had identified the approach to be used in general.

Notice that in his statement "*particles of bodies . . . are repelled and recede from each other,*" we have the first hint of Newton's acceptance of an atomic hypothesis and particles moving in a void.

It is fascinating to find, about three hundred years after Newton wrote those words, Nobel Prize–winner and expositor supreme Richard Feynman writing in *The Feynman Lectures*:

> If, in some cataclysm, all of scientific knowledge were to be destroyed, and only one sentence passed on to the next generation of creatures, what statement would contain the most information in the fewest words? I

believe it is the atomic hypothesis (or atomic fact, or whatever you wish to call it) that all things are made of atoms—little particles that move around in perpetual motion, attracting each other when they are a little distance apart, but repelling on being squeezed into one another. In that one sentence, you will see, there is an enormous amount of information about the world, if just a little imagination and thinking are applied.[1]

Newton's speculations have become today's orthodoxy.

4.2. PUBLICATION AND PAYING TRIBUTE TO HALLEY

It is heartening to see that Newton next gives a glowing and unequivocal thank-you to Edmond Halley for his part in giving us the *Principia*:

> In the publication of this work the most acute and universally learned Mr. Edmund[2] Halley not only assisted me with his pains in correcting the press and taking care of the schemes [geometrical figures], but it was to his solicitations that its becoming public is owing; for when he had obtained of me my demonstrations of the figures of the celestial orbits, he continually pressed me to communicate the same to the Royal Society, who afterwards, by their kind encouragement and entreaties, engaged me to think of publishing them.

Newton goes on to explain how turning to other problems and delaying publication allowed him to broaden his work into the comprehensive masterpiece it finally became:

> But after I had begun to consider the inequalities of the lunar motions, and had entered upon some other things relating to the laws and measures of gravity and other forces; and the figures that would be described by bodies attracted according to given laws; and the motion of several bodies moving among themselves; the motion of bodies in resisting mediums; the forces, densities, and motions, of mediums; the orbits of comets, and such like, we deferred that publication till I had made a search into those matters, and put forth the whole together.

In the previous chapter, I mentioned that certain people have criticized the order of some things in the *Principia* and that even Newton was humble enough to recognize that there are a few organizational problems:

> *What relates to the lunar motions (being imperfect), I have put all together in the corollaries of Proposition 66, to avoid being obliged to propose and distinctly demonstrate the several things there contained in a method more prolix than the subject deserved, and interrupt the series of the several [other] propositions. Some things, found out after the rest, I chose to insert in places less suitable, rather than change the number of propositions and the citations.*

Presumably it was just too much trouble to renumber everything to neatly insert the very final discoveries.

A Final Plea

The last sentence gives us a picture of an unusually humble Newton:

> *I heartily beg that what I have here done may be read with candour; and that the defects in a subject so difficult be not so much reprehended as kindly supplied, and investigated by new endeavours of my readers.*

This may refer in part to the organizational defects he mentioned, but for me it seems to entail more than that. Newton must have been aware that some of his ideas (such as gravitational forces attracting bodies over large distances without any specified underlying mechanism) were certain to encounter criticism. Knowing his dislike of disputes and controversy, it seems likely that he was making a plea for a fair hearing and a thorough evaluation of his ideas.

4.3. DISCUSSION

Newton's *Preface* tells us what to expect in the *Principia* and introduces approaches and concepts that became the basis for science from then on. In that sense, it is a remarkable document.

But what of the style and feel of the *Preface* itself?

Newton's *Preface* is very matter-of-fact and almost impersonal. There is no dedication or recognition of the personal struggles that must have been involved in writing the *Principia*. Perhaps the character traits outlined in chapter 1 should tell us not to expect such things. A dedication of the type *To the Glory of God* might have been appropriate, given his deep religious feelings, but again the secretive and private nature of his beliefs might have stood in the way. In fact, it is surprising that God is not mentioned in the third-edition *Principia* until the final *General Scholium*, in which we find a major and forceful statement thereon.

We do glimpse a human side of Newton when he speaks about Edmond Halley. However, in contrast to many of today's books, there is no mention of others who have given him support and helped with the development of his ideas. Unsurprisingly, his rival and sparring partner Robert Hooke is not mentioned, but then neither are people like Christopher Wren, Robert Boyle, John Flamsteed, and Christiaan Huygens, nor Isaac Barrow and his old teachers and others living at Trinity College. However, knowing the complex nature of the man, as discussed in chapter 1, perhaps this is just the kind of preface we should expect him to have written.

4.4. FURTHER READING

Garrison, James W. "Newton and the Relation of Mathematics to Natural Philosophy." *Journal of the History of Ideas* 48 (1987): 609–27.

Gjertsen, Derek. *The Newton Handbook*. London: Routledge and Kegan Paul, 1986. [an invaluable reference for all things Newtonian—good sections on the *Principia* and Newton's *Preface*]

4.5. APPENDIX: MECHANICS, GEOMETRY, THEORY, AND APPLICATIONS

We now come to a somewhat technical matter. Newton talks about accuracy and what we might term the theory-application distinction. He introduces the importance of geometry and its relevance for mechanics:

> But as artificers do not work with perfect accuracy, it comes to pass that mechanics is so distinguished from geometry, that what is perfectly accurate is called geometrical; what is less so, is called mechanical. But, the errors are not in the art, but in the artificers. He that works with less accuracy is an imperfect mechanic; and if any could work with perfect accuracy, he would be the most perfect mechanic of all, for the description of right lines and circles, upon which geometry is founded, belong to mechanics.

Newton is now on to the subtle point that geometry tells us about ideal geometrical figures, but it does not require us to actually produce those ideals. This is the link between practical matters and the ideal theory that was discussed by Plato and, much later, was covered by Hermann von Helmholtz in his famous essay "On the Origin and Significance of Geometrical Axioms." Einstein also discussed this when he suggested that geometry can be taken as the first example of theoretical physics. The problem of constructing actual geometrical figures is part of practical mechanics, but the theory tells us what to expect in those figures, as Newton explains:

> Geometry does not teach us to draw these lines, but requires them to be drawn; for it requires that the learner should first be taught to describe these accurately before he enters upon geometry; then it shows how by these operations problems may be solved. To describe right lines and circles are problems, but not geometrical problems. The solution of these problems is required from mechanics, and by geometry, the use of them, when so solved, is shown.

As we get farther into the *Principia*, we will see Newton setting out the problems of dynamics in geometric form and then showing how the methods of geometry lead to solutions.

5

FUNDAMENTALS

Some readers might expect the *Principia* to begin with the famous laws of motion. However, before setting out the laws, Newton needs to deal with some **fundamental conceptual matters**, including the still-puzzling nature of space and time. In introducing these and in the subsequent discussions, Newton actually gives us his (and now our) theory of motion. The mathematical treatment follows later in *Book I*, with the applications in *Books II* and *III*.

The *Principia* opens with *Definitions*, and, over twelve pages, Newton introduces basic things needed for dynamics and discusses their nature and importance. Aspects of some of these matters can be found in the writings of Kepler, Galileo, Descartes, and Huygens, but it was Newton who finally produced the clear basics for dynamics. The strength of his achievement can be gauged by the fact that these matters are just as relevant—and in parts as controversial—today as they were when Newton first wrote them down. Newton begins with what has turned out to be one of the cornerstones of modern physics: mass.

5.1. MASS

The *Principia* opens with:

> DEFINITION I. *The quantity of matter is the measure of the same, arising from its density and bulk conjointly* [density multiplied by volume].

In the discussion that follows, Newton says, "*It is this quantity that I mean hereafter everywhere under the name of body or mass.*" Today we just say, "mass." The 1979 *Concise Dictionary of Physics* (from Pergamon Press) still echoes Newton: the definition of *mass* is listed as "the quantity of matter in a body."[1]

Often, Newton has been criticized for the circularity of that definition; how can the density be used when it is found using the mass of a given volume of matter? But reading on (and did the critics actually read the little discussion?), it is clear that the definition is framed so that Newton can establish two important properties of quantity of matter or mass. He goes on to say:

> Thus air of a double density, in a double space, is quadruple in quantity; in a triple space, sextuple in quantity. The same thing is to be understood of snow, and fine dust or powders, that are condensed by compression or liquefaction, and of all bodies that are by any causes whatever differently condensed.

Thus the mass is linear in density and volume, so increasing either by a given factor will give a body with the mass increased by that factor. Also, we may change the density by processes like compression and liquefaction, and the mass does not change. Even changing the state of matter, from solid to liquid, for example, does not change the mass. These are key properties of mass.

Newton is using an atomic theory of matter, as he makes clear when he adds, "*I have no regard in this place to a medium, if any there such is, that freely pervades the interstices between the parts of bodies.*" There is no contribution from an aether. The particles of matter may be rearranged and moved closer together or farther apart without changing the value of the mass. (For Newton, there are particles or bodies in space; he considers space later.)

Now Newton comes to the question of how mass is to be measured:

> And the same is known by the weight of each body; for it is proportional to the weight, as I have found by experiments on pendulums, very accurately made, which shall be shown hereafter. [The details are in Book III of the Principia.]

There is a connection between mass and weight. **After just a few lines, Newton tells us about an idea, and its experimental verification,**

that was to have a profound effect on Einstein in setting the course of modern physics. I will return to this in a moment and also later, when discussing pendulum theory in *Book II* and experimental results in *Book III*.

5.2. MOMENTUM

Having introduced mass, Newton can now use it to define what we call momentum:

> DEFINITION II. *The quantity of motion* [momentum] *is the measure of the same, arising from the velocity and quantity of matter conjointly* [velocity multiplied by mass].

If there is more than one particle involved, then "*the motion of the whole is the sum of the motions of all the parts.*" Increasing either mass or velocity increases momentum or, as Newton says, "*Therefore in a body double in quantity, with equal velocity, the motion is double; with twice the velocity, it is quadruple.*"

Momentum is a key quantity in physics, and it remains so even with relativistic or quantum modifications.

5.3. INERTIA

The concept of inertia and its role in mechanics was troublesome for those who first tried to come to grips with the nature of motion, its creation, and its destruction (see chapter 2). Newton gives the following:

> DEFINITION III. *The vis insita, or innate force of matter, is a power of resisting, by which every body, as much as in it lies, endeavours to persevere* [continues] *in its present state, whether it be of rest, or of moving uniformly forwards in a right* [straight] *line.*

Personally I do not like the use of the word *force* in this definition; it would better be *property* or *characteristic*. Because this concept had caused so much confusion, it is interesting to read Newton's explanation:

This force is always proportional to the body whose force it is and differs nothing from the inactivity of the mass, but in our manner of conceiving it. A body, from the inactivity of matter, is not without difficulty put out of its state of rest or motion. Upon which account this vis insita may, by a most significant name, be called vis inertiae [inertia] or force of inactivity. But a body exerts this force only when another force, impressed upon it, endeavours to change its condition.

Thus we are talking about **inertia**, which is related to the **mass** of the body and tells us how hard it is to change its state of motion. Notice that mass is now linked with inertia. In his 2008 book about the origin of mass, *The Lightness of Being*, Nobel-laureate physicist Frank Wilczek defines *mass* in his glossary as "a property of a particle or system, which is a measure of its inertia (that is, a particle's mass tells us how difficult it is to change the particle's velocity)."[2]

Surely that is a neat paraphrase of what Newton was intimating. We now speak about **inertial mass**. Newton has already mentioned weight, which we measure using the gravitational pull of the Earth. This leads to the concept of **gravitational mass**. It is the consideration of these two masses that so influenced Einstein (more of which, later).

The sophistication of Newton's thinking is evident in his final discussion point:

Resistance is usually ascribed to bodies at rest, and impulse to those in motion; but motion and rest, as commonly conceived, are only relatively distinguished; nor are those bodies always truly at rest, which commonly are taken to be so.

Newton is already introducing ideas of relativity as he recognizes fundamental matters concerning motion.

5.4. FORCES

We are now at the point where Newton introduces the concept that allows him to move beyond the essentially kinematic treatment of motion given by Kepler and Galileo to a full-blown theory of dynamics. In the discussions, Newton explains how motion works, paving the way for the formal laws and mathematical treatment in *Book I*, then their application in *Books II* and *III*. Newton gives:

> DEFINITION IV. *An impressed force is an action exerted upon a body, in order to change its state, either of rest, or of moving uniformly forward in a right* [straight] *line.*

Newton now makes it clear that a force acts on a body, after which it has a new state of motion:

> *This force consists in the action only, and remains no longer in the body when the action is over. For a body maintains every new state it acquires, by its inertia only. Impressed forces are of different origins, as from percussion, from pressure, from centripetal force.*

Together with what has been said about motion and inertia, **Newton has now contradicted the Aristotelian viewpoint: a force will change the state of motion, but then there is no need of a cause for maintaining that motion.** He next elaborates the concept a little more.

5.4.1. Centripetal Force

The kind of force of greatest importance is what we now refer to as a **central force,** but Newton uses the term *centripetal* (allegedly in honor of Huygens, who used the term *centrifugal force* when discussing the outward acceleration associated with circular motion).

> DEFINITION V. *A centripetal force is that by which bodies are drawn or impelled, or any way tend, towards a point as to a centre.*

Newton immediately points out that gravity is a centripetal force pulling things toward the center of the Earth and that the planets are also controlled by such a force. He uses the (now rather archaic-sounding) example of a stone in a sling that is kept whirling around by the force exerted through the string and that flies off when the user releases the sling. Generally:

> *The same thing is to be understood of all bodies, revolved in any orbits. They all endeavour to recede from the centre of their orbits; and were it not for the opposition of a contrary force which restrains them to, and detains them in their orbits, which I therefore call centripetal, would fly off in right [straight] lines, with an uniform motion.*

Similarly, a projectile is continuously drawn away from a straight-line path by the centripetal force of gravity. Newton now links projectile motion and orbits in a superb **thought experiment**. A diagram helps to follow the discussion, and figure 5.1 reproduces the one from Newton's "popular" version of *Book III* (see chapter 27).

> *If a leaden ball, projected from the top of a mountain by the force of gun-powder, with a given velocity, and in a direction parallel to the horizon, is carried in a curved line to the distance of two miles before it falls to the ground; the same, if the resistance of the air were taken away, with a double or decuple velocity, would fly twice or ten times as far. And by increasing the velocity, we may at pleasure increase the distance to which it might be projected, and diminish the curvature of the line which it might describe, till at last it should fall at the distance of 10, 30 or 90 degrees, or even might go quite round the whole earth before it falls; or lastly, so that it might never fall to earth but go forward into the celestial spaces, and proceed in its spaces in infinitum.*

Newton is linking the idea of projectile motion on Earth with orbits around the Earth (see figure 5.1) and even the modern concept of escape velocity, which ensures that a projectile or rocket will escape the Earth's gravity and fly off into space.

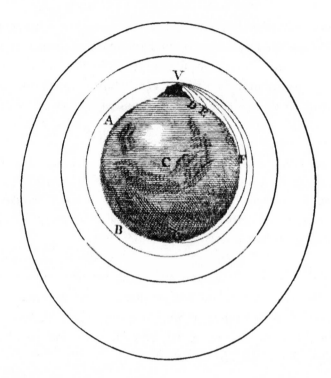

Figure 5.1. The fate of cannonballs shot off with varying speeds. (From Newton, *System of the World, Book III* of the *Principia*, in *The* Principia, trans. Andrew Motte [Amherst, NY: Prometheus Books, 1995].)

With these thoughts, Newton has now explained how a centripetal force can give orbital motion; at last, the old ideas of some power sweeping or pushing the planets around their orbits has been replaced. Newton has challenged the vortices theory of his rival Descartes and his *Principia Philosophiae* statements:

24. That the heavens are fluid
25 That the heavens carry with them all the bodies which they contain.

Newton uses the example of the Moon, telling us that it would fly off in a straight line were it not for the Earth continually pulling it back into Earth's orbit. All of this leads to following the challenge (answered, of course, in the *Principia*):

It is necessary, that the force be of a just quantity, and it belongs to the mathematicians to find the force that may serve exactly to retain a body in a given orbit with a given velocity; and vice versa, to determine the curvilinear way into which a body projected from a given place, with a given velocity, may be made to deviate from its natural rectilinear way, by means of a given force.

5.4.2. Measures of Force

Newton's theory of dynamics, in stark contrast to Descartes's, was designed to lead to a mathematical formalism and quantitative results, so he needs to say how "*the quantity of any centripetal force may be considered.*" He gives three more definitions:

DEFINITION VI. The absolute quantity of a centripetal force is the measure of the same, proportional to the efficacy of the cause that propagates it from the centre, through the spaces round about.

DEFINITION VII. The accelerative quantity of a centripetal force is the measure of the same, proportional to the velocity which it generates in a given time.

DEFINITION VIII. The motive quantity of a centripetal force is the measure of the same, proportional to the motion which it generates in a given time.

Newton discusses these with reference to magnetic and gravitational forces, and of course they are to be treated more mathematically when the laws of motion are given and used. During the course of these discussions, Newton makes clear his mathematical approach:

For I here design only to give a mathematical notion of those forces, without considering their physical causes and seats.

And again, a little later, even more emphatically:

I likewise call attractions and impulses, in the same sense, attractive and motive; and use the words attraction, impulse, or propensity of any sort

*towards a centre, promiscuously, and indifferently, one for another; consid-
ering those forces not physically, but mathematically: wherefore the reader
is not to imagine that by those words I anywhere take upon me to define the
kind, or manner of action, the causes or the physical reason thereof, or that
I attribute forces, in a true and physical sense, to certain centres (which are
only mathematical points); when at any time I happen to speak of centres as
attracting, or as endued with attractive powers.*

One can almost feel Newton struggling to counter the (inevitable)
criticism that he does not give physical mechanisms for the transmission
and effect of his forces—something he famously returns to in the very final
General Scholium of the *Principia*. **Remember that his aim in the first part
of the book is to present a mathematical formalism that later can be
applied to physical problems.**

Much has been written about Newton's definitions of force and the
subtleties involved; I return to this topic in chapter 6.

5.5. SPACE AND TIME

It might seem that we are ready to move on to the laws and their
applications, but in defining things as we have above, we have implicitly
made certain assumptions; for example, we need to know about position and
its change with time in order to discuss velocity. It is a measure of Newton's
thoroughness that he realizes this and adds a *Scholium*, which begins:

*Hitherto I have laid down the definition of such words as are less known, and
explained the sense in which I would have them to be understood in the following discourse. I do not define time, space, place, and motion, as being well
known to all.*

However, if those things are to be used in a theory of dynamics, there
are some further, more-subtle points that should be considered. The
discussion Newton gives is one of the most quoted and analyzed parts
of the *Principia*. It is also an indication of the sophistication and depth

of his thinking. Thus I must make some comments about this *Scholium*, but I refer the interested reader to the references given in the "Further Reading" section at the end of the chapter for greater detail.

The quote just given continues:

> *Only I must observe that the vulgar* [ordinary people] *conceive those quantities under no other notions but from the relation they bear to sensible objects. And thence arise certain prejudices, for the removing of which it will be convenient to distinguish them into absolute and relative, true and apparent, mathematical and common.*

I think it is fair to say that most people are happy to work using the everyday time and position specifications that are an integral part of our lives. Most scientists do that most of the time too (although, from the viewpoint of certain parts of advanced physics and philosophy, a deeper analysis is required). You may be one of those people, and that includes many scientists, who share the sentiment expressed long ago by essayist Charles Lamb: "Nothing puzzles me more than time and space; and yet nothing puzzles me less, for I never think about them."[3]

In that case, you may like to find out a little more about just what the puzzle entails, which we'll discuss in the following sections. A few more comments on connected practical matters will come in the next chapter.

5.5.1. Time

We all may feel that there is some overall time that "marches on," and we all measure our own local or private time using watches and clocks. Newton's words still seem convincing today:

> *Absolute, true and mathematical time, of itself, and from its own nature, flows equably without regard to anything external, and by another name is called duration: relative, apparent, and common time, is some sensible and external (whether accurate or unequable) measure of duration by the means of motion, which is vulgarly* [commonly] *used instead of true time; such as an hour, a day, a month, a year.*

It is our local measures of time that we use when studying motion (like Galileo's pulse counting, or his water clocks for balls rolling down inclined planes). In certain modern experiments, the passing of time may be measured with exquisite accuracy.

We do seem to have some natural "feel" for time, but for many of us, Saint Augustine's famous words probably still ring true: "What, then, is time? If no one ask of me, I know; if I wish to explain to him who asks, I know not."[4]

5.5.2. Space

Newton begins with what has become one of his most famous statements:

> *Absolute space, in its own nature, without relation to anything external, remains always similar and immovable. Relative space is some movable dimension or measure of the absolute spaces; which our senses determine by its position to bodies; and which is vulgarly [commonly] taken for immovable space; such is the dimension of subterraneous, an aereal, or celestial space determined by its position in respect of the earth.*

Why is Newton concerned with a possible **absolute space** when it is the **relative space** that we all use? Already in the above definitions and discussions, Newton has introduced the ideas of inertia and rest or, equivalently, the idea of motion in a straight line when no forces act (*"but motion and rest, as commonly conceived, are only relatively distinguished"*). As he makes even clearer when the laws of motion and their consequences are introduced (see the next chapter), the theory of mechanics uses what today we call an inertial frame of reference in which those properties of bodies to be at rest or in uniform motion are apparent. But are there any such reference frames (or relative spaces, as Newton might say)? One response is to say that we do use mechanics and that it works, so operationally we can use suitable reference coordinates; I will expand on that approach in the next chapter. Newton too is very clear on this point and its necessity:

But because the parts of space cannot be seen, or distinguished from one another by our senses, therefore in their stead we use sensible measures of them. For from the positions and distances of things from any body considered as immovable, we define all places; and then with respect to such places, we estimate all motions, considering bodies as transferred from some of those places into others. And so, instead of absolute places and motions, we use relative ones; and that without any inconvenience in common affairs.

The whole scheme might work well, but what if we think a little more deeply? Newton continues the above quote about using everyday measurements:

but in philosophical disquisitions, we ought to abstract from our senses, and consider things themselves, distinct from what are only sensible measures of them. For it may be that there is no body really at rest, to which the places and motions of others may be referred.

Newton is saying that there are (in our modern terminology) suitable and useable inertial frames of reference that allow us to use his theory of mechanics, but is there any independent way of being certain that such reference frames do truly exist? Before tackling that, it is worthwhile noting that Newton understands relative motion and makes it clear to the reader.

5.5.3. Relative Motion

Newton gives an example of dealing with relative motion involving a sailor on a moving ship and recognizing that, from some larger viewpoint, the Earth is also moving (see figure 5.2). Newton explains:

If the body [the sailor] *moves also relatively to the ship, its* [his] *true motion will arise, partly from the true motion of the earth* [remembering that larger viewpoint] *in immovable space, and partly from the relative motions as well of the ship on the earth, as of the body in the ship; and from these relative motions will arise the relative motion of the body on the earth.*

And to make things clear, Newton goes on with an example (see figure 5.2):

> *As if that part of the earth, where the ship is, was truly moved towards the east, with a velocity of 10010 parts; while the ship itself with a fresh gale, and full sails, is carried towards the west, with a velocity expressed by 10 of those parts; but a sailor walks in the ship towards the east, with 1 part of the said velocity; then the sailor will be moved truly in immovable space towards the east, with a velocity of 10001 parts, and relatively on the earth towards the west with a velocity of 9 of those parts.*

Figure 5.2. Relative velocities of sailor, ship, and Earth.

5.5.4. Absolute Motion and the Bucket Experiment

We still have the problem of the **immovable space**, although Newton raises the possibility of the so-called fixed stars providing the reference frame (something we still do today). However, "*absolute rest cannot be determined from the position of bodies in our regions.*" But although we cannot detect absolute space, we can look at motions and discuss their **relative** and **true** forms. To this end, Newton now gives an example in a simple but highly thought-provoking experiment. Basically, a bucket of water is made to rotate by suspending it with a twisted cord and the state of the water is observed. Here is Newton's famous description:

> *If a vessel, hung by a long chord, is so often turned about that the chord is strongly twisted, then filled with water, and held at rest together with the water; after, by the sudden action of another force, it is whirled about the contrary way, and while the chord is untwisting itself, the vessel continues for some time in this motion; the surface of the water will at first be plain [flat], as before the vessel began to move; but the vessel, by gradually communicating its motion to the water, will make it begin sensibly to revolve, and recede little by little from the middle, and ascend to the sides of the vessel, forming itself into a concave figure (as I have experienced), and the swifter the motion becomes, the higher will the water rise, till at last, performing its revolutions in the same times with the vessel, it becomes relatively at rest in it.*

Newton is identifying the three stages illustrated in figure 5.3. Stage (a) has the bucket at rest and the water lying in it with a flat surface; in (b), the untwisting cord makes the bucket rotate, and the water surface initially remains flat; in (c), the water and the bucket are both rotating together, and the water has formed a concave surface.

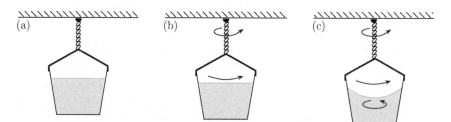

Figure 5.3. Stages in the rotating-bucket experiment.

What should we make of this? Newton points out that in both stage (a) and stage (c), **the water is at rest relative to the bucket,** but we can observe a quite **different** surface. He concludes:

> *This ascent of the water shows its endeavour to recede from the axis of its motion; and the true and absolute circular motion of the water, which is here*

directly contrary to the relative, discovers itself, and may be measured by this endeavour.

The water is not rotating relative to the bucket, but we have detected that it is finally rotating around some axis in space; we have learned about "*the true and absolute circular motion of the water.*" Newton continues with a long discussion, and interested readers may wish to study it and come to their own conclusions. This bucket experiment is discussed in many books and papers, and at the end of the chapter, I give references to some more-recent comprehensive analyses of the experiment and the types and validity of any conclusions that may be drawn from it. One thing is clear: this is an incredibly ingenious experiment, the sort of experiment that only an Isaac Newton would suggest!

5.5.5. The Globes Thought Experiment

In the bucket experiment, we may think about the motion in a larger sense, with the fixed stars proving a suitable reference. But what if they were not there? Would we still get the same rotation phenomena? These sorts of problems were raised by the philosopher and scientist Ernst Mach (1838–1916), and they influenced Einstein as he struggled to produce his general theory of relativity.

Newton does not explicitly discuss such questions, but he does consider the case where all extraneous matter is taken away. How can he do that? While the bucket experiment is one that we, like Newton, can actually try, there is another class of "experiments" known as *thought experiments* (see the "Further Reading" section at the end of this chapter) in which the experiment is imagined. Newton sets the scene with this summary of the situation:

It is indeed a matter of great difficulty to discover, and effectually to distinguish, the true motions of particular bodies from the apparent; because the parts of that immovable space, in which those motions are performed, do by no means come under the observation of our senses. Yet the thing is not altogether desperate; for we have some arguments to guide us, partly from the apparent motions, which are the differences of the true motions; partly from the forces which are the causes and effects of the true motions.

If we accept the ideas about forces and motion that Newton has presented, then we may use them to interpret examples and make deductions. Then comes the thought experiment:

For instance, if two globes, kept at a distance from one another by means of a chord that connects them, were revolved about their common centre of gravity, we might, from the tension of the chord, discover the endeavour of the globes to recede from the axis of their motion, and from thence we might compute the quantity of their circular motions.

Newton elaborates a little more and concludes that:

We might find . . . the determination of this circular motion, even in an immense vacuum, where there was nothing external or sensible with which the globes could be compared.

Newton takes the discussion a little further, but I leave that for interested readers to follow as they wish. This experiment too has provoked great discussion (see the "Further Reading" section at the end of this chapter), and once again we must recognize the inventive genius at work in the *Principia*.

5.6. MODERN NOTATION

Throughout the book, an outline of Newton's theory will also be given in its modern form. I begin with the basic definitions and notations. Newton used classical geometry where today we use a symbolic, algebraic form. I encourage everyone, even readers with less-developed mathematical skills, to read over these sections, because I believe you will gain both a feeling for the modern theory and, often, a better idea of what Newton was describing.

5.6.1. Kinematics

We begin with the basic problem of specifying a body's position in space and how that varies with time. (In calculus, differentiating a quantity with respect to time gives us its derivative, which measures how that quantity is varying with time. Thus we go from position to velocity, and from velocity to acceleration.)

In modern notation, we begin with a reference frame, often taken to give a set of Cartesian axes x, y, and z. A "body" or "particle" is assumed small enough (the idealized "point particle") so that its position in space is given by the position vector $\mathbf{r} = (x, y, z)$. Where necessary, and for larger bodies, we may sum or integrate over the collection of smaller particles forming that body, a procedure Newton uses to great effect.

The position vector $\mathbf{r}$ may be represented by an arrow from the coordinate-system origin to the particle, as in figure 5.4. It has a magnitude r (length of the arrow), and its direction may be described by the unit vector $\hat{\mathbf{r}}$. (A unit vector has a length or magnitude of one.) The algebraic form $\mathbf{r} = (x, y, z)$ indicates that the coordinates of the body, x, y, and z, are the components of $\mathbf{r}$.

The velocity vector $\mathbf{v}$ and the acceleration vector $\mathbf{a}$ of the particle are given by differentiation with respect to time t. If we use Newton's dot notation (which is neat and concise and does still appear in mechanics texts, although not widely in most other mathematics and science writing) so that a dot over a symbol indicates the derivative,

$$\dot{x} = \frac{dx}{dt},$$

then

$$\mathbf{v} = \dot{\mathbf{r}} = (\dot{x}, \dot{y}, \dot{z}) \text{ and } \mathbf{a} = \dot{\mathbf{v}} = \ddot{\mathbf{r}} = (\ddot{x}, \ddot{y}, \ddot{z}). \tag{5.1}$$

The vector $\mathbf{v}$ is tangent to the particle's path, and its magnitude gives the speed (see figure 5.4).

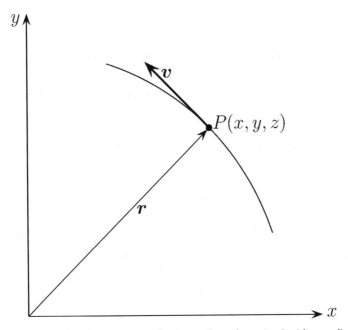

Figure 5.4. Position and velocity vectors for a particle at the point *P* with coordinates *x, y,* and *z*. The velocity vector *v* is tangent to the particle's path. In the illustrated example, motion is confined to the *x–y* or *z* = 0 plane.

If our coordinate system is attached to an **inertial frame**, then a particle initially (time $t = 0$) at the origin, $\mathbf{r} = (0, 0, 0,)$, say, and moving with speed s in the x direction will have the position vector $\mathbf{r} = (st, 0, 0)$ for all future times **if no forces are acting on it**. Similarly, if it was initially at rest at the origin, it would remain there until such time as a force acted on it.

5.6.2. Introducing Dynamical Quantities

A body with mass m has momentum $\mathbf{p} = m\mathbf{v}$. Momentum, like velocity, is a vector quantity; it has both magnitude and direction. (Ignoring that directional property caused problems in some early studies in dynamics; Descartes slipped up in this way.) If more than one body is present, we use subscripts to label the relevant vectors, so $\mathbf{r}_2$ is the position vector for particle 2. Then for two particles, Newton defines the following:

$$\text{total momentum} = \mathbf{p}_1 + \mathbf{p}_2 = m_1\mathbf{v}_1 + m_2\mathbf{v}_2.$$

A force is represented by the vector $\mathbf{F}$. The magnitude F of $\mathbf{F}$ tells us the size of the force, and the direction of $\mathbf{F}$ indicates the direction in which the force acts on a particle. For Newton's special case of a centripetal force acting on a particle,

$$\mathbf{F} = F\hat{\mathbf{r}}, \tag{5.2}$$

where $\hat{\mathbf{r}}$ is the unit vector pointing away from the origin along the position vector $\mathbf{r}$ (see figure 5.4). The center of force has been chosen as the origin for the position vector $\mathbf{r}$. If $F = 6$, say, the force has magnitude 6 in some units and is pointing away from the origin, so it is repelling the particle. If a minus sign is introduced, we have $\mathbf{F} = -6\hat{\mathbf{r}} = 6(-\hat{\mathbf{r}})$, so now we have a force of magnitude 6 acting in the $-\hat{\mathbf{r}}$ direction, which is back toward the origin. This is an attractive force.

5.6.3. Moving between Reference Frames

Consider two reference frames (such as the ship and the Earth in Newton's example) in which the x axes are in relative motion. Let us call the second coordinate $x_{obs,1}$ so that it could be an observer watching a passenger using the coordinate x on a train (as in Einstein's favorite example). If the train has constant speed s, then particle 1 has passenger and observer coordinates linked by

$$x_{obs,1} = x_1 + st, \tag{5.3}$$

assuming that the origins coincide when $t = 0$. In the two reference frames, the particle speeds $\dot{x}_1$ and $\dot{x}_{obs,1}$ are related by

$$\dot{x}_{obs,1} = \dot{x}_1 + s. \tag{5.4}$$

This is the sort of equation Newton used when working out relative speeds for the sailor on the moving ship.

Notice that if we have a second particle, then the distance between particles is the same in both coordinate systems:

$$x_{obs,2} - x_{obs,1} = x_2 - x_1. \qquad (5.5)$$

The differentiating equation (5.4) shows that the particle accelerations are also measured to be the same in the two coordinate systems or frames of reference, as long as the relative speed s remains constant:

$$\ddot{x}_{obs,1} = \ddot{x}_1 \qquad \ddot{x}_{obs,2} = \ddot{x}_2. \qquad (5.6)$$

5.7. DISCUSSION

It is remarkable that in just the first five pages, Newton has defined the key quantities to be used in mechanics and indicated how inertia and forces combine to explain how motion works (for a summary, see figure 5.5).

There are now two challenges. The first challenge is to develop these ideas into a mathematical formalism that may be applied to a variety of problems, and that is what Newton does in the rest of the *Principia*. The second challenge is to find the mechanisms behind the forces (or risk being accused of using occult notions). Newton does not respond to that in the *Principia*; his position is made clear both here (*"for I design only to give a mathematical notion of those forces, without considering their causes and seats"*) and at other points later in the book.

It is amusing to compare the words used three hundred years later by Murray Gell-Mann after his Nobel Prize–winning work on quarks, the entities forming and locked inside protons, neutrons, and other "elementary particles":

I referred to such quarks as "mathematical", explaining carefully what I meant by the term, and contrasted them with what I called "real quarks", which would be capable of emerging so that they could be detected singly. The reason for the choice of language is that I didn't want to face arguments with philosophically inclined critics demanding to know how I could call quarks "real" if they were always hidden.[5]

Newton's fundamental ideas remain as the basis for classical mechanics. Some books (see this chapter's "Further Reading" section for details) give an introductory discussion something like that of Newton's; for example, see Thomson (Lord Kelvin) and Tait (writing in 1888 that "we cannot do better, at all events in commencing, than to follow Newton somewhat closely"[6]), or the 1942 Sommerfeld classic, or the 1966 Kibble text, or the 2004 Thornton and Marion book. Others are much less forthcoming, and Goldstein's highly acclaimed and sophisticated book opens thusly:

> Basic to any presentation of mechanics are a number of fundamental physical concepts, such as space, time, simultaneity, mass and force. For the most part, however, these concepts will not be critically analyzed here; rather, they will be assumed as undefined terms whose meanings are familiar to the reader.[7]

Perhaps even more remarkable is the fact that Newton has already identified the puzzle of mass, with its involvement in two apparently distinct roles (inertia and weight or gravitation), and the need for an analysis of the part played in physics by time and space. These deep and most-fundamental problems remained puzzling for over two hundred years, until Einstein appeared on the scene. (For many people, they remain puzzling!)

Einstein's theory of special relativity uses inertial frames of reference (just as Newton understood them) but incorporated other laws (those describing electrodynamics) into the physics and mechanics. However, Einstein also introduced the fact that the speed of light is the same in all inertial frames of reference, no matter what their relative motions are. That led to a new approach to time and simultaneity. It also means that the transformations between descriptions in different frames of reference must be changed, so the rules used by Newton in his sailing-ship example and used in equation (5.4) must be modified. However, it is vitally important to note that the introduced corrections depend on the speed s of the particles being studied through the ratio s^2/c^2, where c is the speed of light, equal to 3×10^5 kilometers per second. Although s might be large in some elementary-particle-physics processes, in everyday use it is very small relative to c. For example, for a speed s of 1,000 kilometers per hour, s^2/c^2 is about 10^{-12}, or one part in a million million; the corrections are negligible, and Newton's mechanics can be used with great accuracy.

Figure 5.5. Giving the Fundamentals

Mass: characterizes a body—measures **quantity of matter** in a body independent of configuration; linked to **weight**

Momentum: a characteristic of a body's motion given by the product of mass and velocity: **total momentum** is the **vector sum** of individual momentums when several bodies are present

Inertia: a characteristic of a body whereby it resists being put into motion from rest or resists changes to its state of motion; measured by a body's **mass**

Force: an external agency that causes a body to change its state of motion

Centripetal force: acts along the line from a body to a fixed **force center**

Measure of force: given by the change in velocity or momentum that it causes

Time and **position:** measured by local, sensible methods; calculated for **reference frames in relative motion**

Relative and **absolute time and space:** there are questions about the existence of these; their relationship and importance for dynamics can be seen in the famous **bucket experiment**; the use of **thought experiments**

In 1916 Einstein went on to present his general theory of relativity. The great clue for Einstein was that puzzling equality of inertial and gravitational masses. He produced a whole new theory of gravity linked to the geometry of space and time. Newton has bodies moving in a three-dimensional space and described by Euclidean geometry. Einstein has matter influencing space-time, controlling its geometry, and, hence, the way particles move in it. Matter, space, and time are inextricably linked. Newton's puzzle with mass might be removed, but the conceptual questions around space still stir debate to this very day (see the "Further Reading" section at the end of this chapter). Finally we should note that, once again, for everyday situations (and even far beyond them), approximations may be made in Einstein's theory, and the result is that we are led back to Newton's mechanics.

5.8. FURTHER READING

For the history of this subject, consult the books listed at the end of chapters 2 and 3. Highly recommended for history and analysis are the articles by DiSalle, Cohen, and Smith in:

The Cambridge Companion to Newton (Cambridge: Cambridge University Press, 2002).

There are a great many books on classical mechanics. Here is a short personal-favorites list of both old and recent books:

Barger, V. D., and M. G. Olsson. *Classical Mechanics: A Modern Perspective.* 2nd ed. New York: McGraw-Hill, 1995.
Goldstein, Herbert, C. Poole, and J. Safko. *Classical Mechanics.* 3rd ed. 1950. Reprint, New York: Addison-Wesley, 2002.
Kibble, T. W. B. *Classical Mechanics.* New York: McGraw-Hill, 1966.
Sommerfeld, Arnold. *Mechanics: Lectures on Theoretical Physics.* Vol. 1. 1942. Reprint, New York: Academic Press, 1964.
Thomson, Sir William (Lord Kelvin), and P. G. Tait. *Principles of Mechanics and Dynamics.* 1888. Reprint, New York: Dover Press, 1962.
Thornton, S. T., and J. B. Marion. *Classical Dynamics of Particles and Systems.* 5th ed. Belmont, CA: Thomson, 2004.

Recent, very approachable books on the origin of mass and time, their histories, and their measurement include the following:

Callender, C., and R. Edney. *Introducing Time*. Royston, UK: Icon Books, 2004.

Jones, Tony. *Splitting the Second: The Story of Atomic Time*. Bristol, UK: IoP Publishing, 2000.

Wilczek, Frank. *The Lightness of Being: Mass, Ether, and the Unification of Forces*. New York: Basic Books, 2008.

Two books about thought experiments are listed below. (Both take Newton's bucket experiment as a thought experiment, even though he clearly writes *"as I have experienced."*)

Brown, J. R. *The Laboratory of the Mind: Thought Experiments in the Natural Sciences*. 2nd. ed. New York: Routledge, 2011.

Cohen, Martin. *Wittgenstein's Beetle and Other Classic Thought Experiments*. Malden, MA: Blackwell, 2005. [a "popular" book]

The material in this chapter is linked to the theory of relativity; books on that subject cover and debate the issues involved. The first two books listed below are classics by masters of the subject and do not require the reader to have much background therein. The third is a modern example of a nontechnical book. The next three are technical treatments (although some of the discussion sections are not too difficult). The final book (highly recommended and very readable) assesses the subject and surveys the experimental confirmation of the accuracy of the various theories of dynamics.

Einstein, Albert. *Relativity: The Special and General Theory*. 15th ed. New York: Crown Publishers, 1952.

Eddington, Sir Arthur. *Space, Time and Gravitation: An Outline of the General Relativity Theory*. Cambridge: Cambridge University Press, 1920. [later reprinted by other publishers]

Stannard, Russell. *Relativity: A Very Short Introduction*. Oxford: Oxford University Press, 2008.

Rindler, Wolfgang. *Essential Relativity*. New York: Van Nostrand Reinhold, 1969.

Weinberg, Steven. *Gravitation and Cosmology*. New York: John Wiley, 1972.

Lambourne, R. J. A. *Relativity, Gravitation and Cosmology*. Cambridge: Cambridge University Press, 2010.

Will, Clifford. *Was Einstein Right?* 2nd. ed. New York: Basic Books, 1993.

There is an enormous literature on the topic of space and time. Appendix V, "Relativity and the Problem of Space," in the above book by Einstein contains some of his final thoughts on the subject. For some history and a range of viewpoints, see:

Grant, Edward. *Much Ado about Nothing: Theories of Space and Vacuum from the Middle Ages to the Scientific Revolution*. Cambridge: Cambridge University Press, 1981.

Huggett, Nick. *Space from Zeno to Einstein: Classic Readings with a Contemporary Commentary*. Cambridge, MA: MIT Press, 1999.

Here are references to Mach's original criticism of Newton's ideas, two recent papers discussing the *Scholium*, and a list of some of the very useful entries in the *Stanford Encyclopedia of Philosophy* (*SEP*) (http://plato. stanford.edu):

DiSalle, Robert. *Space and Time: Inertial Frames*. SEP.

Huggett, Nick, and Carl Hoefer. *Absolute and Relational Theories of Space and Motion*. SEP.

Laymon, Ronald. "Newton's Bucket Experiment." *Journal of the History of Philosophy* 16 (1978): 399–413.

Mach, Ernst. *The Science of Mechanics: A Critical and Historical Account of Its Development*. 1893. Reprint, La Salle, IL: Open Court Publishing, 1960.

Rynasiewicz, Robert. "By Their Properties, Causes and Effects: Newton's Scholium on Time, Space, Place and Motion." *Studies in the History and Philosophy of Science* 26 (1995): 133–53.

———. *Newton's Views on Space, Time and Motion*. SEP.

6

NEWTON'S LAWS OF MOTION AND THEIR IMMEDIATE CONSEQUENCES

With those preliminaries taken care of, Newton is now ready to set out his famous laws of motion and to identify some immediate consequences. As you might expect, there is an enormous interest in the details and history of this subject; some references are given later in this chapter, in the "Further Reading" section. My approach is to follow Newton's text and his discussions without too much attention to origins and the particular nuances that so fascinate some people. I concentrate on what Newton bequeathed to us in the final, third edition of the *Principia*, and I use his own words as much as possible.

This part of the *Principia* extends over sixteen pages and comprises the laws of motion, six *Corollaries*, and a final *Scholium* in which Newton discusses justifications and examples of the laws in action. It gives the last of the basics to be put in place before Newton begins the mathematical treatment of dynamics in *Book I*.

6.1. THE LAWS OF MOTION

Newton headed this part of his book "*Axioms, or Laws of Motion.*" The word *axioms* reminds us of texts on Euclidean geometry, and we know from his preface that Newton had great respect for that subject and its use both as a tool and as a model for his treatment of mechanics.

6.1.1. The First Law

> LAW I. *Every body perseveres* [remains] *in its state of rest, or of uniform motion in a right* [straight] *line, unless it is compelled to change that state by forces impressed thereon.*

Newton gives the example of a projectile that would follow a straight-line path but for the effects of air resistance (which destroys the uniformity) and gravity (which bends the path). Obviously Newton is considering motion with respect to an inertial frame of reference, and he discusses that a little later.

6.1.2. The Second Law

> LAW II. *The alteration* [change] *of motion is ever proportional to the motive force impressed; and is made in the direction of the right line in which that force is impressed.*

Notice that Newton uses the word *proportional* and he does not write down an equation. (More on that in the next chapter.) Also, he uses the word *motion*, implying that he is talking about the change in momentum, which is what the second law is really all about. (See section 6.6 for the modern formulation and interpretation.)

In his discussion, Newton shows us the thoroughness and care he takes in his formulation as he notes two important facets of this law. First, the change is linear in the force, therefore:

> *If any force generates a motion, a double force will generate double the motion, a triple force triple the motion* [and so on].

Second, the force may change an existing motion, but only in the direction of its application:

> *And this motion (being always directed the same way with the generating force), if the body moved before, is added to or subtracted from the former*

motion, according as they directly conspire with or are directly contrary to each other; or obliquely joined when they are oblique, so as to produce a new motion compounded from the determination of both.

Today this is probably just taken as a property of the vector nature of force and motion (hopefully to be appreciated by the student at some stage!) rather than explicitly spelled out as Newton does here.

The second law is the key to Newton's mathematical formulation of mechanics, and so it is worthwhile to look carefully at what he is saying. Notice that Newton talks about "*the alteration* [change] *of motion,*" whereas today we tend to begin with the rate of change (again, see section 6.6 for the modern formulation). Newton is talking about an impulse (as when two bodies collide), and his first comment on *Law II* given above continues, "*whether that force be impressed altogether and at once, or gradually and successively.*" We shall see later, in chapter 8, that Newton uses a limiting process to deal with cases where the force acts continually. He certainly does understand such cases (and the role of mass), as this quote from the discussion of *Proposition XXIV* in *Book II* indicates:

For the velocity which a given force can generate in a given matter in a given time is as the force and the time directly, and the matter [mass] inversely. The greater the force or the time is, or the less the matter, the greater the velocity will be generated. This is manifest from the second Law of Motion.

In the *Scholium* coming up shortly, he gives the example of gravity, which may be taken as constant locally near the Earth:

When a body is falling, the uniform force of its gravity acting equally, impresses, in equal intervals of time, equal forces upon that body, and therefore generates equal velocities; and in the whole time impresses a whole force, and generates a whole velocity proportional to the time.

This seems to be a case where Newton could have been more explicit (less "slipshod," as we saw Whiteside complaining in section 3.4) and perhaps saved some confusion and debate around his second law by immediately expanding his comment on forces *"gradually and successively"* instead of leaving it to later applications.

6.1.3. The Third Law

> *LAW III. To every action there is always opposed an equal reaction; or, the mutual actions of two bodies upon each other are always equal, and directed to contrary parts.*

As homely illustrations, Newton points out that *"if you press a stone with your finger, the finger is also pressed by the stone."* This law also tells us about forces transmitted by the rope used when a horse pulls a stone.

The enunciation of the third law is one of Newton's greatest achievements. It is crucial for moving beyond cases of a single body moving under the action of given forces to those dealing with several interacting bodies. The simplest case concerns the collision between two bodies, and Newton comments:

> *If a body impinge upon another and by its force change the motion* [momentum] *of the other, that body also (because of the equality of the mutual pressure) will undergo an equal change, in its own motion, towards the contrary part. The changes made by these actions are equal, not in the velocities but in the motions of bodies; that is to say if the bodies are not hindered by any other impediments. For, because the motions* [momentums] *are equally changed the changes of the velocities made towards contrary parts are reciprocally proportional to the bodies* [their masses]. [In other words, the magnitudes of the changes in the bodies' momentums are the same, but because momentum is mass multiplied by velocity, the velocity changes depend on the masses.] *This law takes place also in attractions, as will be proved in the next SCHOLIUM.*

6.2. MORE THAN ONE FORCE

There will be cases in which more than one force acts simultaneously on a body; for example, the Moon is subject to gravitational forces from both the Earth and the Sun, and a projectile is subject to gravity and the force produced by air resistance. Newton spells out what today is commonly referred to as the parallelogram rule. He first deals with effects and then with manipulation of the forces themselves.

COROLLARY 1. *A body, by two forces conjoined will describe the diagonal of a parallelogram in the same time that it would describe the sides by those forces apart* [separately].

If a body in a given time, by the force M *impressed apart in the place* A, *should with an uniform motion be carried from* A *to* B, *and by the force* N *impressed apart in the same place, should be carried from* A *to* C, *complete the parallelogram* ABCD, *and, by both forces acting together, it will in the same time be carried in the diagonal from* A *to* D. [The forces in this example are impulses acting at A.]

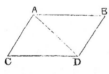

Figure 6.1. Newton's diagram for discussing the first *Corollary*. (From *The Principia*, trans. Andrew Motte [Amherst, NY: Prometheus Books, 1995].)

COROLLARY 2. *And hence is explained the composition of any one direct force* AD, *out of any two oblique forces* AC *and* CD; *and, on the contrary, the resolution of any one direct force* AD *into two oblique forces* AC *and* CD: *which composition and resolution are abundantly confirmed from mechanics.*

The second *Corollary* tells us how to combine forces and also how to resolve a given force into two components (which Newton will need to do later in his studies of the motion of the Moon).

Today a student reading a mechanics text probably accepts the parallelogram law, usually expressed, or even hidden, in the vector addition of forces, without any further thought. However, notice Newton's comment: "*abundantly confirmed from mechanics.*" He appears to appreciate that these properties of forces do not follow from the laws of motion but require an independent statement and justification. Arnold Sommerfeld considered this to be so important that, in his classic 1942 text, he suggests the parallelogram of forces should be the fourth law. The assumption is that **forces act independently**, so they may be **linearly superposed**. This physical property of forces is vitally important for work in mechanics, although it is not universally true in all branches of physics (see references given in this chapter's "Further Reading" section).

Newton ends his discussion of the second *Corollary* by noting its widespread importance in practical systems:

> For from hence are easily deduced the forces of machines, which are compounded of wheels, pullies, levers, cords, and weights, ascending directly or obliquely, and other mechanical powers; as also the force of the tendons to move the bones of animals.

6.3. CONSEQUENCES

Newton is now ready to give us two of the great general results in mechanics, results that are of major significance physically and that also allow us to take simplifying steps in the mathematical theory. These results lean heavily on the third law and show the transition from one particle being acted on by given forces to a system of particles mutually interacting through their various forces.

> COROLLARY 3. *The quantity of motion* [momentum], *which is collected* [found] *by taking the sum of the motions directed towards the same parts, and the difference of those that are directed to contrary parts, suffers no change from the action of bodies among themselves.*

For a system of particles with no external forces acting on them, the total momentum is constant. This is the **conservation of momentum**. Newton shows how this follows from his third law; in section 6.6, the argument is given in modern notation.

Notice that momentum is a vector quantity so we must carefully use appropriate signs to take account of directions. This is something Descartes appeared to have trouble with. It is also interesting to see that, for Newton, this conservation law follows mathematically from the laws of motion, whereas for Descartes (in his *Principles of Philosophy*), "God is the primary cause of motion; and that He always maintains an equal quantity of it."[1]

Newton considers collisions of two bodies in his discussion of *Corollary 3*, as the total momentum must be the same before and after the collision. In an impressive measure of his thoroughness, he notes that the case where the collision is not "head-on" (*"moving in different right lines, impinge obliquely one upon the other"*) can be dealt with by examining how the impact occurs and using the resolution result presented in *Corollary 2*. You can find the theory of oblique collisions in modern textbooks, for example, see Thornton and Marion or Barger and Olsson.

There is also something that can be said about the motion of the whole system:

> COROLLARY 4. *The common centre of gravity of two or more bodies does not alter its state of motion or rest by the actions of the bodies among themselves; and therefore the common centre of gravity of all bodies acting upon each other (excluding outward [external] actions and impediments) is either at rest, or moves uniformly in a right [straight] line.*

For the cases considered here, we can equivalently use the center of mass (the average position of the bodies weighted by their masses). Newton gives an argument for this *Corollary* based on the laws, and in section 6.6, I give the modern derivation. This property of the center of mass will be vitally important when dealing with planetary motion and the solar system (see chapter 23).

6.4. REFERENCE FRAMES AND RELATIVITY

The final two *Corollaries* very clearly set out Newton's ideas about what today we might call relativity principles.

COROLLARY 5. *The motions of bodies included in a given space are the same amongst themselves, whether that space is at rest, or moves uniformly forwards in a right line without any circular motion.*

Newton discusses this in terms of his laws and then gives the physical picture:

A clear proof of which we have from the experiment of a ship; where all motions happen after the same manner, whether the ship is at rest, or is carried uniformly in a right line.

This example was given by Galileo (with rather more colorful descriptions of life in the ship's cabin). By the time of Einstein, it was common to use trains instead of ships, and today, airplanes take over! In the previous chapter, we saw how Newton used the motion of a ship to explain how relative velocities are to be calculated when comparing these different situations. That allows us to move between the inertial frames of reference to which this *Corollary* is linked.

Consideration of physics in different and uniformly moving reference frames is the business of relativity theory, and *Corollary 5* is a statement of the classic theory that Einstein modified to produce his special theory of relativity, as discussed earlier in section 5.7.

6.4.1. A Remarkable Extension

Einstein went on to consider frames of reference related in other ways, and that led to his general theory. It is a measure of Newton's genius that he too thought about other situations and gave a result that would be used in *Book III* when discussing planets and their moons, a key topic in establishing his theory of gravitation (see chapter 22). His result is:

COROLLARY 6. *If bodies, any how moved among themselves, are urged in the direction of parallel lines by equal accelerative forces, they will continue to move among themselves, after the same manner as if they had not been urged by those forces.*

Newton adds an explanation that also makes clear how the term *equal accelerative forces* means forces proportional to the mass of the body acted upon (obviously with gravitation in mind). Then:

> *For these forces acting equally (with respect to the quantities of the bodies to be moved), and in the direction of parallel lines, will (by Law II) move all the bodies equally (as to velocity), and therefore will never produce any change in the positions or motions of the bodies among themselves.*

Here is a case then where the *"excluding outward* [external] *actions and impediments"* caveat, as in *Corollary* 4, may be removed and a useful result still obtained. (It is amusing to note that Einstein considered locally parallel gravitational forces in his famous laboratory in a rocket-ship thought experiment leading up to general relativity!)

6.5. NEWTON'S *SCHOLIUM*

It is hardly surprising to find that such a comprehensive basis for mechanics is followed by a discussion section that contains both justifications and illustrations of the above material. What is surprising is that Newton begins by saying that Galileo used the first two laws in his work on projectiles. This untypical generosity seems to be stretching it a bit; Galileo still retains vestiges of perfect circular motion in his ideas about inertia and does not introduce forces or use anything like Newton's second law. It might have been more appropriate to mention Descartes, but Newton was not too well disposed toward him (and, in fact, parts of the *Principia* seem to be designed specifically to discredit Descartes's theories).

Newton shows how his thinking gives the results for projectile motion and, furthermore,

> *On the same Laws and Corollaries depend those things which have been demonstrated concerning the times of vibration of pendulums, and are confirmed by daily experiments of pendulum clocks.*

Newton moves on to collisions as it is there that the third law reveals its importance. He acknowledges the work of Christopher Wren, John Wallis, and Christiaan Huygens, as well as the colliding-pendulum experiments of Frenchman Edme Mariotte. Newton carried out his own refined experimental program using pendulums with both equal and unequal bobs. In order to obtain the greatest accuracy, he devised a method for correcting for air resistance. Figure 6.2 reproduces Newton's diagram, and the marks show how he measured swings to make the necessary corrections in the experiments in which the pendulum bobs A and B collide. He took great care with these experiments, using ten-foot pendulums; the *Scholium* contains a detailed report. In addition to hard balls, like steel and glass, he also investigated collisions using a variety of softer materials, like balls of wool and cork. In effect, he introduced what today we call the coefficient of restitution as a measure of the change in relative velocities—just another passing touch of brilliance. (See the modern texts listed in section 6.9.1, "To Help as We Read the *Principia*.") After everything, he concludes:

> *And thus the third Law, so far as it regards percussions and reflections, is proved by a theory exactly agreeing with experience.*

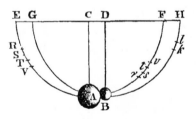

Figure 6.2. Newton's figure for colliding pendulums, showing how he marked the points of successive swings in order to correct for air resistance. (From *The* Principia, trans. Andrew Motte [Amherst, NY: Prometheus Books, 1995].)

He next tackles the case of attractive forces, beginning with another thought experiment. Two bodies, A and B, are attracting one another, and an object is placed between them so that the bodies now press on the intervening obstacle. Newton argues that unless the attractions are equal and opposite, the whole system will "*go forward in infinitum with a motion perpetually accelerated; which is absurd and contrary to the first Law.*"

He concludes that the bodies must press equally on the obstacle and *"be equally attracted one by the other."* He states that he also *"made the experiment on the loadstone* [magnet] *and iron. If these, placed apart in proper vessels, are made to float by one another in standing water, neither of them will propel the other; but, by being equally attracted, they will sustain each other's pressure, and rest at last in equilibrium."*

Obviously Newton wanted to be sure that his major innovation, the third law, was correct, so he gives a further (somewhat curious and not altogether convincing) example involving planes cutting opposite, symmetrical sections of the Earth. He argues that, unless the attractions are equal, *"the whole Earth floating in the nonresistant aether would give way to the greater weight, and, retiring from it, would be carried off in infinitum"*!

The *Scholium* concludes with a discussion of the importance of his laws of motion for understanding clocks, balances, pulleys, screw presses, mallets, and other practical devices. It is clear that Newton did not see his work as just some abstract theoretical notion, but that its physical validity could be perceived in all the mechanical devices around us.

6.6. MODERN FORMULATION

This is a continuation of the basis set out in section 5.6. The task is to put the laws of motion and their consequences into algebraic form using simple ideas from calculus.

Asked what is Newton's law of motion, a great many people will respond "$F = ma$" and, using the notation introduced in section 5.6, we can indeed write for a particle of mass m

$$m\mathbf{a} = \mathbf{F} \tag{6.1}$$

as a modern form of Newton's second law. If we recall that acceleration $\mathbf{a}$ is the time derivative of the velocity $\mathbf{v}$, and that $m\mathbf{v}$ is the momentum $\mathbf{p}$, we can rewrite equation (6.1) as

$$m\frac{d\mathbf{v}}{dt} = \mathbf{F} \quad \text{or} \quad \frac{d\mathbf{p}}{dt} = \mathbf{F}. \tag{6.2}$$

(I have dropped Newton's dot notation for derivatives to use the more conventional notation.) The first form for equation (6.2) also assumes the mass m is a constant. This is not really the law as Newton gave it because equation (6.2) expresses the **rate** of change of the momentum. To go to Newton's form, we assume a constant force, say $\mathbf{F}_0$, then the rate of change of $\mathbf{p}$ is constant, and equation (6.2) implies that

the change in the particle's momentum
= $\mathbf{F}_0$ multiplied by the time it is acting.

Newton gives the limiting form of this for an impulse and then generalizes to continually acting forces, as we saw above.

In the general case, by integrating equation (6.2), we find the momentum $\mathbf{p}(t)$ at times t_1 and t_2 is given by

$$\mathbf{p}(t_2) - \mathbf{p}(t_1) = \int_{t_1}^{t_2} \mathbf{F}\,dt. \tag{6.3}$$

For the constant force $\mathbf{F}_0$, this becomes

$$\mathbf{p}(t_2) - \mathbf{p}(t_1) = (t_2 - t_1)\mathbf{F}_0 \text{ or } \mathbf{p}(t_2) = \mathbf{p}(t_1) + (t_2 - t_1)\mathbf{F}_0. \tag{6.3a}$$

This is the algebraic expression for Newton's general change of momentum result.

When two forces, say $\mathbf{F}_1$ and $\mathbf{F}_2$, are acting on a particle, the total force is

$$\mathbf{F}_{total} = \mathbf{F}_1 + \mathbf{F}_2,$$

which is just the algebraic-vector form of the diagrammatic parallelogram rule for combining forces. Newton's diagram is just the geometric representation of the algebraic formula for the addition of two vectors.

6.6.1. The Third Law and Conservation of Momentum

If we have two interacting particles, the second law equations of motion are

$$\frac{d\mathbf{p}_1}{dt} = \mathbf{F}_{12} \quad \text{and} \quad \frac{d\mathbf{p}_2}{dt} = \mathbf{F}_{21}, \tag{6.4}$$

where $\mathbf{F}_{12}$ is the force exerted by particle 2 on particle 1, and $\mathbf{F}_{21}$ is the force exerted by particle 1 on particle 2 (see figure 6.3). Now, by Newton's third law, those two forces are equal in magnitude but opposite in direction, so $\mathbf{F}_{21} = -\mathbf{F}_{12}$ and equations (6.4) can be written as

$$\frac{d\mathbf{p}_1}{dt} = \mathbf{F}_{12} \quad \text{and} \quad \frac{d\mathbf{p}_2}{dt} = -\mathbf{F}_{12}. \tag{6.4a}$$

If we add these two equations, we get

$$\frac{d\mathbf{p}_1}{dt} + \frac{d\mathbf{p}_2}{dt} = \mathbf{0} \quad \text{or} \quad \frac{d(\mathbf{p}_1 + \mathbf{p}_2)}{dt} = \mathbf{0}. \tag{6.5}$$

This tells us that if particles 1 and 2 interact through forces obeying Newton's third law, then their total momentum does not vary with time. Thus, we have conservation of momentum:

$$\mathbf{p}_1 + \mathbf{p}_2 = \text{a constant.} \tag{6.6}$$

This derivation can be extended to cover any number of particles interacting among themselves.

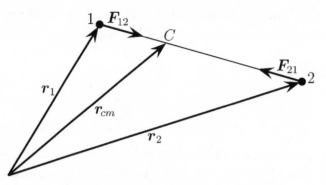

Figure 6.3. Position vectors for bodies 1 and 2 and their center of mass C.

6.6.2. The Center-of-Mass Theorem

For two particles, the position vector for the center of mass $\mathbf{r}_{cm}$ is defined by

$$\mathbf{r}_{cm} = \left(\frac{m_1}{m_1 + m_2} \right) \mathbf{r}_1 + \left(\frac{m_2}{m_1 + m_2} \right) \mathbf{r}_2. \tag{6.7}$$

See figure 6.4. Taking the time derivative of that equation (and recalling that $md\mathbf{r}/dt = m\mathbf{v} = \mathbf{p}$), we find

$$\frac{d\mathbf{r}_{cm}}{dt} = \left(\frac{m_1}{m_1 + m_2} \right) \frac{d\mathbf{r}_1}{dt} + \left(\frac{m_2}{m_1 + m_2} \right) \frac{d\mathbf{r}_2}{dt}$$

$$= \left(\frac{1}{m_1 + m_2} \right) m_1 \frac{d\mathbf{r}_1}{dt} + \left(\frac{1}{m_1 + m_2} \right) m_2 \frac{d\mathbf{r}_2}{dt}$$

$$= \left(\frac{1}{m_1 + m_2} \right) (m_1 \mathbf{v}_1 + m_2 \mathbf{v}_2)$$

$$= \left(\frac{1}{m_1 + m_2} \right) (\mathbf{p}_1 + \mathbf{p}_2).$$

However, equation (6.6) tells us that $(\mathbf{p}_1 + \mathbf{p}_2)$ is constant, so $d\mathbf{r}_{cm}/dt$, the velocity with which the center of mass moves, is also a constant (which I call $\mathbf{v}_{cm0}$). Thus the position of the center of mass is given by

$$\mathbf{r}_{cm} = \mathbf{v}_{cm0}t + \mathbf{r}_{cm0}. \qquad (6.8)$$

We have shown that the center of mass moves with a uniform velocity $\mathbf{v}_{cm0}$ or remains at rest at the time $t = 0$ position, $\mathbf{r}_{cm0}$, as Newton showed. The derivation can be readily extended to any number of particles.

6.6.3. Frames of Reference

Newton's laws are to be used with inertial frames of reference. If no force is acting on a particle, equation (6.1) tells us that

$$\mathbf{a} = 0, \text{ implying that } \mathbf{v} = \mathbf{v}_0 \text{ and } \mathbf{r} = \mathbf{v}_0 t + \mathbf{r}_0. \qquad (6.9)$$

From equation (6.9), we see that the particle is moving with constant velocity $\mathbf{v}_0$, or it is at rest $(\mathbf{v}_0 = 0)$ at $\mathbf{r} = \mathbf{r}_0$, as the first law requires. If a second inertial reference frame is moving relative to the first one with constant velocity $\mathbf{w}_0$, then an observer using that second frame of reference will report

$$\mathbf{v} = \mathbf{v}_0 + \mathbf{w}_0 \text{ and } \mathbf{r} = (\mathbf{v}_0 + \mathbf{w}_0)t + \mathbf{r}_0,$$

and the first law is still satisfied.

If particles are interacting through forces depending on only the distance between them, then transforming to a different reference frame still gives the same equations for their internal motion since, as we saw in section 5.6, both those interparticle distances and accelerations remain unchanged when different inertial reference frames are used.

I will return to the case with particular external forces, as in *Corollary* 4, in a later chapter.

6.7. DISCUSSION

Newton has now set out the formalism to be used for mechanical problems, essentially the same formalism we use today. (Perhaps rather than simply "mechanics," we should say "point mechanics," because some problems with composite bodies require summation or integration over the small bodies forming them; there are theory extensions as made later by Leonard Euler and others—see chapter 28.) Newton has also established some general consequences; that is, he has shown how the laws of motion lead directly to certain results that do not depend on specific force details. That process has become a standard part of theoretical physics, and conservation of momentum remains an essential part of the foundations of physics.

We have also begun to see Newton's mathematical style, which might be called geometrical and rhetorical rather than algebraic (as in the modern formulation just given). The *Scholium* demonstrates that Newton also recognizes the importance of experiments for justifying and confirming his theoretical approach. We see that Newton himself was a skilled and innovative experimentalist, a natural extension of his childhood toys and model-making enterprises.

In keeping with his commitment to the mechanical philosophy (see chapters 2 and 4), Newton emphasizes the importance of particles directly interacting through collisions, and he refers to both his own and others' experimental and theoretical work in support of his formalism. He was probably unaware of or chose not to use Huygens neat use of different reference frames for solving the problem of the collision of two equal bodies. Observer 1 sees body 1 with speed s approaching a second body at rest and wishes to know what will be the result of the collision. To solve this problem, Huygens imagines an observer 2 moving toward body 2 with speed $s/2$ relative to observer 1; now for observer 2, the bodies are each moving with speed $s/2$, as shown in figure 6.4. (Huygens imagines one observer in a moving boat and the other on the river bank.) For observer 2, the situation is symmetrical and obviously the bodies bounce off each other with equal speeds. This final state will be viewed by observer 1 as a stationary first body and the second one now moving off with speed s.

Notice that, for observer 2, the center of mass is at rest (at all times, of course), and this use of "the center-of-mass frame" has become an important technique for analyzing problems in modern physics. This is a beautiful example of Newton's *Corollaries* 3, 4, and 5 in operation.

Observer 1	O → s	O	body 1 with speed s approaching body 2
Observer 2	O → s/2	s/2 ←O	bodies approaching with equal speeds
Observer 2	OO		bodies collide
Observer 2	s/2 ← O	O → s/2	after the collision
Observer 1	O	O → s	after the collision

Figure 6.4. Outline of Huygens's collision theory.

6.7.1. The Meaning of the Laws

There has been much analysis of and debate about Newton's laws and his struggles toward the formulation of the theory of mechanics (see this chapter's "Further Reading" section). I will deal with only one important point, which concerns the status of the laws and definitions of mechanical quantities. The second law involves force and mass, and some people see philosophical or methodological problems here: What are those two things; and is the second law just a definition of force? I follow the lead of physicists. Rudolf Peierls was one of the greatest physicists in the twentieth century, and here he is in *The Laws of Nature*:

> People sometimes argue whether Newton's second law is a definition of force or of mass, or whether it is a statement of an objective fact. It is really a mixture of all these things; and as this is a very typical situation for a physical law, it is worth making this point quite clear.[2]

First, I consider mass. Suppose we have a force, magnitude F acting in the x direction, and we measure the accelerations given to two particles by that same force. According to equation (6.1), we will find

$$m_2 a_2 = m_1 a_1 \text{ or } m_2 = (a_1/a_2)m_1. \qquad (6.10)$$

This means that once we choose a reference mass m_1, all other masses may be determined in terms of it using equation (6.10). One may object that arranging such a force is virtually impossible, but if we let the two particles interact, then the third law tells us that they are both acted on by the same force but with opposite directions. For this case, equation (6.10) becomes

$$m_2 a_2 = -m_1 a_1 \text{ or } m_2 = -(a_1/a_2)m_1. \qquad (6.10a)$$

Thus, in principle, at least, the laws tell us how to measure the inertial mass of a particle. We may go on to deduce properties of the inertial mass—see the Kibble source, for example. (Of course, as mentioned in chapter 5, once we know that inertial and gravitational masses are the same, it is easy to find mass by weighing. Newton revisits this point in *Book III*.)

This approach to laws and the concepts involved in them is important; it is worth reading the strong statement made by physicist Taylor and theoretical physicist Wheeler in their acclaimed book *Spacetime Physics*:

> All the laws of physics have this deep and subtle character, in that they both define for us the needful concepts and make statements about these concepts. Contrariwise, the absence of some body of theory, law, and principle deprives us of a means properly to use or even define concepts. How far out of date is that view of science which used to say, "Define your terms before you proceed"! The truly creative nature of any step forward in human knowledge is such that theory, concept, law, and method of measurement—forever inseparable—are born into the world in union.[3]

That is the "operational" approach taken by the physicist. However, to use Newton's theory, we also need to have a way of specifying forces since, otherwise, as Einstein stressed, the theory is logically incomplete. We should take a lead from master expositor and physics Nobel laureate Richard Feynman:

> The real content of Newton's laws is this: that the force is supposed to have some independent properties, in addition to the law $F = ma$; but the specific independent properties that the force has were not completely described by Newton or by anybody else, and therefore the physical law $F = ma$ is an incomplete law.[4]

Finding the appropriate form for the force is a part of physics. Newton did it for the gravitational force and made progress toward finding the force due to air resistance. Charles-Augustin de Coulomb found the form of the force to be used for the electrical interaction between charged particles. Johannes Diderik van der Waals discovered a formula for modeling the force between molecules in order to study properties of a gas (something Newton alluded to in his *Preface*). For each new physical situation under consideration, the scientist must find the form of the force to use in $F = ma$, and that is an independent piece of physics, as Feynman explained. Of course, before they are used to study other phenomena, suggested force types are checked by using $F = ma$ and seeing how well the results fit the observational data. This is really what Newton's program as set out in the *Preface* is all about: *"for all the difficulty of philosophy seems to consist in this—from the phenomena of motions to investigate the forces of nature, and then from these forces to demonstrate the other phenomena."*

We can also substitute various mathematical forms for the force in the equations of motion and then explore theoretically what sorts of results they produce. In that way, we can have ready a set of results against which experimental data can be compared in order to pick out the most suitable force type. This is exactly what Newton does in the *Principia*.

6.7.2. A Great Piece of Physics Is Put in Place

The second law is one of Newton's great (if not his greatest) contributions to physics. Recall that before Newton, the work of people like Kepler and Galileo was essentially kinematics and, to repeat (from chapter 3) Einstein's words:

We have to realize that before Newton there existed no self-contained system of physical causality which was somehow capable of representing any of the deeper features of the empirical world.[5]

The equation for the second law has, on one side, quantities characterizing the moving body (mass and position and change of position with time), and, on the other, a quantity (force) that describes how the world acts on that body to give its dynamical path. Thus the planetary paths described by Kepler are related to the force of gravity or to a *"deeper feature of the empirical world."* This use of forces is Newton's great step forward, and his third law tells us how to use forces when they result from the interaction of two bodies.

We are so familiar with Newton's laws of motion that it is easy to miss the remarkable step forward in physics they represent. In recent times, that obvious form, "$F = ma$," has been questioned by certain astrophysicists. The result has been a modified Newtonian dynamics (MOND) in which the right-hand side changes from the simple ma form to something else when very small accelerations are involved. MOND has been introduced in response to astronomical observations that seem to require the presence of large amounts of "dark matter" for their explanation. MOND offers an alternative approach in which force laws are unchanged but the basic dynamics is varied. There are now experimental tests of Newton's second law (see the "Further Reading" section at the end of this chapter), and so far, the form suggested by Newton seems to be holding its own.

Those who found the concept of force troublesome and questioned the logical use of it in physics were relieved when two hundred years after Newton, Einstein introduced his general theory of relativity for which forces are not required. Nevertheless, as always, it is worth repeating that Newton's formulation remains useful and acceptably accurate in all but the most extreme situations. Chapter 28 expands on these topics.

6.8. HIGHLIGHTS SUMMARY

The material in this chapter is of such profound importance in the subject of dynamics that it is worth making a few summary points. A diagrammatic summary is given in figure 6.5.

In formulating his three laws of motion, Newton gave the final statement on inertial motion, showed how to introduce forces and how to add and resolve them, and explained how the equal-action-and-reaction third law allowed us to move to a system of interacting particles. **Newton set out the framework for classical mechanics**, a framework that remains valid today. Newton took what was previously known and incorporated it into one **coherent system**.

Next, Newton demonstrated a most powerful approach for the theorist to follow: **he showed how mathematical manipulations of the framework leads to general dynamical results**, like the conservation of momentum law and the result for the motion of the center of mass. Today we still follow that quest for such general results using the approach that Newton pioneered (see chapter 28).

In all of that, we saw the care and thoroughness of Newton's working, and that carried over into the complementary experimental work, both in thought experiments and in physical demonstrations. Newton's work with colliding pendulums showed his attention to detail in the corrections for air resistance and in the investigation of a range of materials for the collisions. With little fanfare, he showed how to deal with inelastic collisions and oblique collisions.

Only a scientist of Newton's stature could have produced this great part of the *Principia*.

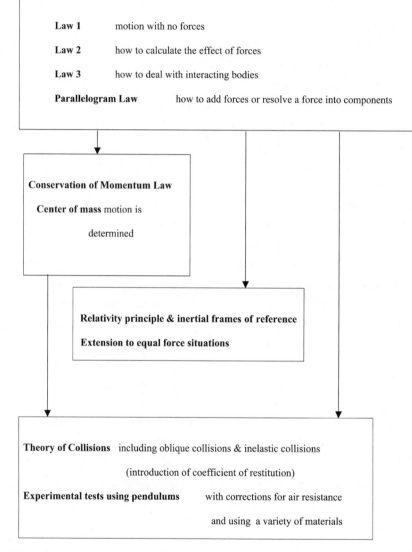

Figure 6.5. Newton's Laws and consequences.

6.9. FURTHER READING

As might be expected, there is an enormous literature on the topic of this chapter. Apart from the books listed in section 6.9.1, see those by Cohen, Bertoloni Meli, and Westfall referenced at the end of chapter 2, as well as M. J. Crowe, *Mechanics from Aristotle to Einstein* (Santa Fe, NM: Green Lion Press, 2007).

Crowe gives an excerpt from Huygens's paper on colliding bodies; the full paper can be read at Christiaan Huygens, "The Motion of Colliding Bodies," trans. J. Blackwell *Isis* 68 (1977): 574–97.

Other interesting and very readable articles on this topic are:

Cook, I. "Newton's 'Experimental' Law of Impacts." *The Mathematical Gazette* 70 (1986): 107–14.
Erlichson, H. "The Young Huygens Solves the Problem of Elastic Collisions." *American Journal of Physics* 65 (1997): 149–54.
Hall, A. Rupert. "Mechanics and the Royal Society, 1688–70." *British Journal for the History of Science* 3 (1966): 24–38.
Smith, George E. "The Vis Viva Dispute: A Controversy at the Dawn of Dynamics." *Physics Today* 59 (October 2006): 31–36.

A small selection of other relevant papers includes:

Lange, Marc. "Why Do Forces Add Vectorially? A Forgotten Controversy in the Foundations of Classical Mechanics." *American Journal of Physics* 79 (2011): 380–88.
Marquit, E. "A Plea for a Correct Translation of Newton's Law of Inertia." *American Journal of Physics* 58 (1990): 867–70.
Pourciau, Bruce. "Newton's Interpretation of Newton's Second Law." *Archive for the History of the Exact Sciences* 60 (2006): 157–207.
———. "Is Newton's Second Law Really Newton's?" *American Journal of Physics* 79 (2011): 1015–22.
Whitrow, G. J. "The Laws of Motion." *British Journal for the History of Science* 5 (1971): 217–34.

The references for Peierls, Feynman, and Taylor and Wheeler are:

Feynman, R. P., R. B. Leighton, and M. Sands. *The Feynman Lectures on Physics*. Reading, MA: Addison-Wesley, 1963.

Peierls, R. E. *The Laws of Nature*. London: George Allen and Unwin, 1955.

Taylor, Edwin F., and John Archibald Wheeler. *Spacetime Physics*. San Francisco: W. H. Freeman, 1963.

On modifications of Newton's second law, see:

Gundlach, J. H. "Laboratory Test of Newton's Second Law for Small Accelerations." *Physical Review Letters* 98 (2007): 150801.

Haycan, Shahen. "What Does It Mean to Modify or Test Newton's Second Law?" *American Journal of Physics* 77 (2009): 607–609.

6.9.1. To Help as We Read the *Principia*

For this and the following chapters, valuable commentaries on the *Principia* may be consulted in:

Brackenridge, J. Bruce. *The Key to Newton's Dynamics*. Berkeley: University of California Press, 1995.

Chandrasekhar, S. *Newton's* Principia *for the Common Reader*. Oxford: Clarendon Press, 1995. [do not be fooled by the title; this is a highly mathematical book]

Cohen, I. Bernard. "A Guide to Newton's *Principia*." In *Isaac Newton: The Principia*, by I. Bernard Cohen and Anne Whitman. Berkeley: University of California Press, 1999.

Cohen, I. Bernard, and George E. Smith, eds. *The Cambridge Companion to Newton*. Cambridge: Cambridge University Press, 2002.

Densmore, Dana. *Newton's* Principia: *The Central Argument*. Santa Fe, NM: Green Lion Press, 2003.

Gjertsen, Derek. *The Newton Handbook*. London: Routledge and Kegan Paul, 1986.

Guicciardini, Niccolo. *Reading the* Principia. Cambridge: Cambridge University Press, 1999.

Smith, George E. *Newton's* Philosophiae Naturalis Principia Mathematica. In the *Stanford Encyclopedia of Philosophy* (http://plato.stanford.edu).

Westfall, Richard S. *Never at Rest: A Biography of Isaac Newton.* Cambridge: Cambridge University Press, 1980.

In particular, to find Newton's mathematical working explained in considerable detail, see the books by Chandrasekhar and Densmore, also those by Brackenridge and Guicciardini.

There are many books giving the modern treatment of classical mechanics. Here are some personal favorites. (Fowles, Goldstein, and Thornton may be taken as standard modern references; Moulton is an excellent older text.)

Barger, V. D., and M. G. Olsson. *Classical Mechanics: A Modern Perspective.* 2nd ed. New York: McGraw-Hill, 1995.

Fowles, G. R. *Analytical Mechanics.* 4th ed. Philadelphia: Saunders College Publishing, 1986.

Goldstein, Herbert, C. Poole, and J. Safko. *Classical Mechanics.* 3rd ed. 1950. Reprint, New York: Addison-Wesley, 2002.

Kibble, T. W. B. *Classical Mechanics.* New York: McGraw-Hill, 1966.

Moulton, F. R. *An Introduction to Celestial Mechanics.* 2nd ed. 1914. Reprint, New York: Dover, 1970.

Sommerfeld, Arnold. *Mechanics: Lectures on Theoretical Physics.* Vol. 1. 1942. Reprint, New York: Academic Press 1964.

Thornton, S. T., and J. B. Marion. *Classical Dynamics of Particles and Systems.* 5th ed. Belmont, CA: Thomson, 2004.

MATHEMATICAL METHODS

Some readers might find the title of this chapter a little threatening. However, most of it is a discussion about the role and style of mathematics in the Principia, and about how that impacted Newton and his contemporaries. Section 7.3 deals with the mathematical base set out by Newton in the Principia, and section 7.4 presents an example of Newton's working alongside a modern version, so that the methods and styles may be compared. Readers averse to mathematical details might gloss over those two sections (although section 7.3.5 on Newton's use of ratios rather than equations and section 7.4.2 on what we learn from the example contain material that is useful to know when reading the Principia).

To move beyond the essentially kinematic approach of Galileo and the world of constant accelerations, two things were needed: some more physics and a mathematical formalism. In the last three chapters, we saw how Newton gave the philosophical, conceptual, and physical foundations. To complete the preliminaries, he now turns to the required mathematics.

While those conceptual foundations were given separately, the mathematical methods are introduced as the first of the fourteen sections forming *Book I* of the *Principia*. Interestingly, this approach is also taken in two other major groundbreaking writings. James Clerk Maxwell begins his *Treatise of Electricity and Magnetism* with a thirty-one-page *preliminary* giving the mathematical methods he intends to use. Albert Einstein's "The Foundation of the General Theory of Relativity" (the great successor to Newton's theory in the *Principia*) comprises fifty-four pages, twenty-two of which form part B, "Mathematical Aids," which comes after a discussion of basic physical matters much on the *Principia* pattern. (Incidentally, Einstein knew that his theory must contain Newton's remarkably successful theory, so his section 21 explains "Newton's Theory as a First Approximation." In fact, as we saw earlier, it is a very good first approximation in most cases.)

The role of mathematics in the *Principia* is a complex topic, and in this chapter, I enlarge on the brief comments made in chapter 3. Some mathematically inclined readers might see even this as inadequate, in which case I refer them to the enormous literature devoted to the subject. References are given in the "Further Reading" section at the end of this chapter, but I particularly recommend (and have leaned on) the works of Guicciardini and Whiteside.

7.1. MATHEMATICS FOR THE *PRINCIPIA*

Before dealing with the mathematical methods themselves, there are four questions that might be asked.

7.1.1. Why Does Newton Need Mathematics?

Newton followed in Galileo's footsteps, recognizing the need for mathematics in science, and we saw him writing in his *Preface* that "*I have endeavoured to subject the phenomena of nature to the laws of mathematics, I have in this treatise cultivated mathematics as far as it regards to philosophy.*" The key point is this: mathematics is essential for exploring his formulation of physics. We shall see that, by using mathematics, Newton leads us inexorably from his laws to the theory of the solar system and much about the Earth and motion on it. This "system of physical causality," as Einstein put it,[1] can be manipulated and developed only by using a mathematical framework.

This contrasts with Descartes's *Principia Philosophiae*, with its claim that vortices in the material of the heavens carry the planets around the

Sun, but then giving no accompanying mathematics to show how such a theory can lead to things like Kepler's laws. Newton was scathing on this point and Descartes's approach in general:

> But if without deriving the properties of things from Phaenomena you feign hypotheses and think by them to explain all nature you may make a plausible system of Philosophy for getting your self a name, but your systeme will be little better than a Romance.[2]

7.1.2. Why Does Newton Need a New Type of Mathematics?

The work of Galileo and Huygens concerns uniform motion, or motion with constant acceleration, with an emphasis on straight-line motion (as for falling bodies, either freely or on inclined planes) and circular motion (as in the constrained motion of a pendulum's bob). Newton is going much further with his objective (see his *Preface*) to produce "*the science of motions resulting from any forces whatsoever.*" He needs mathematics to deal with that generality and to allow for particles moving on any sort of curved path.

A plot of speed versus time as used by Galileo for constant acceleration (equivalent to a constant force) is a straight line. The distance traveled is given by the area under that line and so requires only knowledge of the area of a triangle. Newton will need to relate results to the areas under other types of curves and under even general curves (see section 7.4 for an example). The velocity of a particle moving along a circular path is in the direction of the tangent, which was well known to Euclid, but Newton needs to know the properties of a particle's motion along more complex curves.

Those two requirements can be met only by introducing ideas in the calculus. Thus, Newton is not just introducing a whole new approach to the physics of motion, he is also insisting on a whole new branch of mathematics for formulating and exploring that approach. Einstein too had to introduce a new mathematical formalism (tensor analysis), but while he could draw on the work of people like Gauss, Riemann, and Christoffel, for Newton it was a matter of also developing the mathematical formalism almost from scratch (see section 7.5).

7.1.3. What Guidance Does Newton Give to the Reader?

The *Principia* is in many ways an uncompromising book, and Newton seems to have made little attempt to guide the reader through it. It might have helped if, right at the start, Newton had advised his readers as he does at the start of *Book III*: "*It is enough if one carefully reads the Definitions, the Laws of Motion, and the first three sections of the first book. He may then pass on to this book, and consult such of the remaining Propositions of the first two books, as the references in this, and his occasion, shall require.*" As if even that wasn't a daunting enough prospect for virtually all readers!

Newton did give some individual advice to Richard Bentley:

> *Next after Euclid's Elements the Elements of ye Conic sections are to be understood. And for this end you may read either the first part of ye Elementa Curvarum of John De Witt, or De la Hire's late treatise of ye conick sections, or Dr Barrow's epitome of Apollonius.*
>
> *For Algebra read first Bartholin's introduction & then peruse such Problems as you will find scattered up & down in ye Commentaries on Carte's Geometry and other Algebraical writings of Francis Schooten. I do not mean you should read over all those Commentaries, but only ye solutions of such problems as you will here & there meet with.*[3]

Clearly Newton was assuming a considerable background for the times.

7.1.4. What Are the Implications of All of This?

The use of mathematics, and a new type of mathematics at that, meant reading the *Principia* was extremely difficult for all but a few experts, people like Christiaan Huygens, Gottfried Leibnitz, Pierre Varignon, Abraham de Moivre, and Richard Cotes. Others could check with those experts that everything did indeed make sense, or simply give up. (I return to such matters in section 7.5.)

It is interesting to recall that when Einstein's theory of general relativity was published, it too contained radical new ideas and unfamiliar mathematics and few people could understand it in any detail. One of those few people was Sir Arthur Eddington, and there is a lovely story of his being asked whether it was true that only three people could understand Einstein's work; after a short silence and on being asked not to be so modest, Eddington finally replied that he was merely wondering who the third person might be!

A modern reader, even one familiar with mathematics, might still feel baffled when confronting the *Principia* for reasons I will make clear in a moment. However, we should note that it is not so difficult to appreciate what Newton's *Propositions* are about and to follow his exposition of classical mechanics (the main aim of this book), even if we need to consult an expert to unravel his mathematical proofs. (Of course, for modern readers, a presentation in analytical form is relatively simple, and that is why I include it in this book.)

7.2. NEWTON'S MATHEMATICAL CHOICES

Newton had to decide on a basic framework for his mathematical developments. It is one of the great strengths of mathematics that things can be approached in a variety of equivalent ways, each of which may be particularly useful or apt for a given application. For example, conic sections may be related to slices through a cone, seen as curves generated by focus-directrix rules or by projecting a circle onto various planes or as the result of plotting curves representing second-order polynomial equations in x and y. This last case belongs to the analytical or coordinate geometry pioneered by Fermat and Descartes.

Newton studied analytical geometry but also came to appreciate the elegance and power of Euclidean geometry—recall his *Preface* statement that "*it is the glory of geometry that from those few principles, brought from without, it is able to produce so many things.*" He had now given "*those few principles*" for mechanics and wished to show how the methods of geometrical synthesis could extract so much from them too. (I return to that point in section 7.5.)

Today the analytical or Cartesian approach seems the natural setting for calculus, but Newton wished to present it in a more classical geometric form. Clearly, a mixture of approaches can be discerned in the *Principia* (see sections 7.4 and 7.5), but Newton sets out his mathematical base in *Book I* in the geometric mode.

7.3. SETTING OUT THE MATHEMATICAL METHODS

Book I begins, then, with *Section I*: "*The Method of First and Last Ratios of Quantities, by the Help whereof We Demonstrate the Propositions That Follow.*"

Before getting to the details, I note that any impression that what follows is all that is needed is somewhat off the mark. Also, mathematical interludes come elsewhere in the *Principia*. For example, *Sections IV* and *V* present stunning geometrical results for fitting orbits in thirteen *Lemmas* and eleven *Propositions*, such as *Proposition XXIII*: "*To find a conic that shall pass through four given points, and touch a given right line.*" The amazing *Lemma XXVII*, assessing the possibility of solving a particular transcendental equation, will be discussed when we come to *Section VI*. In *Section II* of *Book II*, we shall find *Lemma II*, where Newton slips in three pages of details of his approach to calculus.

I will discuss in detail three of the eleven lemmas forming *Section I* as examples. See section 7.3.4 for a summary in modern terms.

7.3.1. The Idea of a Limit

LEMMA I. *Quantities and the ratios of quantities, which in any finite time converge continually to equality, and before the end of that time approach nearer the one to the other than by any given difference, become ultimately equal.*

Newton adds further explanation in the *Scholium* for the case of ratios:

For those ultimate ratios with which quantities vanish are not truly the ratios of ultimate quantities, but limits towards which the ratios of quantities decreasing without limit do always converge; and to which they approach nearer than by any given difference, but never go beyond, nor in effect attain to, till the quantities are diminished in infinitum.

It is remarkable how close Newton goes to the modern "ε - δ" form for the definition of a limit (see the paper by mathematics historian Bruce Pourciau for a full discussion). It is typical of Newton's writing that no simple example is given to help the reader. He might have asked about

$$\text{the limit as } x \to 2 \text{ of } \frac{(2x^2 - x - 6)}{(x - 2)}$$

and, given a table of results showing what happens as x gets closer and closer to 2:

x	$(2x^2 - x - 6)$	$(x - 2)$	$\dfrac{(2x^2 - x - 6)}{(x - 2)}$
2.1	0.72	0.1	7.2
2.01	0.0702	0.01	7.02
2.001	0.007002	0.001	7.002
2.0001	0.00070002	0.0001	7.0002

Thus we see "*quantities decreasing without limit*" while the "*ratios converge*" to 7, in this case. Such discussions are readily found in modern calculus texts.

7.3.2. Areas

LEMMA II. *If in any figure AacE, terminated by the right lines Aa, AE, and the curve acE, there be inscribed any number of parallelograms Ab, Bc, Cd &c, comprehended under equal bases AB, BC, CD, &c, and the sides*

Bb, Cc, Dd, *&c, parallel to one side* Aa *of the figure; and the parallelograms* AKbl, bLcm, cMdn, *&c, are completed: then if the breadth of those parallelograms be supposed to be diminished, and their number to be augmented in infinitum, I say, that the ultimate ratios which the inscribed figure* AKbLcMdD, *the circumscribed figure* AalbmcndoE, *and the curvilinear figure* AabcdE, *will have to one another, are ratios of equality.*

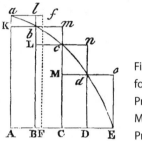

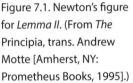

Figure 7.1. Newton's figure for *Lemma II*. (From *The Principia*, trans. Andrew Motte [Amherst, NY: Prometheus Books, 1995].)

The figure shows the area under the curve *abcdE* approximated by rectangles (although Newton uses the general term *parallelogram*), which over- and underestimate it. In the limit that the number of rectangles is increased "*in infinitum*," the area will be given exactly. Of course, this is the type of limiting process found in many calculus textbooks to introduce the concept of the integral as a way to find the area under curves. *Lemmas III* through V specify various properties of areas under curves, or integrals, as we might now say. Thus *Lemmas II* through V give a version of what today we call integral calculus.

Newton here does not describe curves using a formula with y given as some function of x, so he does not give the conventional integral $\int y(x)dx$. However, he did know how to evaluate such integrals when required; an example is given in section 7.4.

7.3.3. Tangents to Curves

LEMMA VI. *If any arc* ACB, *given in position, is subtended by its chord* AB, *and in any point* A, *in the middle of the continued curvature, is touched by a right line* AD, *produced both ways; then if the points* A *and* B *approach one another and meet, I say, the angle* BAD, *contained between the chord*

and the tangent, will be diminished in infinitum, and ultimately will vanish.
[See figure 7.2.]

This is the usual limit process by which tangents are found; in calculus textbooks, such a process may be used to find the derivative dy/dx for a curve $y = y(x)$. The other *Lemmas* introduce similar differential-calculus results (see in the summary given below).

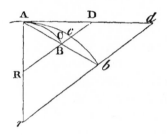

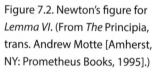

Figure 7.2. Newton's figure for *Lemma VI*. (From *The* Principia, trans. Andrew Motte [Amherst, NY: Prometheus Books, 1995].)

7.3.4. Modern Forms for Newton's *Lemmas*

After a detailed analysis, Bruce Pourciau suggests that Newton's *Lemmas* expressed in their geometric form may be related to modern calculus as follows:

Calculus	Lemmas
Definition of limit	*I*
Limit of a difference	
Definition of the integral	*II, III*
Existence of the integral	
Vertical expansion property	*IV*
Horizontal expansion property	*V*
Translation invariance property	
Definition of the first derivative	*VI, IX*
Derivative of sine	*VII, VIII*
Definition of the second derivative	*XI, X*
Fundamental Theorem of Calculus	*IX*

The interested reader is referred to Pourciau's very readable paper.

7.3.5. Ratios and Equations

There are two discussions or *Scholiums* in *Section I*. In the second, longer one, from which I quoted earlier in section 7.3.1, Newton explains more about limits and how they are to be conceived. The first, much shorter *Scholium* makes clear what Newton means by ratios, and after reading it, we can see the step (which he did not take in most cases) to the equations we are used to today.

SCHOLIUM

If in comparing indetermined quantities of different sorts one with the other, any one is said to be as any other directly or inversely, the meaning is, that the former is augmented or diminished in the same ratio with the latter, or with its reciprocal. And if any one is said to be as any other two or more, directly or inversely, the meaning is, that the first is augmented or diminished in the ratio compounded of the ratios in which the others, or the reciprocals of the others, are augmented or diminished.

And to make matters clear, this time Newton does go on to give an example:

As, if A is said to be as B directly, and C directly, and D inversely, the meaning is, that A is augmented or diminished in the same ratio as $B \times C \times \frac{1}{D}$, *that is to say that A and* $\frac{BC}{D}$ *are to each other in a given ratio.* [Or they are proportional, as we might say.]

Notice that Newton mentions "*quantities of different sorts,*" and this may hark back to earlier mathematicians who worried about equations containing different types of things. For example, in the quadratic expression $x^2 + x + 6$, the x would be taken as a length and the x^2 as an area, so that they would worry about what it means to combine a length and an area in that way. Today we introduce appropriate constants. For example, Boyle's law may be taken in Newton's terms as saying that the pressure P in a gas changes in the same ratio as the temperature T directly and the volume V inversely; whereas we would write the equation

$$P = K\left(\frac{T}{V}\right).$$

Now K is a constant with dimensions chosen so that the whole equation is dimensionally correct.

Many of the results in the *Principia* are given in terms of ratios, which we now routinely convert into equations. In fact, in the modern presentation, we come directly to that form, as we derive results by manipulating equations to produce them in equation form.

7.4. OBSERVING NEWTON AT WORK

I will now give a reasonably simple example of Newton's working, with a modern formulation alongside for comparison. This one example will give you an idea of the difficulties faced by readers of the *Principia*, both at publication time and now. For further examples, see section 10.8 in the Cohen and Whitman translation of the *Principia*; Densmore devotes her book to taking readers through Newton's working; the books by Guicciardini and Brackenridge are exemplary studies of Newton at work in the *Principia* and in the papers leading up to it.

7.4.1. The Example of *Proposition XC, Problem XLIV*

This is a relatively easy problem to come to grips with, and it lets us appreciate the combination of methods Newton uses. I will give all the details (for this particular example), so you can use it as a complete case for comparative study (if you wish to do so!).

(i–a) Newton Poses the Problem

PROPOSITION XC. PROBLEM XLIV.

If to the several points of any circle [actually a circular disk] *there tend equal centripetal forces, increasing or decreasing in any ratio of the distances; it is required to find the force with which a corpuscle is attracted, that is, situated anywhere in a right line which stands at right angles to the plane of the circle at its centre.*

(i–b) A Modern Setting

A circular disk of radius R lies in the x–y plane with center at the origin (see figure 7.3). A particle P is at a height h above the disk on the z axis. If each part of the disk exerts an attractive central force $\mathbf{F}(s)$ per unit area on the particle, where s is the distance from the disk to the particle, what is the total force exerted on it by the whole disk?

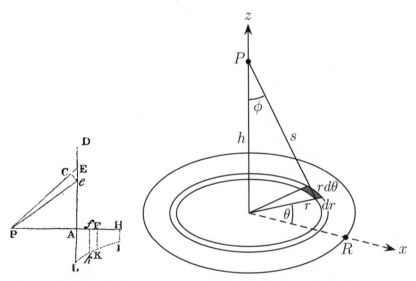

Figure 7.3. A diagram in the *Principia* (*left*) and a diagram for the modern formulation (*right*). (Left image from *The* Principia, trans. Andrew Motte [Amherst, NY: Prometheus Books, 1995].)

(ii–a) Newton's Solution

Suppose a circle to be described about the centre A *with any interval* [radius] AD *in a plane to which the right line* AP *is perpendicular; and let it be required to find the force which a corpuscle* P *is attracted towards the same. From any point E of the circle, to the attracted corpuscle P, let there be drawn the right line PE. In the right line PA take PF equal to PE and make a perpendicular FK, erected at F, to be as the force with which the point E attracts the corpuscle P. And let the curved line IKL be the locus of the point K. Let that curve meet the plane of the circle in L. In PA take PH equal to PD, and erect the perpendicular HI meeting that curve in I; and the attraction of the corpuscle P towards the circle [disk] will be as the area AHIL multiplied by the altitude AP. Q.E.I. [Quod erat inveniendum—what was to be found.]*

(ii–b) Modern Solution

The total force F_{tot} exerted on particle P is directed toward the origin and has magnitude

$$F_{tot} = 2\pi h \int_h^S F(s)\, ds, \quad \text{where} \quad S = \sqrt{h^2 + R^2}\ .$$

It should be clear that Newton too has given the solution in terms of this integral, but in his case, it is expressed in the equivalent form of an area under a curve. We next see how Newton obtains this solution.

(iii–a) Newton's Proof

For let there be taken in AE a very small line Ee. Join Pe, and in PE, PA take [both] PC, Pf equal to Pe. And because the force, with which any point E of the annulus described about the centre A with interval [radius] AE in the aforesaid plane attracts to itself the body P, is supposed to be as FK; and, therefore, the force with which that point attracts the body P towards A is as $\frac{AP \times FK}{PE}$; and the force with which the whole annulus attracts the body P towards A is as the annulus and $\frac{AP \times FK}{PE}$ conjunctly; and that annulus also is as the rectangle under the radius AE and the breadth Ee, and this rectangle (because PE and AE, Ee and CE are proportional) is equal to the rectangle PE × CE or PE × Pf; the force with which that annulus attracts the body P towards A will be as PE × Ff and $\frac{AP \times FK}{PE}$ conjunctly; that is, as the content under Ff × FK × AP or as the area FKkf multiplied by AP. And therefore the sum of the forces with which all the annuli, in the circle [disk] described about the centre

A with radius AD, attract the body P towards A, is as the whole area AHIKL multiplied by AP. Q.E.D. [Quod erat demonstrandum—which was to be demonstrated]

(iii–b) The Modern Proof

If we use polar coordinates (r, θ) in the plane of the disk, then a ring in the disk has radius r and infinitesimal thickness dr as shown in figure 7.3. An infinitesimal element in the ring has area $r \times dr \times d\theta$ and is at a distance s from the particle P as shown in figure 7.3. That element exerts a force $F(s) \times r \times dr \times d\theta$ where $F(s)$ is the magnitude of $\mathbf{F}(s)$. The component of that force on P in the direction of the origin is $\cos(\phi) \times F(s) \times r \times dr \times d\theta$, where ϕ is the angle shown in the figure and its cosine is h/s. All elements in that ring have the same angle ϕ and so contribute the same to the force on the particle directed toward the origin. The components of the forces perpendicular to the z axis balance out or cancel by symmetry.

The total force is given by adding up or integrating over all the rings that form the disk, and so

$$F_{tot} = \int_0^{2\pi} \int_0^R \cos(\phi) F(s)\, r\, dr\, d\theta = 2\pi \int_0^R \cos(\phi) F(s)\, r\, dr.$$

The integral over θ is simply done because ϕ and s do not depend on it, so neither does the whole integrand. If we now note that

$$\cos(\phi) = h/s \quad \text{and} \quad s^2 = h^2 + r^2 \quad \text{so that} \quad sds = rdr,$$
$$\text{and} \quad s = h \quad \text{when} \quad r = 0,$$
$$\text{and} \quad s = S = \sqrt{h^2 + R^2} \quad \text{when} \quad r = R,$$

the formula becomes

$$\mathbf{F}_{tot} = 2\pi \int_h^S \left(\frac{h}{s} \right) F(s) s ds = 2\pi h \int_h^S F(s) ds \quad \text{where} \quad S^2 = h^2 + R^2,$$

which is the required answer.

Newton is basically doing the same thing but in his own geometric style.

(iv–a) Example: Power Law Forces—Newton's Working

Newton now gives the answer for some special cases, those where the force depends inversely on some power of the distance between interacting particles so that $F(s)$ is inversely proportional to s^n. Again I give details so you can compare old and new approaches.

> *Cor. 1. Hence if the forces of the points decrease as the square of the distances, that is, if FK be as $\frac{1}{PF^2}$, and therefore the area AHIKL as $\frac{1}{PA} - \frac{1}{PH}$; the attraction of the corpuscle P towards the circle will be as*
>
> $$1 - \frac{PA}{PH}; \text{ that is, as } \frac{AH}{PH}.$$
>
> *Cor. 2. And universally if the forces of the points at the distance D be reciprocally [inversely] as any power D^n of the distances; that is, if FK be as $\frac{1}{D^n}$ and therefore the area AHIKL as $\frac{1}{PA^{n-1}} - \frac{1}{PH^{n-1}}$, the attraction of the corpuscle P towards the centre of the circle will be as $\frac{1}{PA^{n-2}} - \frac{PA}{PH^{n-1}}$.*

Notice that Newton gives no details about how he moves from his general result to this specific example; he never tells us how he evaluates the integral involved.

(iv–b) Example: Power Law Forces—Modern Working

Let $F(s) = g/s^n$, where g is a constant. Substituting into the formula gives

$$F_{tot} = 2\pi h \int_h^S \left(\frac{g}{s^n}\right) ds = \left(\frac{2\pi hg}{(n-1)}\right)\left[\frac{-1}{s^{n-1}}\right]_h^S$$

$$= \left(\frac{2\pi hg}{(n-1)}\right)\left(\frac{1}{h^{n-1}} - \frac{1}{S^{n-1}}\right).$$

For the special case $n = 2$,

$$F_{tot} = 2\pi hg\left(\frac{1}{h} - \frac{1}{S}\right) = 2\pi g\left(1 - \frac{h}{S}\right).$$

(v–a) Newton's Limiting Case

The answer takes on a simple form when the disk becomes very large. Here is Newton's version:

> Cor. 3. And if the diameter of the circle [disk] be increased ad infinitum, and the number n be greater than unity; the attraction of the corpuscle P towards the whole infinite plane will be reciprocally [inversely] as PA^{n-2}, because the other term $\frac{PA}{PH^{n-1}}$ vanishes.

(v–b) Newton's Limiting Case in Modern Form

As the disk becomes very large, $R \to \infty$ and $S = \sqrt{h^2 + R^2} \to \infty$. To avoid an infinity, we take $n > 1$ and obtain the result

$$F_{tot} = \left(\frac{2\pi hg}{(n-1)} \right) \left(\frac{1}{h^{n-1}} - \frac{1}{S^{n-1}} \right) \rightarrow \left(\frac{2\pi gh}{(n-1)} \right) \left(\frac{1}{h^{n-1}} - 0 \right) = \frac{2\pi g}{(n-1)h^{n-2}}.$$

7.4.2. What Do We Learn from the Example?

You will draw your own conclusions, but I find four points worthy of note. First, to our eyes, Newton's methods are unfamiliar and ponderous. We need to continually and carefully refer to the diagram, much as we do when reading Euclid's *Elements*. However, we see the same underlying approach in both Newton's form and the modern form: set up the problem using infinitesimal areas and then sum, or integrate, to get the required answer. Newton tends to present his answers using ratios, whereas today we use equations with appropriate constants, like the g used in the above equations.

Second, Newton tackles the problem with full generality. No details of the force are initially needed, and a general integral is produced (in terms of an area to be found). Then a general power-law force is assumed and a special case ($n = 2$) is evaluated as an example, one obviously related to gravity.

Third, we see that Newton leaves out the details of vital steps (especially calculus steps, like evaluating an integral), and this made life even more difficult for his contemporary readers.

Fourth, we see Newton the sophisticated mathematician squeezing everything he can out of the general answer by taking the limiting case of an infinitely large disk. The use of such special cases is perhaps second nature for a modern theorist, but this was a man writing three hundred years ago.

7.5. CONTROVERSIES

If you read about Newton's personality traits in chapter 1, you will not be surprised to learn that various confusions and controversies surround the topic of mathematics and the *Principia*. We can be sure about the earlier influences on Newton, as set out in 1728 by Henry Pemberton (who was involved in the editing of the third edition of the *Principia*):

> Sir Isaac Newton has several times particularly recommended to me Huygens style and manner. He thought him the most elegant of any mathematical writer of modern times, and the most just imitator of the ancients. Of their taste and form of demonstration Sir Isaac always professed himself a great admirer: I have heard him even censure himself for not following them yet more closely than he did; and speak with regret of his mistake at the beginning of his mathematical studies, in applying himself to the works of Des Cartes and other algebraic writers before he had considered the elements of Euclide with that attention, which so excellent a writer deserves.[4]

This supports Newton's own statements and shows his respect for ancient geometry and Huygens's use of it. But despite that, Newton did need more than classical geometry, and the *Principia* has elements of the calculus blended into the geometry. We saw that in the example given above and authoritative writers past (Marquis de l'Hôpital in 1696: "*Principia* is almost wholly of this calculus"[5]) and more recent (Clifford Truesdell in 1968: "a book dense with the theory and application of infinitesimal calculus"[6]) are examples of a general opinion (see also chapter 17).

That all seems clear, but in later life, Newton claimed that he also used the method of analysis (methods related to Descartes's work and algebra). Here is Newton himself writing in 1715:

By the help of the new Analysis Mr. Newton found out most of the Propositions in his Principia Philosophia: *but because the Ancients for making things certain admitted nothing into Geometry before it was demonstrated synthetically, he demonstrated the Propositions synthetically, that the System of the Heavens might be founded upon good Geometry. And this makes it now difficult for unskilled men to see the Analysis by which those Propositions were found out.*[7]

Expert opinion (Cohen and Whiteside, for example) suggests that this is not correct, and not a single piece among the voluminous papers he left behind supports Newton's contention. However, as Guicciardini has stressed, Newton used a variety of mathematical methods, sometimes obviously so in the *Principia* and sometimes not. Historian Herman Erlichson presents evidence of Newton's use of calculus. We know that Newton was a complex and devious man, but why this claim? The answer, at least in part, has to do with his battle with Leibnitz over the discovery of calculus. Newton kept his early work on calculus secret. One of several examples is his work on infinite series, calculus, and integration, which was recorded in *De analysi per aequationes numero terminorum infinitas*, a document known to close colleagues by 1669 but not published until 1711. The *Methodis fluxionum* was written around 1670 but not published until 1736, after Newton had died! Leibnitz, on the other hand, had published his mathematical discoveries and could point to those documents in support of his priority claims.

So it was that Newton wished to show that his claims were supported by his use of calculus in the *Principia*, which was not easy, given the way his new methods were woven together with his geometric approach. He is often far from explicit about the use of calculus, as we saw in the example in section 7.4 where he slips in the result of an integration without giving any details. It seems that Newton's secretive and suspicious nature had finally caught up with him; he was left struggling to use his great *Principia* as a priority-claim weapon instead of being able to point to a string of regular papers on mathematics.

There is also the question of Newton's ego problems and his hatred of challenges and disputes. We already saw in chapter 3 that, later in life, there was his claim that *"to avoid being baited by little Smatteres in Mathematics . . . he designedly made his* Principia *abstruse."* Was the *Principia* deliberately made difficult to read? The introduction to *Book III*, "*System of the World*," contains a revealing statement:

> *Upon this subject I had, indeed, composed the third book in a popular method that it might be read by many; but afterwards, considering that such as had not sufficiently entered into the principles could not easily discern the strength of the consequences, nor lay aside the prejudices to which they had been many years accustomed, therefore, to prevent the disputes which might be raised upon such accounts, I chose to reduce the substance of the book into the form of Propositions (in the mathematical way), which should be read by those only who had made themselves masters of the principles established in the preceding books.*

Three points need comment. First, Newton did write such a book "*in a popular method*" (called "*System of the World*"), but it was not published (in Latin and English) until 1728. (More on that in the part 7 of this book.) Second, it is a legitimate point that without a strong mathematical underpinning the "*strength of the consequences*" might not be apparent and could be doubted. Third, the mathematical difficulty of the *Principia* did deter many readers and possibly avoided some disputes for Newton, although there were plenty of those, as we will see in later chapters.

It is amusing (and maybe just a little kinder to Newton) to note that Descartes had similar priority and readership troubles when he published *La Géométrie*. In a letter to mathematician Marin Mersenne, he wrote:

> I have omitted a number of things that might have made it [*La Géométrie*] clearer, but I did this intentionally, and would not have it otherwise. The only suggestions that have been made concerning changes in it are in regard to rendering it clearer to readers, but most of these are so malicious that I am completely disgusted with them.[8]

7.6. CONCLUSIONS

The theory of mechanics presented in the *Principia* did require new mathematical techniques, and Newton invented many of them. It was partly the lack of those mathematical skills that held back Hooke and allowed Newton to overshadow him. There is no doubt that dressing up ideas from calculus in geometrical form made it virtually impossible for many to read the *Principia*. Perhaps the best and most graphic summary was given by William Whewell in his 1837 *History of the Inductive Sciences*:

> The ponderous instrument of synthesis, so effective in Newton's hands, has never since been grasped by one who could use it for such purposes; and we gaze at it with admiring curiosity, as on some gigantic implement of war, which stands idle among the memorials of ancient days, and makes us wonder what manner of man he was who could wield as a weapon what we can hardly lift as a burden.[9]

We shall find that, while conceptually on the right track, Newton reached the limits of his technical ability in several problems. It was the use of analytical methods that allowed those scientists who followed Newton to extend the theory of dynamics into the form we use today, as I discuss in chapter 28.

7.7. FURTHER READING

A great deal has been written on this topic. Here are some references I found particularly useful.

Densmore, Dana. *Newton's* Principia: *The Central Argument.* Santa Fe, NM: Green Lion Press, 2003.

Erlichson, H. "Evidence That Newton Used the Calculus to Discover Some Propositions in His *Principia*." *Centaurus* 39 (1996): 253–66.

Gjertsen, Derek. *The Newton Handbook.* London: Routledge and Kegan Paul, 1986.

Guicciardini, Niccolo. "Analysis and Synthesis in Newton's Mathematical Work." In *The Cambridge Companion to Newton.* Edited by I. Bernard Cohen and George E. Smith. Cambridge: Cambridge University Press, 2002.

————. "'Gigantic Implements of War': Images of Newton as a Mathematician." In *The Oxford Handbook of the History of Mathematics*. Edited by Eleanor Robson and Jacqueline Stedall. Oxford: Oxford University Press, 2009.

————. *Isaac Newton on Mathematical Certainty and Method*. Cambridge, MA: MIT Press, 2011.

————. *Reading the* Principia: *The Debate on Newton's Mathematical Methods for Natural Philosophy from 1687 to 1736*. Cambridge: Cambridge University Press, 1999.

Newton, Isaac. *Isaac Newton: The* Principia. Translated by I. Bernard Cohen and Anne Whitman. Berkeley: University of California Press, 1999.

Pourciau, Bruce. "Newton and the Notion of Limit." *Historia Mathematica* 28 (2001): 18–30.

————. "The Preliminary Mathematical Lemmas of Newton's *Principia*." *Archive for the History of the Exact Sciences* 52 (1998): 279–95.

Whiteside, D. T. *The Mathematical Principles Underlying Newton's* Principia Mathematica. 9th lecture in the Gibson Lectureship in the History of Mathematics. Glasgow, UK: University of Glasgow, 1970.

For the dispute about calculus between Newton and Leibnitz, see *Isaac Newton on Mathematical Certainty and Method* by Guicciardini. Books on the subject (a recent popular-style one and a standard reference) are:

Bardi, Jason. *The Calculus Wars*. London: High Stakes Publishing, 2006.

Hall, A. Rupert. *Philosophers at War*. Cambridge: Cambridge University Press, 1980.

Readers wishing to see a comprehensive, detailed account of the development of mathematics around Newton's time may consult:

Whiteside, D. T. "Patterns of Mathematical Thought in the Later Seventeenth Century." *Archive for the History of the Exact Sciences* 1 (1961): 179–388.

PART 3

DEVELOPING THE BASICS OF DYNAMICS

Newton is now ready to begin developing his theory of dynamics. His first basic problem is how to describe the motion of a single body when acted upon by a centripetal (central) force. In these chapters, he gives a complete solution for that problem, giving general results for a variety of forces, but also emphasizing many specific results for the inverse-square-law case.

Reading plans: This is where Newton develops the vital results to be used in his theory of the solar system and other things in Book III. If you are keen to get on to those things, you can probably manage by reading chapters 8, 9, and 10, plus the summarizing chapter 18 in part 4, before moving on to part 6. That fits in with advice Newton himself gave. Of course, for a full understanding, you should try to appreciate what is in all these chapters. The results in chapters 12, 13, and 14 will show you Newton at his brilliant best.

8

DOWN TO BUSINESS

ecall that after some preliminary material, Newton divided the *Principia* into three books: *Book I* develops the basic theory of mechanics; *Book II* extends that theory to resisted motion and continuous systems; *Book III* has the great applications to the Earth and the solar system. We are now into *Book I*.

Newton has dealt with the preliminaries—methodological matters in the *Preface*, definitions, the laws of motion, and useful mathematical methods (as discussed in the previous four chapters). He now begins to set out the theory for dynamics, gradually increasing the level of complexity over the next thirteen sections of *Book I* (having used *Section I* for describing his mathematical methods). There are ninety-eight *Propositions*, each also labeled as a *Theorem* or a *Problem*, usually accompanied by *Corollaries* giving examples or extensions of the work just covered. Interspersed among the *Propositions* are twenty-nine *Lemmas* giving required extra mathematical background details. There are also *Scholiums* discussing the *Propositions* and their applications.

Please do not panic at this point! I do not intend to go through all that material and the relevant mathematical details as Newton gives them. If I did, then you would soon give up, just as many readers of the original *Principia* gave up. My approach is to

(1) pick out the most important *Propositions* and accompanying results,
(2) explain the significance of those *Propositions* for the development of mechanics, and
(3) show how this material appears in a modern formulation of mechanics.

171

Occasionally I will give a few details or comments on Newton's mathematical proofs, but you may skip over those details as you wish. (In fact, Newton himself suggested readers might just cover the first seventeen *Propositions* and return to others as they felt the need when reading *Book III*.)

In this chapter, I will briefly outline the whole plan for *Book I* and then discuss the first two *Propositions*, which are among the most celebrated in the whole *Principia*. As an example of Newton at work, in section 8.2.1, I gently guide you through his proof for the first *Proposition*. I will also set out the modern approach to those *Propositions*, and we will be able to compare Newton's style of thinking with our own.

8.1. THE PLAN FOR BOOK *I*

So that you can see where we are going, I will briefly outline the contents of *Book I* in figure 8.1 (and no doubt many of Newton's readers wished he had done the same!).

8.2. THE FIRST STEP

Newton begins by finding a general property of the motion of a moving body acted on by a centripetal force. This is one of the great fundamental steps in mechanics (as I will explain), and this first *Proposition* is quite rightly one of the most celebrated and analyzed in the whole *Principia*. It will also allow us to see a reasonably simple example of Newton's way of working.

PROPOSITION I. THEOREM I.

The areas which revolving bodies describe by radii drawn to an immovable centre of force do lie in the same immovable planes, and are proportional to the times in which they are described.

BOOK I. OF THE MOTION OF BODIES

SECTION

I. Of the method of first and last ratios . . .

> [**This is the mathematical methods section** discussed in the previous chapter.]

II. Of the invention of centripetal forces

III. Of the motion of bodies in eccentric conic sections

> [**First steps in the theory of dynamics: the "one-body problem"—motion of a single body under the influence of a centripetal (central) force.** This is the simplest place to start.]

IV. Of the finding of elliptic, parabolic and hyperbolic orbits, from the focus given

V. How the orbits are to be found when neither focus is given

> [**These are basically mathematical sections giving results that might be used to fit orbits specified in various mathematical ways.** They just seem to clutter up *Book I*—and bewilder the reader! According to I. Bernard Cohen, Newton had completed them well before writing the *Principia*, which became a convenient place to publish these results.]

VI. How the motions are to be found in given orbits

VII. Concerning the rectilinear ascent and descent of bodies

> [**Application of the theory to various dynamical problems involving bodies revolving in orbits and moving "vertically."**]

VIII. Of the invention of orbits wherein bodies will revolve, being acted upon by any sort of centripetal force

IX. Of the motion of bodies in movable orbits; and of the motion of the apsides

> [**More complicated orbital problems. For some centripetal forces and combinations of forces, the orbit is not closed. The body may move in a nearly closed orbit that slowly rotates.**]

X. Of the motion of bodies in given superficies; and of the reciprocal motion of funependulous bodies

> [**Motion when there is a constraint; for example, the body is confined to a smooth surface or suspended by a string as in a pendulum.**]

XI. Of the motion of bodies to each other with centripetal forces

> [**This is a major step: Newton shows how to go beyond the one-body problem and how to handle a system of two or more interacting bodies.**]

XII. Of the attractive forces of spherical bodies

XIII. Of the attractive forces of bodies which are not of a spherical figure

> [**Another major step: going beyond the "point-particle" and dealing with the interaction and motion of large, composite bodies, like the Earth.**]

XIV. Of the motion of very small bodies when agitated by centripetal forces tending to the several parts of any very great body

> [**Solution of some problems related to a mechanical theory of light.**]

Figure 8.1. An outline of the contents of *Book I*.

No doubt you will immediately be reminded of Kepler's second law: in any equal time intervals, a line from the planet to the sun will sweep out equal areas. Of course, in linking *Proposition I* with Kepler's law, we must assume that the Sun is providing the centripetal force. I will return to this (as Newton does) when we get to *Book III*.

8.2.1. The Proof

Figure 8.2 is Newton's diagram that shows the motion broken down into steps made when we "*suppose the time to be divided into equal parts.*" The center of force is S. Newton's strategy is to consider those steps and then to take the limit as the time intervals tend to zero and a continuous curve for the motion is produced. For each step, the motion is a combination of straight line or inertial motion and a change imposed by the centripetal force; the two are combined using the parallelogram rule introduced along with the *Laws* (and discussed in section 6.2).

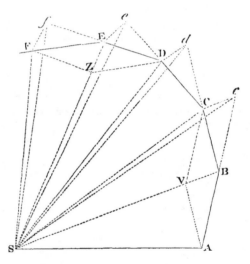

Figure 8.2. Newton's diagram for discussing *Proposition I*. (From *The* Principia, trans. Andrew Motte [Amherst, NY: Prometheus Books, 1995].)

This proof is a good introduction to Newton's methods because it uses only the simple geometry of triangles (reviewed in figure 8.3 in case your Euclidean geometry is a little rusty). The particular triangles involved are extracted from Newton's figure, figure 8.2, and shown separately in figure 8.4.

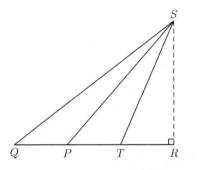
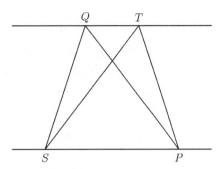

(a) If $PQ = PT$, then $\triangle QPS = \triangle PTS$ (same perpendicular height SR and equal bases)

(b) $\triangle PSQ = \triangle PST$, if lines QT and SP are parallel

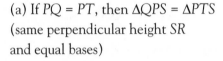

Figure 8.3. Triangle area properties. Column (a) accompanies the diagram on the left; column (b) accompanies the diagram on the right.

Here then is Newton's proof of *Proposition I*, based on figure 8.2, mostly in his own words but sometimes in my words and with extra detail inserted to help you keep track of the geometry.

(1) "*For suppose the time to be divided into equal parts, and in the first part of that time let the body by its innate force* [inertia] *describe the right line* AB. *In the second part of that time the same would (by Law I), if not hindered, proceed directly to* c, *along the line* Bc *equal to AB.*"

(2) Using triangle area property (a) as in figure 8.3, we observe that $\triangle SAB$ and $\triangle SBc$ have equal area, $\triangle SAB = \triangle SBc$, because they have equal bases (from step [1]) and the same height p, as shown in equation (a) in figure 8.4.

(3) Newton now introduces the effect of the centripetal force: "*But when the body is arrived at* B, *suppose that the centripetal force acts at once with a great impulse, and, turning aside the body from the right line* Bc *compels it afterwards to continue its motion along the right line* BC.

"*Draw* cC *parallel to* BS *meeting* BC *in* C; *and at the end of the second part of the time, the body (by Cor. I of the Laws* [parallelogram law]) *will be found in* C, *in the same plane with the triangle* ASB.*"

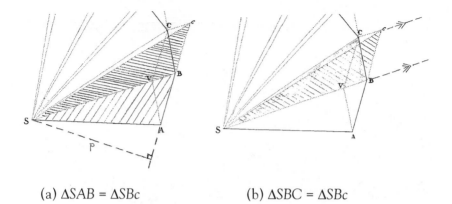

(a) $\Delta SAB = \Delta SBc$ (b) $\Delta SBC = \Delta SBc$

Figure 8.4. The lower part of Newton's figure extended to show triangles of interest. Equation (a) accompanies the diagram on the left; equation (b) accompanies the diagram on the right. (Originally from *The* Principia, trans. Andrew Motte [Amherst, NY: Prometheus Books, 1995]; modified by Annabelle Boag.)

(4) Join C to the force center, that is, draw in SC. We now consider ΔSBC and ΔSBc. Note that by construction of the parallelogram in step (3), the lines SB and Cc are parallel. We can now use triangle area property (b) as in figure 8.3; ΔSBC and ΔSBc are on the same base, SB, and between the parallel lines SB and Cc as shown in triangle area property (b) as in figure 8.4. Thus we have $\Delta SBC = \Delta SBc$.

(5) From steps (2) and (4), we have

$\Delta SAB = \Delta SBc$ and $\Delta SBC = \Delta SBc$, so we conclude that $\Delta SAB = \Delta SBC$.

With that result, Newton has shown that, in the two motions A to B and B to C, the same area is swept out.

(6) It remains to point out that the same thing happens in the following steps in the motion, and then we can take a limit to move to the continuous case. Here is how Newton puts it:

By the like argument, if the centripetal force acts successively in C, D, E &c.,
and makes the body, in each single particle of time, to describe the right lines
CD, DE, EF, &c., they will all lie in the same plane; and the triangle SCD
will be equal to the triangle SBC, and SDE to SCD, and SEF to SDE. And
therefore, in equal times, equal areas are described in one immovable plane:
and, by composition, any sums SADS, SAFS, of those areas, are one to
other as the times in which they are described.

Now let the number of those triangles be augmented, and their breadth
diminished in infinitum; and (by Cor. 4, Lem. III) their ultimate perimeter ADF
will be a curve line: and therefore, the centripetal force, by which the body is
perpetually drawn back from the tangent of this curve, will act continually; and
any described areas SADS, SAFS, which are always proportional to the times of
description, will, in this case also, be proportional to those times. Q.E.D.

What a clever, simple, elegant and powerful proof! The steps are all
simple, but of course, it takes the genius of Newton to build them together
to reach the desired result. (However, I should warn you that the rigor
and style of Newton's work has come in for extensive discussion—see the
"Further Reading" section at the end of the chapter.) This proof provides
a good demonstration of how Newton deals with a continuously acting
force through his limit approach.

8.2.2. The *Corollaries*

Newton follows the proof of *Proposition I* with six *Corollaries*. *Corollaries*
2 to *5* concern geometrical results related to the force, and Newton uses
them when he comes to *Proposition VI* about finding forces (see chapter
9). Of great interest is *Corollary 1*, which translates the area result into one
about how the velocity of the body in motion varies with distance from
the force center. (Note that although Newton says "velocity" it is really
just the magnitude of the velocity, or the speed, that is involved.)

COR. 1. The velocity of a body attracted towards an immovable centre, in
spaces void of resistance, is inversely as the perpendicular let fall from that
centre on the right line that touches the orbit.

The proof follows directly from the triangles result in the main proof:

For the velocities in those places A, B, C, D, E, *are as the bases* AB, BC, CD, DE, EF, *of equal triangles; and these bases are inversely as the perpendiculars let fall upon them.*

The situation using a more modern notation is illustrated in figure 8.5. Two sections of the path, A_1A_2 and B_1B_2, are covered in the same time so that the shaded areas are equal. Obviously the speed along the path is greater for the first, longer section. The above *Corollary* tells us how to find the speed: the velocity vector **v** is tangent to the curve and the magnitude v of **v**, that is the speed, remains always inversely proportional to the perpendicular distance p from the force center S to the tangent, as shown in figure 8.5. Today we write this in the form of an equation:

$$vp = a \text{ constant}. \tag{8.1}$$

This is a version of the modern conservation of angular-momentum law, as we will see in section 8.5.2.

In the final *Corollary*, Newton shows his thoroughness and emphasizes the generality of his results by pointing out that they hold in any inertial frame of reference:

COR. 6. *And the same things do all hold good (by* COR. 5. *of the Laws) when the planes in which the bodies are moved, together with the centres of force which are placed in those planes, are not at rest, but move uniformly forward in right lines.*

8.3. AND CONVERSELY: PROPOSITION II

PROPOSITION II. THEOREM II.

Every body that moves in any curve line described in a plane, and by a radius, drawn to a point either immovable, or moving forwards with an uniform rectilinear motion, describes about that point areas proportional to the times, is urged by a centripetal force directed to that point.

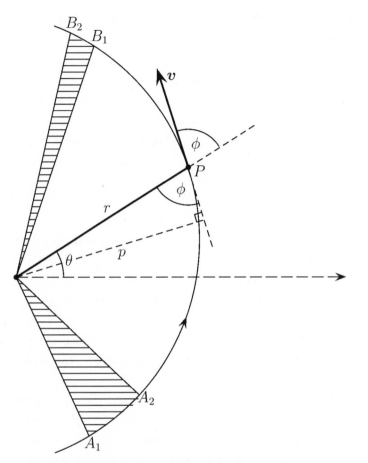

Figure 8.5. A body at point *P* on its orbit has velocity vector *v* and the perpendicular distance from *v* to the force center *S* is *p*. The polar coordinates are *r* and *θ* with the origin at the force center. According to *Corollary 1*, the magnitude of *v*, the speed, is inversely proportional to *p*.

Thus *Proposition I* tells us that a centripetal force implies the equal-areas-swept-out-in-equal-times law, and now *Proposition II* tells us that the converse is also true: if we find evidence for that equal-areas law, then a centripetal force is acting on the body in motion. We are seeing the sophistication and high level of Newton's mathematical working, the impetus for which I will come to in a moment.

8.4. DISCUSSION: APPRECIATING THIS FIRST STEP

Newton extends the theory in *Proposition III* to the case where two bodies are interacting. That seems a little out of place here, and I will return to it later when discussing how we move beyond the one-body problem.

There are (at least) five important points that should be made about this first, monumental development in orbital mechanics embodied in those first two *Propositions*.

(1) Newton has now discovered **a general property of motion under the influence of a centripetal force, any centripetal force;** the result is quite general. The speed of a body so moving will vary with its position on its path, but area swept out in a given time period will always be the same, no matter where on the path the body is located. *Corollary 1* tells us how to translate this property into a formula for the body's speed if that is required. This is one of the great conservation laws of physics, and I discuss its modern form in section 8.5.3.

(2) This is one of the first (if not the first) great examples of how modern science proceeds by combining physical knowledge and laws with a mathematical formalism to deduce new results. This is so standard today that we rarely stop to appreciate what a wonderful and powerful method we are using.

(3) Because Newton has so carefully presented this as an if-and-only-if result (*Propositions I* and *II* together say that equal areas are swept out in equal times **if and only if** the force is centripetal), it may be used to explore the nature of a force under investigation. And remember, we are in *Section II*, which is called "*The Invention of Centripetal Forces*"—more on that point in the next chapter. Newton makes this clear in two ways. The first way is as a *Corollary* to *Proposition II* outlining what happens if the area law is not found to be valid:

In nonresisting spaces or mediums, if the areas are not proportional to the times, the forces are not directed to the point in which the radii meet, but deviate therefrom in consequentia, or towards the parts to which the motion is directed, if the description of the areas is accelerated; but in antecedia [i.e., away from the parts], if retarded.

The second way is in a short *Scholium* where he suggests how we may use the area law:

Because the equable description of areas indicates that a centre is respected by that force with which the body is most affected, and by which it is drawn back from its rectilinear motion, and retained in its orbit, why may we not be allowed, in the following discourse, to use the equable description of areas as an indication of a centre, about which all circular motion is performed in free spaces?

(4) We begin to see the care taken in, and the comprehensive nature of, Newton's work. He stresses that these results hold in any inertial frame of reference and that there is to be no air or other resistance influencing the motion (and in *Corollary 2* to *Proposition II*, he even comments on what happens when that is not the case). Perhaps even more remarkable is his little *Scholium* following *Proposition II*, in which he states that if the centripetal force acts in a given plane, then another force acting perpendicular to that plane will not destroy the area property. Today this is dealt with in terms of separable forces or potentials, and equations of motion are independently solved for the relevant coordinates.

The fifth important point for discussion is so significant for the theory of mechanics that it should be highlighted in its own subsection.

8.4.1. A Major Advance in the Theory of Orbital Motion

The position of a body on its path in a plane is given by specifying the Cartesian coordinates (x, y) or the polar coordinates (r, θ) (as shown in figure 8.5). For motion under centripetal forces, the polar coordinates are more suitable. Thus, at time t, the body is at point $(r[t], \theta[t])$. We could say that the orbit curve is given in terms of the parameter t.

The equal-areas law given above allows us to replace time with area swept out, and essentially we may find how r varies as the angle θ changes. (The mathematical details will be given in the next chapter.) This means that we find the result for $r = r(\theta)$, which is the equation of the body's path. For example, we may find that

$$r = 3\theta,$$

which is the equation of a spiral. (We shall see later that Newton gives an example of motion on a different type of spiral path.) It is also possible to use the equal-areas law to move from angle θ to time and, hence, then calculate where on the orbit the body may be found at a given time.

This may be a little hard to grasp at this point, but basically **the equal-areas law allows us to split the difficult, full dynamical problem into two simpler parts:**

(1) **first, find the equation for the body's path or orbit,**
(2) **then find the time at which the body reaches a given point on that orbit.**

Newton too understood how to use the equal-areas result in this way and its importance for the theory of mechanics. I will show this more explicitly in the next chapter.

8.5. COMPARING APPROACHES, OLD AND NEW

This is a good point to examine the way Newton has tackled this first step and to compare it with the way we proceed today.

8.5.1. Reviewing Newton's Approach

Newton considers a small step in the motion of a body, and he uses the idea that the total movement can be resolved into an inertial part along the tangent to the path and a part directed toward the force center. The two parts are combined using the parallelogram law. It is then a matter of straightforward geometry to see that a series of such steps maintains the equal-areas-swept-out-in-equal-times law. A remarkable fact is involved here: the only property of the force used in the proof is that it is a centripetal (central) force, that is, one always directed toward (or away from) a fixed force center (S in figure 8.2).

Newton analyzes what happens at points along the path and finally uses a limiting process to argue for the continuous case.

8.5.2. The Modern Approach

In the modern theory, our ideal ultimate objective is to find a formula for the position vector $\mathbf{r}(t)$, which specifies where the body is on its path at time t. To that end, we use the given force and Newton's second law to produce a differential equation—the *equation of motion*—that we can then solve to find $\mathbf{r}(t)$. I return to this in the next chapter. However, we also need to do the equivalent of Newton's first step, and doing so highlights three differences in the approaches. (Further discussion of this point and the modern approach is given in chapter 11, particularly in section 11.3.3.)

First, we are looking for a way to establish a property of the motion at every point on the body's path, and this is done in different ways. In the modern approach, we manipulate the equation of motion to discover, or use it to prove, the validity of the sought-after property. We use a mathematical analysis (and, in fact, one way of formulating classical mechanics is called analytical mechanics). This is extremely powerful, but it can hide the physics. Newton shows the physics very explicitly in his geometrical proof.

Second, Newton's limiting process leading to a continuous path has already been done in the modern approach since it uses the continuous path and derivatives from the differential calculus. Thus we use the velocity vector $\mathbf{v} = \dfrac{d\mathbf{r}}{dt}$ and its derivative, the acceleration $\mathbf{a} = \dfrac{d\mathbf{v}}{dt}$.

Newton's second Law $ma = F$ then tells us about the variation of $\mathbf{v}$:

$$m\frac{d\mathbf{v}}{dt} = \mathbf{F} \text{ and so for a centripetal force } \frac{d\mathbf{v}}{dt} = \left(\frac{1}{m}\right)F(r)\hat{\mathbf{r}}. \quad (8.2)$$

The force has center at the origin and magnitude $F(r)$. This is the modern form for relating the change in velocity to the force and is thus the equivalent of step (3) in Newton's proof as given above. Equation (8.2) is the modern analytical form, but it still retains the obvious interpretation that the change is made only in the direction of the force as it acts at that particular point on the body's path. I suspect that in retrospect, many people will feel that in his step (3), Newton exhibited the physics of the motion better than we see it through equation (8.2).

Third, where Newton finds something that is always the same throughout the body's motion, we phrase this as establishing a **constant of motion**. We saw that for momentum in chapter 6. In the present case, if we call the constant h, then it is given by Newton's first *Corollary* expressed through equation (8.1),

$$h = vp \qquad\qquad (8.3)$$

If h is a constant of motion, it does not change with time, implying that its derivative with respect to t must be zero. Thus we must have

$$\frac{dh}{dt} = \frac{d(vp)}{dt} = 0. \qquad\qquad (8.4)$$

In the modern formalism, Newton's geometry is replaced by an analytical proof that equation (8.4) is correct. The proof uses the mathematical definitions and dynamics fed in via equation (8.2). The proof is set out in section 8.5.3, if you would like to see how it works.

The constant h multiplied by the body's mass m is called the magnitude of the **angular momentum**. We also show that the direction of the angular-momentum vector $m\mathbf{h}$ is fixed throughout the motion, and that is equivalent to Newton's statement that "*radii drawn to an immovable centre of force do lie in the same immovable planes.*" **Motion under the influence of a centripetal force is confined to a plane.**

8.5.3. Details of the Modern Proof

The modern proof uses the vector $\mathbf{h}$ defined by the vector cross product of the position and velocity vectors:

$$\mathbf{h} = \mathbf{r} \times \mathbf{v} \qquad (8.5)$$

Here $\times$ denotes the vector cross product, which, for us, has the following two important properties:

Property 1: the magnitude of $\mathbf{a} \times \mathbf{b}$ is the magnitude of $\mathbf{a}$ times the magnitude of $\mathbf{b}$ times the sine of the angle between them;

Property 2: the direction of $\mathbf{a} \times \mathbf{b}$ is given by a unit vector perpendicular to both $\mathbf{a}$ and $\mathbf{b}$, that is, by a unit vector perpendicular to the plane containing $\mathbf{a}$ and $\mathbf{b}$.

If two vectors are **parallel**, then the angle between them is zero, the sine of which is zero, and so their cross product is also zero, according to property 1.

Referring to figure 8.5, we can see that $r \sin (\phi) = p$, and so property 1 tells us that the magnitude h of $\mathbf{h}$ is $rv \sin (\phi) = vp$, which agrees with equation (8.3). Next, to satisfy the requirement of equation (8.4), we must show that

$$\frac{d\mathbf{h}}{dt} = \frac{d(\mathbf{r} \times \mathbf{v})}{dt} = \mathbf{0}.$$

The proof follows a set of straightforward steps:

$$\frac{d(\mathbf{r}\times\mathbf{v})}{dt} = \frac{d\mathbf{r}}{dt}\times\mathbf{v} + \mathbf{r}\times\frac{d\mathbf{v}}{dt} \qquad \{\text{derivative of a product rule}\}$$

$$= \mathbf{v}\times\mathbf{v} + \mathbf{r}\times\left(\frac{1}{m}\right)F(r)\hat{\mathbf{r}} \qquad \{\text{using definition of } \mathbf{v} \text{ and equation (8.2)}\}$$

$$= \mathbf{0} + \left(\frac{1}{m}\right)F(r)\mathbf{r}\times\hat{\mathbf{r}} \qquad \{\text{cross product is zero for parallel vectors}\}$$

$$= \mathbf{0} + \left(\frac{1}{m}\right)F(r)\mathbf{0} \qquad \{\text{cross product is zero for parallel vectors}\}$$

$$= \mathbf{0}.$$

Thus h, the magnitude of $\mathbf{h}$, is a constant, and the direction of $\mathbf{h}$ must also be constant. According to property 2 of the cross product, this tells us that $\mathbf{r}$ and $\mathbf{v}$ must always lie in the same plane. This completes the modern approach to Newton's result.

8.6. CONCLUSIONS

Newton has shown how to obtain a general property of motion under the influence of a centripetal (central) force by analyzing that combination of inertial motion and pull of a centripetal force responsible for orbital motion. To do that, he introduces some highly original, clever, and innovative mathematics. This is where Hooke failed; he understood the mechanism for orbital motion (and he may even have been correct in his claims that he told Newton about that mechanism), but he could not translate that into a mathematical formalism and manipulate it to get the required general results. For example, Hooke believed the speed was inversely proportional to the distance from the force center rather than the perpendicular distance p, as in equation (8.1). (Hooke's result is correct for the special case of circular motion, of course.)

Although Newton found a law that is essentially equivalent to our modern conservation of angular momentum, he did not recognize that there is a quantity, called angular momentum, in the way that he dealt with linear momentum (or *quantity of motion*), as discussed in chapters 5 and 6. I will return to this point when discussing the rotation of the Earth and other matters in the chapters discussing *Book III*.

The modern formulation easily produces the conservation of angular momentum for central forces (equivalent to the equal-areas law) by simply using the calculus form of Newton's second law. However, I cannot help but think that many students would easily follow the mathematics in section 8.5.3 but yet struggle to explain physically why it is that a force directed toward a center produces orbital motion rather than one that sweeps the body around. (Put an object in a hollow tube, rotate the tube so that the object moves around—is swept around—and we find that, instead of the body being confined to a circular path, it shoots out the end of the tube.)

We are now ready to move on in the theory with a new tool that allows us to tackle cases other than uniform motion in a circle by *"substituting the equable description of areas in place of equable motion,"* as Newton puts it. As we have seen in section 8.4.1, that allows a crucial step to be made in the modern formulation of mechanics; the next chapter sets out the details.

8.7. FURTHER READING

See the references in section 6.9.1. Further commentary and analysis for this chapter may be found in:

Densmore, Dana. *Newton's* Principia: *The Central Argument*. Santa Fe, NM: Green Lion Press, 2003.

Nauenberg, M. "Hooke, Orbital Motion, and Newton's *Principia*." *American Journal of Physics* 62 (1994): 331–50.

———. "Kepler's Area Law in the *Principia*: Filling in Some Details in Newton's Proof of Proposition I." *Historia Mathematica* 20 (2003): 441–56.

Whiteside, D. T. *The Mathematical Principles Underlying Newton's* Principia Mathematica. 9th lecture in the Gibson Lectureship in the History of Mathematics. Glasgow, UK: University of Glasgow, 1970.

9

WHAT ISAAC DID NEXT
FINDING FORCES

Recall that the main program for this book is to ask at each stage: What did Newton do? How did he do it? How does it fit into the scheme of mechanics and its applications? And how do we carry out such things today? We are now at the point in the *Principia* where Newton must decide how to go beyond the fundamentals and show how his mechanics works and is applied. As planned, I will describe his approach, the results he produces, and the modern response. However, I will put a little less emphasis on the details of how he worked, since we have already seen representative examples of his methods, and to continue with great detail would not gain too much beyond boring many readers. (Those wanting greater depth should consult, as always, the recommendations in the "Further Reading" section at the end of this chapter.)

9.1. PLAN OF ATTACK

If this were a modern mechanics textbook, the next step would be to take a selection of forces (some chosen for their mathematical convenience and simplicity; others, for their physical relevance) and show how to derive the resulting trajectories for bodies moving under their influence. Newton does do that (setting the pattern for that modern work), but that is not his first step. To understand this, we must recall the great driving force for the *Principia*: the paths of the planets are known (the details given in Kepler's laws), but which force will produce those particular paths? Newton set out his approach in his *Preface*:

For all the difficulty of philosophy seems to consist in this—from the phenomena of motions to investigate the forces of nature, and then from these forces to demonstrate the other phenomena.

For Newton, the next step must be to show how to go from a given path to the underlying forces, and that is the subject of this chapter. We shall see that Newton first used the case of motion in a circle to demonstrate his methods and to give some important results; then he goes on to the general case, giving a general formula and a set of examples.

9.1.1. Terminology: Direct and Inverse Problems

Many physical problems can be viewed in two opposite ways, called the direct problems and the inverse problems. Inverse problems are often of great practical importance. For example:

Direct problem: calculate the trajectory of a shell shot from a particular cannon.

Inverse problem: a shell has landed nearby; where did it come from!?

Direct problem: calculate the way in which radio waves scatter off an object like an airplane.

Inverse problem: use RADAR to plot aircraft positions by interpreting scattered waves.

Direct problem: calculate how far a liquid expands up this tube as it gets hotter.

Inverse problem: the liquid has moved to this level; what is the temperature?

For Newton, the direct problem was to find the force leading to a particular orbit, so the inverse problem is to find the orbit given the force. Today we would tend to reverse that nomenclature, and the modern inverse problem (go from orbits to forces) is barely mentioned in most modern mechanics textbooks. (See this chapter's "Further Reading" section.)

9.2. FORCES AND UNIFORM CIRCULAR MOTION

This is the simplest case, and Newton considers it first. For uniform motion in a circle of radius R and with angular velocity ω, the acceleration away from the center—the centrifugal acceleration, as Huygens called it—is given by $R\omega^2$ (a simple fact from dynamics and known in Newton's time). To keep a body of mass m in a circular orbit, the centripetal force F must attract that body toward the circle center with a balance according to

$$F = mR\omega^2. \tag{9.1}$$

If the period (time for one complete circle) is T, then $T = 2\pi/\omega$ (so $\omega = 2\pi/T$), and the force-balance equation becomes

$$F = mR\left(\frac{2\pi}{T}\right)^2 = 4m\pi^2\left(\frac{R}{T^2}\right). \tag{9.2}$$

Thus, balancing the centrifugal acceleration and the force tells us that, for the body to travel with period T in a circular orbit of radius R, **the force must be proportional to the radius divided by the square of the period**. Then, if we know how the period varies as the orbit size changes, how T depends on R, we will know how the force depends on the distance R from the force center.

This is the result Newton obtained using his previous *Propositions* and the one he expressed in *Proposition IV, Theorem IV*:

> The centripetal forces of bodies, which by equable motions describe different circles, tend to the centres of the same circles; and are one to the other as the squares of the arcs described in equal times applied to [divided by] the radii of the circles.

He sets out further results and consequences in the *Corollaries*.

9.2.1. Examples

If we know how the period T changes as the radius of the orbit changes, we can deduce how the force varies with distance R from the force center. In the *Corollaries*, Newton gives the following examples:

COR. 3	if the period T is constant	the force F is proportional to R
COR. 4	if the period T is proportional to $\sqrt{R}$	the force F is constant, that is, independent of R
COR. 5	if the period T is proportional to R	the force F is proportional to $\frac{1}{R}$
COR. 6	if the period T is proportional to $R^{3/2}$	the force F is proportional to $\frac{1}{R^2}$
COR. 7	if the period T is proportional to R^n	the force F is proportional to $\frac{1}{R^{2n-1}}$

Typically Newton has concluded with an impressive general result. He also suggests another generalization: if the concentric circles are replaced by some other shape, then the equal-areas result will allow that situation to be tackled in a similar way (*Corollary* 8).

9.2.2. Discussion

Equation (9.2) has solved Newton's direct problem of finding the force when given circular orbit details. Of course, it also solves the inverse problem: if you know the force at a distance R from the force center, then equation (9.2) lets you calculate the period T of the resulting circular motion.

These are obviously important results, and Newton follows it with a discussion section or *Scholium*. One result stands out, and in Newton's own words:

> *The case of the 6th Corollary obtains in the celestial bodies (as Sir Christopher Wren, Dr. Hooke and Dr. Halley have severally observed); and therefore in*

what follows, I intend to treat more at large of those things which relate to
centripetal forces decreasing in a square ratio of the distances from the centres.

The case where the period T is proportional to $R^{3/2}$ is Kepler's third
law (here assuming circular orbits), and this argument suggesting an
inverse-square law for gravitation was well known. Newton recognizes the
work of others—even Hooke! Of course, while suggestive, this does not
establish that gravity acts through such a law, as that requires reference to
interacting bodies and noncircular orbits, the major task of *Book III*. In
the meantime, Newton says, he will be giving lots of results for inverse-
square-law forces ready for future use.

Newton also mentions that his results (in *Corollary 9*) agree with the
ideas that Christiaan Huygens used in order to measure the local acceleration due to gravity (what today we would call the gravitational constant
g), referring to "*his excellent book De horologio oscillatorio*."

9.3. THE GENERAL PROBLEM

Newton now turns to the general problem, so *Proposition VI* begins

In a space void of resistance, if a body revolves in any orbit about an immove-
able centre . . .

Newton is reminding us that there is no air resistance and we are
to deal with "*any orbit,*" not just a circular one. He is now using all his
ingenuity and technical brilliance to show how the earlier *Propositions* and
their *Corollaries* (such as those for *Proposition I*) can be used to produce
an expression for the force. Newton's diagram reproduced in figure 9.1 is
one of the most famous in the *Principia*, but very few people wish to go
through, in detail, the arguments based on it—I will follow suit here, but
see the "Further Reading" section at the end of this chapter if you wish
to do so. However, I will indicate how the argument appears in modern
terms.

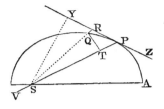

Figure 9.1. Newton's figure accompanying *Proposition VI*. Note that *SY* is perpendicular to the tangent *ZPRY*, and *QT* is constructed to be perpendicular to *SP*. (From *The Principia*, trans. Andrew Motte [Amherst, NY: Prometheus Books, 1995].)

The diagram shows a particle moving from point P to point Q with the usual inertial (PR) and force-related (RQ) contributions. The center of force is S. The line $ZPRY$ is tangent to the orbit at P, and the line SY is perpendicular to that tangent, so we know from *Proposition I, Corollary 1* that the body's speed at P is inversely proportional to the length SY. Using that and other arguments, Newton proves that "*the centripetal force will be reciprocally* [inversely] *as the solid* [that is, the product] $\frac{SP^2 \times QT^2}{QR}$, *if the solid be taken as that magnitude which it ultimately acquires when the points* P *and* Q *coincide*." That is, we are to take the limit as the point Q approaches P.

Newton says in *Corollary 2* that we may use $\frac{SY^2 \times QP^2}{QR}$ in the same way. He mentions other geometrical results based on figure 9.1 that may be applied.

Newton has now shown how to calculate the force producing a given orbit, the solution to his direct problem.

The limit will have to be worked out using the geometrical properties of the particular orbit under consideration, and Newton gives some examples that I will come to shortly. However, before that, it might help some readers and remove some of the mystery if I sketch the derivation of Newton's result in more modern terms. I will also indicate what you will find if you consult a modern textbook on this matter.

Some readers may wish to skim lightly over these next two subsections or skip them altogether; but they are relatively simple and we are talking about one of science's greatest results!

9.3.1. Derivation of Newton's Solution

The complicated diagram and the geometrical jargon tend to hide the simplicity of Newton's approach: consider the inertial and forced components of the motion at a point on the orbit to find an expression for the force, and then use the result in *Proposition I* (what we call the conservation of angular momentum) to eliminate time and leave a totally geometric result. **This gives an expression for the force depending on only the geometric properties of the given orbit.**

Suppose the particle moves from P to Q (as in figure 9.1) with inertial component PR and force-related component RQ, or QR, as Newton writes it. If it does so in time Δt, then

$$PR = v_0 \Delta t \quad \text{and} \quad QR = \left(\frac{F_0}{2m}\right)(\Delta t)^2. \tag{9.3}$$

In equations (9.3), v_0 is the speed at P, F_0 is the force acting at P (directed toward the force center S, of course), and m is the mass of the particle. The first equation is simply the old "distance traveled = speed multiplied by time" formula. The second equation is the old formula (essentially Galileo's result) for distance traveled under the effect of a constant force acting over a given time. We now have an expression for the required force:

$$F_0 = \left(\frac{2m}{(\Delta t)^2}\right) QR. \tag{9.4}$$

It remains to replace the time factor Δt with a geometric factor. We can do that in two ways, both of which use *Proposition I*, as discussed in the previous chapter (note particularly section 8.4.1). The first method uses the first form in equation (9.3) to write Δt as

$$\Delta t = \frac{PR}{v_0} = \left(\frac{PR}{\left(h/SY\right)}\right) = \frac{PR \times SY}{h},$$

where the result from *Proposition I* (the angular-momentum form) expressed in equation (8.3) has been used to replace the speed v_0 with the perpendicular distance SY, as shown in figure 9.1. Recall that h is a constant (in modern terms, the magnitude of the angular momentum). Using the above expression for Δt in equation (9.4) produces

$$F_0 = (2mh^2)\frac{QR}{PR^2 \times SY^2} \cdot \tag{9.5a}$$

Remember that Newton uses proportions, so the constant term $2mh^2$ does not appear in his results.

The second approach is to directly use the concept of equal areas swept out in equal times to replace time with geometrical area, as discussed in section 8.4.1. Here, the area in question is the sector SPQ (see figure 9.1) which, as we must take the Q-approaching-P limit, may be replaced by the area of ΔSPQ, which has area $\frac{1}{2}SP \times QT$. Using that way of thinking in equation (9.4) leads to

$$F_0 = k\left(\frac{QR}{SP^2 \times QT^2}\right), \tag{9.5b}$$

where k is a composite constant. Equations (9.5) contain geometric results as Newton found them (but be careful because, for some reason, he insisted on giving things **inversely proportional** to the force).

9.3.2 Modern Results

The results presented so far are typical in the way they express things in geometric terms, and, in applications, they call for geometric manipulations that are quite unfamiliar to modern scientists, such as Whewell's *"gigantic implements of war"* discussed in section 7.6. Today we tend to look for results given in explicit calculus form (rather than in terms of limits, as Newton does for this *Proposition VI*).

Some older textbooks (for example, in 1914, Lamb's section 85, "The Inverse Problem") give results that seem close to Newton's. The force can be expressed in terms of the radial distance r to the orbit and the perpendicular distance p to the tangent or velocity vector (see figure 8.4):

$$F(r) = \left(\frac{h^2}{p^3}\right)\frac{dp}{dr}. \tag{9.6}$$

Of course, to use that formula, the orbit must be defined in terms of p and r, which is something not at all familiar to modern scientists. The interested reader will find examples in Lamb's book (see the "Further Reading" section at the end of this chapter).

More-recent textbooks, if they mention the modern inverse problem at all—perhaps as a problem as in Fowles's text—resort to the standard equation of motion for the orbit, derived from Newton's second law. That equation, to be discussed in more detail in chapter 11, relates the equation for the orbit in polar coordinates, $r = r(\theta)$, to the controlling central force $F(r)$:

$$\frac{1}{r^2}\frac{d}{d\theta}\left(\frac{1}{r^2}\frac{dr}{d\theta}\right) - \frac{1}{r^3} = \frac{1}{mh^2}F(r) \,. \tag{9.7}$$

While this equation may look formidable to some readers, the only point to note here is that for the **modern direct problem**, we use a given force $F(r)$ in it and then solve to find the orbit $r = r(\theta)$; whereas for the **modern inverse problem**, we substitute the given form for an orbit $r = r(\theta)$ and, hence, find the responsible force $F(r)$. (Readers wishing to relate this procedure and equation [9.7] to Newton's results given above may consult the Whiteside reference in this chapter's "Further Reading" section. The book by Moulton is also good on this topic.)

9.4. EXAMPLES

Newton gives four examples showing how to use his formulas. The first one involves motion in a circle where the force center is not necessarily at the center of the circle. The second example deals with motion in a semicircle when the force center is very far away. The other examples are of more interest, so I will briefly explain them.

The third example deals with the force required to produce motion in a spiral:

PROPOSITION IX. PROBLEM IV.

If a body revolves in a spiral PQS, cutting all the radii SP, SQ, &c., in a given angle; it is proposed to find the law of the centripetal force tending to the centre of that spiral.

Newton's diagram is shown in figure 9.2, with the spiral path extended beyond his original section. Newton's statement that the spiral is "*cutting all the radii SP, SQ, &c., in a given angle*" means that he is dealing with the so-called equiangular spiral, which, in polar coordinates, is given by

$$r = ae^{-b\theta}, \tag{9.8}$$

where a and b are constants. In figure 9.2, r is the distance from the origin S to the orbit point Q and θ is the angle between SP and SQ. (The spiral always cuts a radial line to make an angle whose cotangent is b.)

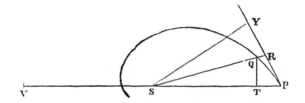

Figure 9.2. Newton's diagram for motion along an equiangular spiral. The spiral is extended beyond the section Newton gave. (Originally from *The* Principia, trans. Andrew Motte [Amherst, NY: Prometheus Books, 1995]; modified by Annabelle Boag.)

Newton shows that for motion along this spiral, the centripetal (central) force directed toward the origin S must follow an inverse cube law,

$$F(r) = -\frac{c}{r^3}, \tag{9.9}$$

where c is a constant. It is easy to give a modern derivation of this result by substituting equation (9.8) into equation (9.7).

It should be noted that even now, more than three hundred years since Newton was working, there is still controversy over and analysis of the methods he used for *Proposition IX*—see the reference in this chapter's "Further Reading" section.

The fourth and final example concerns motion on an ellipse:

PROPOSITION X. PROBLEM V.

If a body revolves in an ellipse; it is proposed to find the law of the centripetal force tending to the centre of the ellipse.

Note that the center of force is at the center of the ellipse and not at a focus, as in the familiar planetary-motion case. Newton's diagram is shown in figure 9.3. If we let the distance CP be r and the angle between CA and CP be θ, then we have polar coordinates (r, θ), and the equation of the ellipse may be written as

$$r^2\{1 + a^2\sin^2(\theta)\} = R^2, \tag{9.10}$$

where a and R are constants.

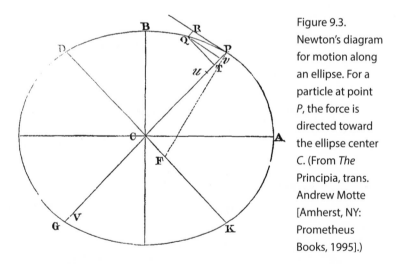

Figure 9.3. Newton's diagram for motion along an ellipse. For a particle at point P, the force is directed toward the ellipse center C. (From *The Principia*, trans. Andrew Motte [Amherst, NY: Prometheus Books, 1995].)

Newton shows that the force is "*directly as the distance* PC," or, in modern terms, $F(r) =$ a constant $\times r$. That result is also given by substituting equation (9.10) into equation (9.7). This linear force will be used often in the *Principia*, as it is in modern textbooks. (See section 11.2.2 for the modern theory.)

9.4.1. A Hidden Gem

The *Corollaries* 2 and 3 attached to *Proposition VII* (about forces giving circular orbits) contain one of those amazing results that Newton seems to just casually add in on the side. If the centripetal force leading to a given orbit is found for one specified force center, then, Newton explains, that result can be used to write down the appropriate force if a different force center is chosen instead. (The interested reader will find that Newton gives some examples of how to use this result; a modern exposition and evaluation is given by Chandrasekhar. I return to this matter in section 10.4.1.)

9.4.2. Newton's Discussion

Newton notes that using his results from here and previous *Propositions*, he can deduce that all such elliptical motions will have the same period. That is one of the nice properties of linear-force motion. He also uses the ellipse result in a clever way to derive Galileo's result that a body moves on a parabolic path when acted upon by a constant gravitational force. Here is Newton's own version, taken from his *Scholium*:

> If the ellipse, by having its centre removed to an infinite distance, degenerates into a parabola; and the force, now tending to a centre infinitely remote, will become equable [constant]. Which is Galileo's theorem.

Newton is saying that over a relatively small vertical distance, the top part of the ellipse is like a parabola and the force is a constant, thus mimicking the dynamical problem studied by Galileo.

9.5. CONCLUSIONS

Once again we see the mathematical genius at work as Newton solves his direct problem and gives us the way to find the centripetal force required to produce motion along a given orbit. The work builds on *Proposition*

I or, for us, the conservation of angular momentum, demonstrating the importance of that basic result, as alluded to in the previous chapter. Just as in a modern textbook, Newton attaches some simple (but important) examples so the reader can see how it all works. Finally, there is a little touch of mathematical extravagance as he shows how to think in terms of a limiting process to obtain one of Galileo's major results.

The orbit-to-forces problem, the modern inverse problem, is not as important today as it was in Newton's time in classical mechanics. But for some decades of the twentieth century, it was a major research problem when cast in terms of quantum mechanics and the nuclear force through which protons and neutrons interact.

Newton now has the tools to tackle one of the key problems driving the *Principia*, and I turn to that in the next chapter.

9.6. FURTHER READING

As usual, the references listed in section 6.9.1 may be consulted. The details of Fowles's book are given there. I. Bernard Cohen's guide (this time, particularly section 10.8) is always worth reading.

Other classical-mechanics references:

Lamb, H. *Dynamics*. Cambridge: Cambridge University Press, 1914. [there are many later editions]

Whittaker, E. T. *A Treatise on the Analytical Dynamics of Particles and Rigid Bodies*. New York: Dover, 1944. [this classic was originally published by Cambridge University Press in 1904; the inverse problem is discussed in sec. 47]

A good discussion by one of the experts is given in Whiteside's lecture:

Whiteside, D. T. *The Mathematical Principles Underlying Newton's* Principia Mathematica. 9th lecture in the Gibson Lectureship in the History of Mathematics. Glasgow, UK: University of Glasgow, 1970.

For Huygens's work on measuring the gravitational constant (or the equivalent thereof), see:

Huygens, Christiaan. *The Pendulum Clock; or, Geometrical Demonstrations concerning the Motion of Pendula as Applied to Clocks.* Translated by Richard J. Blackwell. Ames: Iowa State University Press, 1986. [see sec. 6 of the introduction by H. J. M. Bos]

For a discussion of the geometric details of *Proposition IX*, see the following and the references therein:

Wilson, Curtis. "Newton on the Equiangular Spiral: An Addendum to Erlichson's Account." *Historia Mathematica* 21 (1994): 196–203.

THE INVERSE-SQUARE LAW
TRIUMPH AND CONTROVERSY

Newton is now ready to tackle the great driving problem for the *Principia*: What is the force that produces **elliptical** orbits? It was generally believed that the force followed an inverse-square law, and now Newton can change that belief into a fact. Thus we find *Section III: "Of the Motion of Bodies in Eccentric Conic Sections."*

Notice that Newton is (as ever) generalizing and going beyond ellipses to include all conic sections. For some readers, a brief review of conic sections may be useful before going on to section 10.2, where we see how Newton tackles the full dynamical problem for inverse-square-law forces. We will see Newton at his dazzling best, but also at his most mysterious and irritating.

10.1. CONIC SECTIONS: A BRIEF REVIEW

Early mathematicians defined conic sections as those curves revealed when a double cone is sliced through by a plane, as in figure 10.1. Depending on the angle of the plane to the cone axis, a circle, an ellipse, a parabola or a hyperbola is revealed (and a pair of straight lines, in a very special case). Hence the name *conic sections*.

They also found a logical definition for constructing these curves using a given line, called the **directrix**, and a point, called the **focus** (see figure 10.2). The points P, whose distance from the focus is a constant multiple ε of its distance from the directrix, form a conic section. The value of ε, the **eccentricity**, determines which curve is obtained: $\varepsilon < 1$ gives an ellipse, ε

= 1 gives a parabola, and ε >1 gives a hyperbola. An ellipse has two focal points and a line at each end, which can serve as a directrix. The directrix is a distance D from the focus. The line through the focus and parallel to the directrix cuts the curve a distance p from the focus and $p = \varepsilon D$. (The line of length $2p$ through the focus and joining the upper and lower parts of the curve is called the **latus rectum**; it is important in classical geometry and prominent in details in the *Principia*.)

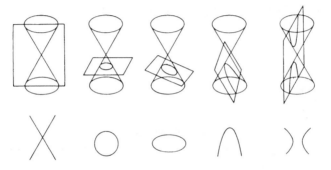

Figure 10.1. Slices through a cone produce the conic sections—lines, circle, ellipse, parabola, and hyperbola.

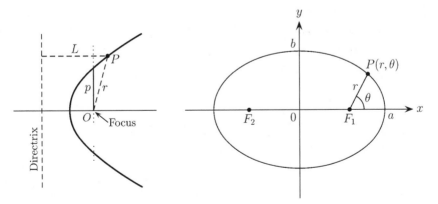

Figure 10.2. *Left*: The focus-directrix approach to conic sections. For a point P, the distance r from the focus and the distance L from the directrix are related by $r = \varepsilon L$, where ε is the eccentricity. For the ellipse (ε < 1), there are two focal points, F and F', each with its own directrix. *Right*: Coordinates for the ellipse with focal points F_1 and F_2. The polar coordinates (r,θ) use an origin at F_1.

We can also describe conic sections using coordinate geometry. If we use the center of the ellipse as the origin and the line through the focal points as the x axis (see figure 10.2), the equation of the ellipse is

$$\frac{x^2}{a^2} + \frac{y^2}{b^2} = 1 \quad \text{or} \quad x^2 + \frac{y^2}{(1-\varepsilon^2)} = a^2. \tag{10.1}$$

I have taken the major axis to have length $2a$, and the focal points are at $x = \varepsilon a$ and $x = -\varepsilon a$. The minor axis has length $2b$, and the eccentricity is given by

$$\varepsilon = \sqrt{1 - \frac{b^2}{a^2}} \quad \text{or} \quad b = a\sqrt{1-\varepsilon^2}.$$

The latus rectum length $2p$ has $p = a(1 - \varepsilon^2)$.

If we use polar coordinates r and θ, with the origin now taken at the right-hand focus (see figure 10.2), the equation of the ellipse is

$$r = \frac{p}{1 + \varepsilon \cos(\theta)} = \frac{a(1-\varepsilon^2)}{1 + \varepsilon \cos(\theta)}. \tag{10.2}$$

The ellipse is characterized by the length p (or a) and the eccentricity ε. (The special case of the circle has $\varepsilon = 0$ and $p = a$, the radius.) If the left-hand focus is taken as origin for polar coordinates, the ε in equation (10.2) must be changed to $-\varepsilon$.

10.2. FORCES RESPONSIBLE FOR CONIC SECTION ORBITS

Newton can now use the formulas given in *Section II* of the *Principia*, as discussed in the previous chapter, section 9.3. Newton's work is based on those formulas and, as always, the consideration of the inertial contribution and force-related contribution to path development at a point, as illustrated in figure 10.3.

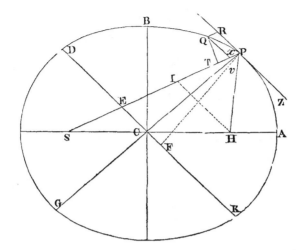

Figure 10.3. Newton's diagram for explaining *Proposition XI*. A body moves from *P* to *Q* on its orbit through a combination of the inertial contribution *PR* and the force-related contribution *RQ*. An inverse-square-law centripetal force has force center *S*, so when acting at *P*, its magnitude is inversely proportional to *SP²*. (From *The* Principia, trans. Andrew Motte [Amherst, NY: Prometheus Books, 1995].)

I will concentrate on Newton's results rather than the usual geometrical argument that he develops—just looking at figure 10.3 will tell you that it is somewhat involved. (We have already seen samples of Newton's geometric approach, and I refer you to the books listed in section 6.9.1, especially those by Chandrasekhar and Densmore, for an explanation of the details of the present case.) Newton begins with the following:

PROPOSITION XI. PROBLEM VI.

If a body revolves in an ellipse; it is required to find the law of the centripetal force tending to the focus of the ellipse.

After some considerable geometric manipulations, he reaches that famous conclusion: "*the centripetal force is inversely as* L × SP², *that is, inversely as the square of the distance* SP. *Q.E.I.*" At last, the question that plagued Hooke, Halley, and Wren had been answered.

Following that result comes a little part of the *Principia* reminding us about some of Newton's attributes and intentions. He tells us, in the section titled *"The Same Otherwise,"* how to get that result in a different, shorter way by using some previous results. Hidden away here is one of Newton's flashes of genius. But rather than detour here, I will consider it separately in section 10.2.1.

Newton then goes on:

> With the same brevity with which we reduced the fifth Problem [Proposition X] to the parabola, and hyperbola, we might do the like here; but because of the dignity of the Problem and its use in what follows, I shall confirm the other cases by particular demonstrations.

Newton is recognizing the importance of this work for the development of his whole book, and so, rather than give further clever methods, he intends to deal with each case from first principles (as we would say now). Yes, I can be clever, but it is important to give the complete conic-sections result, starting from my basic theorems. Thus we find:

PROPOSITION XII. PROBLEM VII.

> Suppose a body to move in an hyperbola; it is required to find the law of the centripetal force tending to the focus of that figure.

PROPOSITION XIII. PROBLEM VIII.

> If a body moves in the perimeter of a parabola; it is required to find the law of the centripetal force tending to the focus of that figure.

Of course, in each case, he comes to the conclusion that the force varies *"inversely as the square of the distance SP. Q.E.I."*

The whole problem of finding the forces giving motion on conic-section orbits with the force center at a focus is now solved!

10.2.1. *"The Same Otherwise"*

> *This is one of those sections showing Newton's supreme cleverness. It may appeal more to mathematically inclined readers, but all are encouraged to read it.*

In *Section II*, when Newton is developing his formula for finding the force required to generate a given orbit, there is one of those stunning theoretical asides. It is almost hidden away by being given in the *Corollaries* to one of his examples, *Proposition VII*. For the case of motion in a circle (see figure 10.4), he writes:

COR.2 *The force by which a body in the circle APTV revolves about the centre of force S is to the force by which the same body P may revolve in the same circle, and in the same periodic time, about any other force centre R . . .*

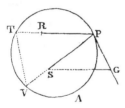

Figure 10.4. Newton's diagram for *Proposition VII, Corollary 2*. (From *The Principia*, trans. Andrew Motte [Amherst, NY: Prometheus Books, 1995].)

Newton gives a formula for relating the two forces and extends the result in *Corollary 3* to other orbits. Thus, knowing the force in one case, he can say what it should be in a different case without resolving the whole problem. Brilliant! (See the "Further Reading" section at the end of this chapter for more details.)

One of Newton's examples in *Proposition XI* concerns motion in an ellipse with the force center at the ellipse center. He can now use that relatively easy result in the method just outlined to find the force when its center is changed (to the ellipse focus!). That is what he means when he says *"The same otherwise"* when dealing with conic-section orbits, as discussed previously. He states the general result in a little *Scholium* at the end of *Section III*.

This is very clever mathematical stuff and, while illustrating Newton's ingenuity, it is probably a good example of dazzling intricacies that befuddled so many readers—that is to say, those who got even this far!

10.2.2. The Modern Solution

The modern approach was discussed in section 9.3.2. Using equation (9.7), we can say that the central force giving an elliptical orbit with focus at the force center is found by

substituting $r = \dfrac{p}{1 + \varepsilon \cos(\theta)}$ into $F(r) = mh^2 \left\{ \dfrac{1}{r^2} \dfrac{d}{d\theta} \left(\dfrac{1}{r^2} \dfrac{dr}{d\theta} \right) - \dfrac{1}{r^3} \right\}.$

Simple differentiations lead to the result

$$F(r) = -\left(\frac{mh^2}{p} \right)\left(\frac{1}{r^2} \right).$$

The minus sign indicates an attractive force; the required force strength is related to the body's mass, its angular momentum h, and the orbit size parameter p; the strength of the force varies inversely as the square of the distance r from the force center.

A similar process gives an inverse-square-law result for parabolic and hyperbolic orbits.

10.3. ORBIT PROPERTIES FOR INVERSE-SQUARE-LAW FORCES

Newton has now established that motion along an ellipse generated by a force centered on a focus requires that force to follow an inverse-square law. The equal-areas-swept-out-in-equal-times property (the modern conservation of angular momentum) must apply, and, in fact, is used by Newton in all his working. Newton now derives two more properties of this motion. The first is geometrical and relates to that equal-areas property, but the second (which actually builds on the first) concerns the orbital period. (Notice that these are new results rather than examples or solutions to given problems, so they are also labeled as theorems.)

PROPOSITION XIV. THEOREM VI.

If several bodies revolve about one common centre, and the centripetal force is inversely as the square of the distance of places from the centre: I say, that the principal latera recta of their orbits are as the squares of the areas, which the bodies by radii drawn to the centre describe in the same time.

The latus rectum is defined in section 10.1.

PROPOSITION XV. THEOREM VI.

The same things being supposed, I say, that the periodic times in ellipses are as the 3/2th power of their greater axes.

In modern form, the greater axis has length $2a$ (section 10.1), so if the period is T, then *Proposition* XV tells us that

$$T \propto (2a)^{3/2} \quad \text{or equivalently} \quad T^2 \propto (2a)^3.$$

We can also write this as

$$T^2 = \kappa a^3,$$

where κ is a constant that depends on the body and force parameters. Some readers will immediately recall Kepler's third law, but Newton does not make such connections until *Book III*. He does realize that this result—relating periodic time and orbit size—does not depend on the eccentricity of the elliptic orbit and so gives the following:

COR. *Therefore the periodic times in ellipses are the same as in circles whose diameters are equal to the greater axes of the ellipses* (2a).

In *Proposition XVII* and its nine *Corollaries*, Newton gives an array (perhaps I should say, a bewildering array) of results linking geometric and dynamical properties of conic-section orbits. One of the simplest results is about the points on the orbit where the body's speed is an extreme: "*the velocities of bodies, in their greatest and least distances from the common focus, are inversely as the distance."* The body is moving fastest when it is at its closest point to the force center at the focus, slowest when farthest away.

10.4. USING THE INITIAL CONDITIONS

Newton has now proved that if a body moves in a conic section under the influence of a force centered at the focus, then that force must have an inverse-square-law dependence on the distance from the center. He has also set out many geometrical and dynamical properties of such orbits. **One obvious question remains: How do we tell which particular conic section will be followed, and how are its defining geometrical properties decided?** (In modern form, we need the eccentricity ε and the size parameter p to be given in equation [10.2].) Mathematically, we ask how the *initial conditions* or the *launching conditions* pick out the correct orbit. Newton answers in his next *Proposition* and its accompanying diagram:

> *PROPOSITION XVII. PROBLEM IX.*
>
> *Supposing the centripetal force to be inversely proportional to the squares of the distances of places from the centre, and that the absolute quantity of that force is known; it is required to determine the line which a body will describe that is let go from a given place with a given velocity in the direction of a given straight line.*

The force center is at S, and the body is projected along the line PR (which must, therefore, be tangent to the orbit at P). Newton shows how to fix the orbit and identifies the conditions leading to an ellipse, a parabola, or a hyperbola. Naturally, his working is geometrical and presented in prose form with no equations, and it rests heavily on results like those given in *Proposition XVI*. Both modern and contemporary readers would struggle with the details, so again I refer you to the "Further Reading" section if you wish to see Newton's working explained in detail.

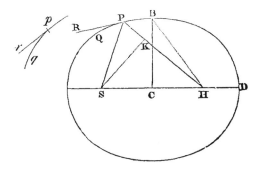

Figure 10.5. Newton's diagram for *Proposition XVII*. (From *The* Principia, trans. Andrew Motte [Amherst, NY: Prometheus Books, 1995].)

However, there is one important point that should be noted. In stating the *Proposition*, Newton uses the words "*it is required to determine the line which a body will describe.*" This sounds quite general ("*the line*"), **but in his working, Newton assumes that the orbit is a conic section.** To be fair, Newton should have replaced "*the line*" with "*the particular conic section.*" You may feel that I am being excessively picky here, but this will become relevant when we delve into a controversy in section 10.5.

10.4.1. Futuristic Theoretical Ideas

Attached to *Proposition XVII* are two *Corollaries* that point to theoretical developments that have become of great importance. It is worth quoting them in full:

> COR. 3 Hence also if a body move in any conic section, and is forced out of its orbit by any impulse, you may discover the orbit in which it will afterwards pursue its course. For by compounding the proper motion of the body with that motion, which the impulse alone would generate, you will have the motion with which the body will go off from a given place of impulse in the direction of a right line given in position.

This is just the situation used when positioning a satellite in orbit around Earth (or indeed in space exploration, around other planets). Small booster rockets may be fired to push the satellite out of its orbit and into the new required orbit—just as Newton explains (see figure 10.6).

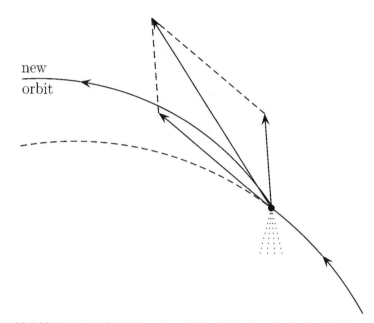

new
orbit

Figure 10.6. Moving a satellite into a new orbit using an impulse from a booster rocket.

The second *Corollary* extends this thinking to a continuous force acting on an orbiting body. Newton obviously has in mind things like the minor influence of planets like Saturn and Jupiter on Earth's orbit, although that does not get mentioned until *Book III*. Newton's ideas are those that underlie perturbation theory, a branch of mathematics that has been used and refined ever since.

> COR. 4. *And if that body is continually disturbed by the action of some foreign force, we may nearly know its course, by collecting the changes which that force introduces in some points, and estimating the continual changes it will undergo in the intermediate places, from the analogy that appears in the progress of the series.*

10.5. DIRECT AND INVERSE PROBLEMS: THE GREAT CONTROVERSY

I cannot leave this review of *Section III*, "*Of the Motion of Bodies in Eccentric Conic Sections*," without discussing one of the strangest and most contentious parts of the *Principia*. Recall from chapter 9 how I explained that Newton set himself the following direct problem: Given that a body moves in a conic section with the force center at the focus, what is the nature of the force involved? The inverse problem is then: given an inverse-square-law force, what is the nature of the resulting orbits? Newton gave a formula for solving the direct problem (see section 9.3), and, in this chapter, we have seen the application to conic-section orbits.

However, after dealing with each conic section in turn in *Propositions XI, XII,* and *XIII*, we find:

> COR. *1. From the last three Propositions it follows, that if any body P goes from the place P with any velocity in the direction of any right line PR, and at the same time is urged by the action of a centripetal force that is inversely proportional to the square of the distance of the places from the centre, the body will move in one of the conic sections, having its focus in the centre of force; and conversely.*

If you read that carefully, you will see that Newton is stating that **both the direct problem and the inverse problem** have been solved—conic section orbits imply inverse-square-law forces, and now, according to *Corollary 1*, inverse-square-law forces produce conic-section orbits. Thus we have one of the great results in the *Principia*.

10.5.1. What Is Wrong?

Nothing—except for the fact that it is not **proven**! Newton has not shown that when the force is inverse-square law, there are no other orbits than those described as conic sections. Because this is such a pinnacle in the *Principia*, there should be a proof. Not to give a proof suggests an error of logic: it is just not true that the solution of a problem implies the solution of its converse.

An example from number theory will make the point. Squaring any odd number and then subtracting 1 produces a number that is a multiple of 8 (e.g., $3^2 - 1 = 8$; $5^2 - 1 = 3 \times 8$; $7^2 - 1 = 6 \times 8$; and so on). However, it is not true that every multiple of 8 can be written as the square of some odd number minus 1; for example, try $4 \times 8 = 32$.

This is an old question in logic, and an early mathematical recognition of its importance is given by Euclid in his *Elements*. Euclid's famous proposition 1.47 is Pythagoras's theorem, but then in proposition 1.48 he proves the converse. If a triangle contains a right angle, then the square on the hypotenuse is equal to the sum of the squares on the other two sides; however, you cannot assume that finding that side's property in a triangle implies that there must be a right angle without giving an independent proof. Someone of Newton's mathematical stature was obviously aware of such things. Furthermore, if you recall, his very first *Proposition* (*Proposition I* about sweeping out equal areas) was followed by its converse and its proof in *Proposition II*.

10.5.2. Newton's Response

Even people like Halley, Hooke, and Huygens did not spot this deficiency in Newton's work, but eventually there were rumblings and Johann Bernoulli recognized the problem. Newton became aware of the defect and made an addition for the second edition. By the time of the third edition (the one discussed in this book), *Corollary 1* as quoted above had been extended:

> . . . *For the focus, the point of contact, and the position of the tangent, being given, a conic section may be described, which at that point shall have a given curvature. But the curvature is given from the centripetal force and velocity of the body being given; and two orbits, mutually touching one the other, cannot be described by the same centripetal force with the same velocity.*

I will not go into details (see this chapter's "Further Reading" section), but Newton is constructing a uniqueness argument—for the given conditions, a conic section orbit can be constructed and there cannot be any other solution. Although debate continued, apparently Johann Bernoulli accepted Newton's argument:

> Gladly I believe what you say about the Addition to Corollary 1, Proposition 13, Book I of your incomparable work the *Principia*, that this was certainly done before these disputes began, nor have I any doubts that the demonstration of the inverse proposition, which you have merely stated in the first edition, is yours.[1]

But it is interesting to read on:

> I only said something against the form of that assertion, and wished that someone would give an analysis that led a priori to the truth of the inverse and without supposing the direct proposition to be already known. This indeed, which I would not have aid to your displeasure, I think was first put forward by me, at least as far as I know at present.[2]

The interested reader may also go to chapter 11 on the modern formulation to find this problem set out in section 11.3.5.

10.5.3. Relevant Parts Elsewhere in the *Principia*

It is interesting to look back at the two examples of the use of the general formula discussed in section 9.4. First, for a spiral orbit, Newton showed that an inverse-cube force is responsible. However, he did not then say that inverse-cube forces **always** lead to such spirals in a general way because he knew that was not correct. Secondly, for the case of motion in an ellipse with the force center at the ellipse center, he found the force to be directly proportional to the distance, and he gives:

> COR. 1. *And therefore the force is as the distance from the centre of the ellipse; and, vice versa, if the force is as the distance, the body will move in an ellipse whose centre coincides with the centre of force, or perhaps a circle into which the ellipse may degenerate.*

Again he has gone from the direct to the inverse problem, apparently without any resulting controversy. In this case, Newton is correct (and I will return to this in modern form in the next chapter).

Clearly Newton appreciated what was going on, but he did not clarify things as he might have done. You might think that he successfully tackled the inverse problem for inverse-square-law forces in *Proposition XVI*, as discussed in section 10.4. However, as I pointed out, in the working, he actually assumed the answer: the orbit takes the form of a conic section.

In chapter 13, I will describe how Newton solved the general problem of how to find the orbit for any given force (the famous *Proposition XLI*). It would be natural to illustrate the use of that result for the vitally important case of an inverse-square-law force. But no, Newton uses an inverse-cube-law force as an example!

It has been suggested that Newton was unable to carry out the mathematical manipulations for the inverse-square-law case, but that is clearly not correct, as the required mathematical steps were given by him elsewhere. So what are we to make of all of this? Newton as perverse, arrogant, playful, weird, teasing, flippant, ignorant, enigmatic . . . ? Take your pick!

10.5.4. Yet Again!

It might seem astounding, but the whole controversy has been replayed some three hundred years after the *Principia* first appeared. It all began with Robert Weinstock's 1982 paper, "Dismantling a Centuries-Old Myth: Newton's *Principia* and Inverse-Square Orbits." There were more papers by Weinstock as well as papers in response by a whole array of historians and mathematicians. Even the eminent Russian mathematician V. I. Arnol'd became involved, and he certainly accepted Newton's version: "Thus, there is no doubt about the uniqueness and Newton correctly proved Kepler's first law."[3] Weinstock has now died, so perhaps there is peace again. See the "Further Reading" section at the end of this chapter to follow up on the details.

10.6. CONCLUSION

We have now learned about one of the great sections in the *Principia*. Newton's major triumph was to show quite generally how conic-section orbits and inverse-square-law forces are linked. This was the kind of sweeping, powerful, and high-level mathematical generalization that people like Hooke were incapable of making. See figure 10.7 for a map of Newton's work.

It is amazing that amid this mathematical brilliance, there lurks a controversial step that, at its worst, would suggest a defect in Newton's logic and mathematical working. However, for Newton, it all seems to be so obvious and his critics *"are deluded,"*[4] as he wrote in the late 1710s when explaining his approach to and use of the orbit-uniqueness property.

It is also interesting to observe the level of mathematical development and sophistication that are evident in the subject matter of that controversy. It is a world away from anything to be found in Galileo's writings or in Descartes's *Principia Philosophiae*; it is already approaching our modern treatments of the subject. (And maybe Newton is being judged a little unfairly using modern levels of rigor and completeness.)

Thinking about progress so far in the development of classical mechanics, as outlined in figure 10.7, we can identify two holes that Newton should next fill in (and he does!). First, if you compare the two flows in figure 10.7, it becomes apparent that on the left (Newton's inverse problem, or the forces-to-orbits problem), there is no box talking about the general solution. Filling this gap will produce some of the outstanding results in the theory of classical mechanics.

Second, there is no mention of time—how do we know where on its orbit a body will be at a specified time? This problem has a general formulation and specific results for inverse-square-law orbits.

Before seeing how Newton meets those challenges (in chapters 12 and 13), I think it will be useful to revisit the modern formulation and see how Newton's results translate into the form we use today.

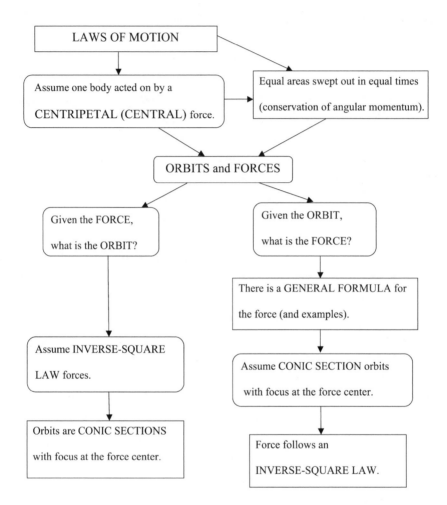

Figure 10.7. A flowchart for the developments in *Principia Sections II* and *III*.

10.7. FURTHER READING

The standard references given in section 6.9.1 should be consulted.

Perturbation methods are discussed by Goldstein and Moulton and are well covered in the following source:

Linton, C. M. *From Eudoxus to Einstein: A History of Mathematical Astronomy*. Cambridge: Cambridge University Press, 2004.

For the controversy over conics and inverse-square laws, see the books by Brackenridge, Cohen (section 6.4), and Guicciardini, as well as references therein. Also see:

Arnol'd, V. I. *Huygens and Barrow, Newton and Hooke*. Basel: Birkhauser Verlag, 1990.

Guicciardini, N. "Johann Bernoulli, John Keill and the Inverse Problem of Central Forces." *Annals of Science* 52 (1995): 537–75.

Nauenberg, M. "The Early Application of the Calculus to the Inverse Square Force Problem." *Archive for the History of the Exact Sciences* 64 (2010): 269–300.

Pourciau, Bruce H. "On Newton's Proof That Inverse-Square Orbits Must Be Conics." *Annals of Science* 48 (1991): 159–72.

Weinstock, R. "Dismantling a Centuries-Old Myth: Newton's *Principia* and Inverse Square Orbits." *American Journal of Physics* 50 (1982): 610–17.

———. "Inverse Square Orbits in Newton's *Principia* and Twentieth-Century Commentary Thereon." *Archive for the History of the Exact Sciences* 55 (2000): 137–62.

For a range of opinions on Newton's work and Weinstock's criticisms, see the May 1994 issue of the *College Mathematics Journal*, which has responses by scientists, historians, and Newton's major biographer, Richard Westfall.

11

INTERLUDE
THE MODERN FORMULATION

Some readers may feel uneasy starting this chapter because it introduces the modern theory of classical mechanics in its algebraic or symbolic form and uses the tools of calculus. I hope that you will still read on, because it is not absolutely essential to appreciate all the details. I believe you can follow the nature of the development and see the form of the equations— which variables are involved and the level of complication. In summary, I believe that even without extensive mathematical skills, you can still get a feel for the modern form of classical mechanics, its triumphs, and its challenges.

Elements of the modern formulation have been given in sections 5.6, 6.6, and 8.5. Recall that we began with kinematics: a particle with mass m is assumed small so that its position is specified by the position vector $\mathbf{r}$ (see figure 5.4), which may depend on the time t, so we write $\mathbf{r}(t)$. A changing position leads to the velocity vector $\mathbf{v} = d\mathbf{r}/dt$, and, in turn, a varying velocity leads to the acceleration vector $\mathbf{a} = d\mathbf{v}/dt$. Dynamical matters are introduced when the particle is characterized by its mass m and the force $\mathbf{F}$ that acts on it.

11.1. LAWS OF MOTION AND CONSEQUENCES

Newton's second law is written as

$$m\mathbf{a} = \mathbf{F} \quad \text{or} \quad m\frac{d\mathbf{v}}{dt} = \mathbf{F} \quad \text{or} \quad m\frac{d^2\mathbf{r}}{dt^2} = \mathbf{F}, \qquad (11.1)$$

and we assume that an inertial frame of reference is being used (see section 6.6.3).

To make contact with Newton's procedures, we may note that a particle at position $\mathbf{r}(t_0) = \mathbf{r}_0$ with velocity $\mathbf{v}_0$ may be tracked to a new position a short time Δt later by substituting equation (11.1) into the first terms in a Taylor series for $\mathbf{r}(t_0 + \Delta t)$:

$$
\begin{aligned}
\mathbf{r}(t_0 + \Delta t) &= \mathbf{r}_0 + \frac{d\mathbf{r}}{dt}\Delta t + \frac{1}{2}\frac{d^2\mathbf{r}}{dt^2}(\Delta t)^2 \\
&= \mathbf{r}_0 + \mathbf{v}_0\Delta t + \frac{1}{2m}\mathbf{F}_0(\Delta t)^2.
\end{aligned}
\tag{11.2}
$$

The first term gives the original position, the second term gives the inertial motion change in the time Δt, and the third term is the effect of the force acting for a time Δt. This is just what we have seen in Newton's working—see figure 9.1, for example.

If we use equation (11.1) together with Newton's third law, we can establish the **conservation of momentum law** and solve for **the motion of the center of mass** (see sections 6.6.1 and 6.6.2). If we assume a single particle and specify that the force acting on it is a centripetal (central) force, then we can establish the **conservation of angular-momentum law** (see section 8.5), equivalent to Newton's result of **equal areas swept out in equal times**.

Those are **general** results in dynamics. We must now turn to the more specific problem: given a force $\mathbf{F}$, find the trajectory or orbit that a particle follows if at time $t = 0$, say, it is launched from a given position with a given velocity. The values for position and velocity at $t = 0$ are called the **initial conditions**.

11.2. THE EQUATION OF MOTION

The third form in equation (11.1) is called the **equation of motion** for the particle. We assume that **F** depends on the position r, so **F** = **F**(**r**) and we have a **second-order differential equation** for **r**(t). (It is second order because it involves the second derivative $d^2\mathbf{r}/dt^2$.) Because we are using vectors, the equation contains details of the components of motion in all directions. If we use (x,y,z) as the components of **r** and resolve (as Newton tells us how to do—see section 6.2) in each direction, we find that there are really three equations:

$$m\frac{d^2x}{dt^2} = F_x \qquad m\frac{d^2y}{dt^2} = F_y \qquad m\frac{d^2z}{dt^2} = F_z. \qquad (11.3)$$

Now the force **F** = (F_x, F_y, F_z), so the component of the force **F** in the x direction is F_x, and so on.

Note that each component may still depend on all position coordinates; we may have $F_x = F_x(x, y, z)$, and only in special cases will we have just $F_x = F_x(x)$. In that way, the three equations in (11.3) are really **coupled together,** and that is often what makes classical mechanics create such a tough problem.

A second-order differential equation, like those in equation (11.3) has a solution that contains **two arbitrary constants**; they may be chosen so that the **initial conditions** are satisfied. Said a different way, the initial position and velocity are used to fix or pick out the **particular** path that the particle will follow. Newton did that in his geometrical way in *Proposition XVII*, which I discussed in section 10.4.

The best way to appreciate all of this is to look at relatively simple examples. I will begin with Galileo's projectile problem (surely the simplest of all meaningful problems) and then use the problem introduced by Newton as an example for finding force laws (see section 9.4).

11.2.1. Projectile Motion

The motion of a projectile in a vertical plane, Galileo's famous problem, provides a simple example. In this case, I take the x and y axes as shown in figure 11.1. The only force operating is downward in the negative y direction with constant magnitude mg, where g is the usual local gravitational constant. Resolving the equation of motion (11.1) in the x and y directions gives

$$m\frac{d^2x}{dt^2} = 0 \quad \text{and} \quad m\frac{d^2y}{dt^2} = -mg . \tag{11.4}$$

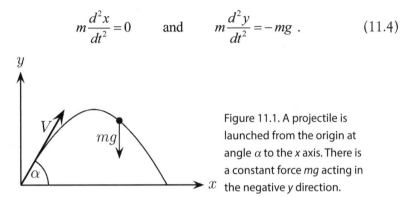

Figure 11.1. A projectile is launched from the origin at angle α to the x axis. There is a constant force mg acting in the negative y direction.

This is a case where the equations are not coupled; the equation for $x(t)$ does not depend on y, and similarly for $y(t)$. These equations of motion are easily solved to give

$$x(t) = At + B \quad \text{and} \quad y(t) = -\tfrac{1}{2}gt^2 + Ct + D. \tag{11.5}$$

As anticipated, we have four constants, $A, B, C,$ and D, in the solution. If we take the projectile to be launched at $t = 0$ from the origin with speed V and at an angle α to the horizontal, as shown in figure 11.1, then those four constants are determined, and

$$x(t) = \{V \cos(\alpha)\}t \quad \text{and} \quad y(t) = -\tfrac{1}{2}gt^2 + \{V \sin(\alpha)\}t. \tag{11.6}$$

This is a case where we have the complete solution to the dynamical problem, with both x and y known as functions of the time t. It is also easy to eliminate time using the fact that

$$t = \frac{x}{\{V \cos(\alpha)\}}, \text{ to find the trajectory path as}$$

$$y = -\left\{\frac{g}{2[V \cos(\alpha)]^2}\right\}x^2 + \tan(\alpha)x. \tag{11.7}$$

This is the parabolic path Galileo discovered, and we can find that the range (the second point where $y = 0$) is a maximum when α is forty-five degrees, as he predicted.

Notice in the y equation in (11.4) that the same m occurs on both sides. This is because I have already assumed that the inertial mass and the gravitational mass are equal, as discussed in section 5.1. I return to this point in a later chapter.

11.2.2. Linear Force Example

We assume an attractive central force (or centripetal force, as Newton would call it) with a magnitude proportional to the distance from the force center (Newton's example as discussed in section 9.4). This is called a linear force. Choosing the center of force as the origin of coordinates gives the force and its resolved components as

$$\mathbf{F} = -k\mathbf{r} \text{ and } F_x = -kx, F_y = -ky, F_z = -kz. \qquad (11.8)$$

Here, k is a positive constant giving the force strength, and the minus sign makes it into an attractive force. Assume motion is in the x–y plane, as shown in figure 11.2.

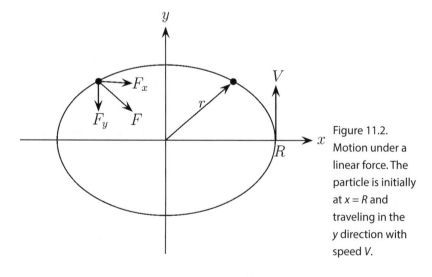

Figure 11.2. Motion under a linear force. The particle is initially at $x = R$ and traveling in the y direction with speed V.

This is a case where the equations in (11.3) are not coupled, and we simply get

$$m\frac{d^2x}{dt^2} = -kx \qquad m\frac{d^2y}{dt^2} = -ky$$

$$\frac{d^2x}{dt^2} = -\omega^2 x \qquad \frac{d^2y}{dt^2} = -\omega^2 y \qquad \text{where} \qquad \omega^2 = \frac{k}{m}. \qquad (11.9)$$

These are relatively simple differential equations—they are linear, so only x occurs in the first one, not x^2 or other powers or more complicated functions of x. Their solutions are

$x(t) = A \cos(\omega t) + B \sin(\omega t)$

$y(t) = C \cos(\omega t) + D \sin(\omega t)$, where A, B, C, and D are constants. (11.10)

As expected for two second-order differential equations, there are four constants in these general solutions, and they may be used to fit any given initial conditions. For example, if at $t = 0$, the particle is crossing the x axis at $x = R$ and traveling in the y direction with speed V, as shown in figure 11.2, we must set

$$A = R \qquad B = 0 \qquad C = 0 \qquad D = V/\omega \qquad (11.11)$$

Thus, the solution is

$$x(t) = R \cos(\omega t) \quad y(t) = (V/\omega)^2 \sin(\omega t) \quad \text{and so} \quad \frac{x^2}{R^2} + \frac{y^2}{(V/\omega)^2} = 1. \, (11.12)$$

The last expression—which follows by using the trigonometric identity $\cos^2(\omega t) + \sin^2(\omega t) = 1$—tells us that the orbit is an ellipse, as shown in figure 11.2. Notice that the time dependence of all possible motions is exactly the same. If the initial conditions are varied, the orbit will change:

if $V = R\omega$, then the orbit will be a circle with radius R;

if $V = 0$, then $y = 0$ and the orbit is part of a straight line—the particle oscillates along the x axis between $x = R$ and $x = -R$.

Notice that, in this case, we have again completely solved the dynamical problem. We have found the **time dependence of the motion, the dependence on the initial conditions, and we have also exhibited the form of the orbit**. That is the ideal solution, but it is rarely found as easily as it is in this case!

11.3. THE GENERAL-CENTRAL-FORCE CASE

I will assume that motion takes place in the x–y plane and take the origin as the center of force. Thus the general force has the form $\mathbf{F} = F(r)\hat{\mathbf{r}}$, where $\hat{\mathbf{r}}$ is the unit vector in the $\mathbf{r}$ direction and r is the magnitude, so that $\mathbf{r} = r\hat{\mathbf{r}}$ (see figure 11.3). Because the force takes this form, the natural coordinates to use are the polar coordinates (r, θ) rather than the Cartesian coordinates x and y.

11.3.1. Kinematics

We also need the unit vector $\hat{\boldsymbol{\theta}}$, which is perpendicular to $\hat{\mathbf{r}}$, as shown in figure 11.3. Then standard manipulations (see the texts listed in "Further Reading" for section 6.9.1) give

$$\text{velocity:} \quad \mathbf{v} = \frac{d\mathbf{r}}{dt} = \frac{dr}{dt}\hat{\mathbf{r}} + r\frac{d\theta}{dt}\hat{\boldsymbol{\theta}} \tag{11.13}$$

The formula for the acceleration is a little more complicated because the polar-coordinate unit vectors also change along the orbit. The result of differentiating equation (11.13) is

$$\text{acceleration:} \quad \mathbf{a} = \frac{d\mathbf{v}}{dt} = \left\{ \frac{d^2r}{dt^2} - r\left(\frac{d\theta}{dt}\right)^2 \right\}\hat{\mathbf{r}} + \left\{ 2\frac{dr}{dt}\frac{d\theta}{dt} + r\frac{d^2\theta}{dt^2} \right\}\hat{\boldsymbol{\theta}}. \tag{11.14}$$

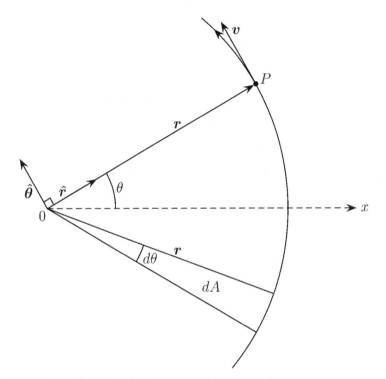

Figure 11.3. A particle *P* has polar coordinates (r, θ).

11.3.2. Equations of Motion

We now take the equation of motion $m\mathbf{a} = \mathbf{F}$, with $\mathbf{F} = F(r)\hat{\mathbf{r}}$, and resolve in the $\hat{\mathbf{r}}$ and $\hat{\boldsymbol{\theta}}$ directions using equation (11.14) for **a**:

$$m\left\{\frac{d^2r}{dt^2} - r\left(\frac{d\theta}{dt}\right)^2\right\} = F(r), \qquad (11.15)$$

$$m\left\{2\frac{dr}{dt}\frac{d\theta}{dt} + r\frac{d^2\theta}{dt^2}\right\} = 0. \qquad (11.16)$$

These are two coupled, second-order differential equations. Solving them will give $r(t)$ and $\theta(t)$, and, hence, the body's position on its orbit at any time t. However, even a casual glance at those equations suggests that is not usually a very simple procedure to follow, and so some further theoretical development is called for.

11.3.3. A Constant of Motion: Angular Momentum

If we multiply equation (11.16) by r, we obtain

$$m\left\{2r\frac{dr}{dt}\frac{d\theta}{dt}+r^2\frac{d^2\theta}{dt^2}\right\} = 0 \quad \text{or} \quad m\left\{\frac{dr^2}{dt}\frac{d\theta}{dt}+r^2\frac{d^2\theta}{dt^2}\right\} = 0.$$

We now recognize that this can be written as a total derivative,

$$m\frac{d}{dt}\left\{r^2\frac{d\theta}{dt}\right\} = 0 \quad \text{or} \quad \frac{d}{dt}\left\{mr^2\frac{d\theta}{dt}\right\} = 0. \quad (11.17)$$

A zero derivative means there is no variation with time, and this tells us that

$$\frac{dh}{dt}=0, \quad \text{where} \quad h=r^2\frac{d\theta}{dt}, \quad \text{so } h \text{ is a constant.} \quad (11.18)$$

This is exactly the conservation law that we discussed in chapter 8 and embodied in equation (8.4). Thus the angular-momentum constant of motion may also be found by examining the equation of motion resolved in the $\hat{\theta}$ direction, that is, in the direction perpendicular to the force.

To complete the link, we can note that in polar coordinates, an element of area dA (see figure 11.3) is given by $(\frac{1}{2})r^2 d\theta$, and we are led to

$$\frac{dA}{dt} = \frac{1}{2}r^2\frac{d\theta}{dt} = \frac{1}{2}h = \text{a constant.}$$

This says that the area is traced out at a constant rate, which is just the equal-areas-swept-out-in-equal-times result, the geometric equivalent of the conservation of angular momentum mh.

11.3.4. Simplifying the Equations of Motion

Recall that in section 8.4.1, I explained how the conservation of angular-momentum law leads to a major advance in the theory, and now I can show that in detail. The fact that h is a constant allows us to take two key steps:

first, replace the θ time derivative using

$$\frac{d\theta}{dt} = \frac{h}{r^2};$$
(11.19)

second, change time t derivatives to θ derivatives by the calculus chain rule

$$\frac{dx}{dt} = \frac{d\theta}{dt}\frac{dx}{d\theta},$$

so that we have

$$\frac{d}{dt} = \frac{h}{r^2}\frac{d}{d\theta}.$$
(11.20)

Using those steps, we can convert equation (11.15) into **the orbit shape equation**

$$\frac{1}{r^2}\frac{d}{d\theta}\left(\frac{1}{r^2}\frac{dr}{d\theta}\right) - \frac{1}{r^3} = \left(\frac{1}{mh^2}\right)F(r).$$
(11.21)

Solving that equation will give us r as a function of θ, which is a formula for the shape of the orbit. We no longer ask where the particle is at any given time, but we find instead the complete orbit or trajectory.

Equation (11.21) involves only r and θ, but it is still a formidable-looking equation. It takes a much simpler form if a variable change is made: instead of r, I introduce the variable u, which is the reciprocal of r (as in, $u = 1/r$). The orbit-shape equation then becomes an equation for $u(\theta)$:

$$\frac{d^2u}{d\theta^2} + u = -\left(\frac{1}{mh^2}\right)\frac{F(1/u)}{u^2}.$$
(11.22)

This is a much friendlier-looking equation, although how easy it is to solve will depend on the form of the force law.

11.3.5. An Important Example

If we consider an attractive inverse-square-law force with strength constant c, we must set

$$F(r) = -\left(\frac{c}{r^2}\right) \quad \text{or} \quad F\left(\frac{1}{u}\right) = -cu^2.$$

Substituting into equation (11.22) gives

$$\frac{d^2u}{d\theta^2} + u = -\left(\frac{1}{mh^2}\right)\frac{(-cu^2)}{u^2} = \frac{c}{mh^2},$$

which can be written as

$$\frac{d^2u}{d\theta^2} + u = K, \qquad \text{where} \quad K = \frac{c}{mh^2}. \qquad (11.23)$$

Equation (11.23) is a simple linear second-order equation with solution

$$u = K + A\cos(\theta) + B\sin(\theta)$$
$$= K + A'\cos(\theta + \beta),$$

where, as expected, there are two constants A and B that may be replaced by constants A' and β. Converting back to r, we get

$$r = \frac{1}{K + A'\cos(\theta + \beta)} = \frac{p}{1 + \varepsilon\cos(\theta + \beta)}, \quad \text{where } p = \frac{1}{K} \text{ and } \varepsilon = \frac{A'}{K}. \quad (11.24)$$

Referring back to section 10.1, we can see that the solution in equation (11.24) is a conic section, and if $\varepsilon < 1$, it is the ellipse-bound orbit. (The extra constant β just sets the orientation of the ellipse relative to the x axis, or $\theta = 0$ axis.) In effect, Newton discusses how to fix the orbit constants in his *Proposition XVII*, as we have seen in section 10.4.

Notice that equation (11.24) gives the complete solution for equation (11.23), and, hence, for the equation of the possible orbits for an inverse-square-law force. By appropriately choosing the constants A and B, all cases are covered. **This is the mathematical answer to Newton's inverse problem: if the force is inverse square, then all possible particle trajectories are conic sections.** This is the missing explicit proof that caused the controversy discussed in section 10.5.

11.3.6. Time and Energy

If we use only the first simplification (equation [11.19]) in the equation of motion (11.15), we get

$$m\left\{\frac{d^2r}{dt^2} - \frac{h^2}{r^3}\right\} = F(r), \tag{11.25}$$

which is the differential equation for $r(t)$. However, writing

$$\frac{d^2r}{dt^2} = \frac{dr}{dt}\frac{d}{dr}\left(\frac{dr}{dt}\right) = \frac{1}{2}\frac{d}{dr}\left(\frac{dr}{dt}\right)^2,$$

and integrating with respect to r changes equation (11.25) into

$$\frac{1}{2}m\left(\frac{dr}{dt}\right)^2 + \frac{1}{2}m\frac{h^2}{r^2} = \int F(r)dr = -V(r) + E.$$

I have introduced the potential $V(r)$ and the constant of integration E. This is usually written as

$$\frac{1}{2}m\left(\frac{dr}{dt}\right)^2 + \frac{1}{2}m\frac{h^2}{r^2} + V(r) = E. \tag{11.26}$$

This is simply kinetic energy plus potential energy equals the total energy E, and we may show that E is truly a constant (independent of time), so that equation (11.26) is a statement of the conservation of energy. Equation (11.26) is now a first-order differential equation, so only one integration is required to find $r(t)$.

11.4. CONCLUSION

The above theory and examples illustrate the modern approach to classical mechanics. For some examples, the complete time solution for particle motion has been given. However, as in the case of Newton's development of the subject outlined in the previous chapter, we are missing the method for finding the time dependence for motion on central-force orbits and a way to give the general solution. Those two problems will be addressed in the next two chapters, where both Newton's and the modern approaches will be given.

Because you may have already thought of them, I mention here two extensions to the above theory that Newton made, and which I will cover in later chapters. The theory presented so far assumes small bodies, the classic "point particle," so we will have to find out how larger, composite bodies may be included. It will also be necessary to go beyond this single-particle case and consider several interacting bodies.

As we move to evermore complex situations, Newton's formulation of classical mechanics becomes cumbersome and, at times, inadequate; therefore, toward the end of the book, in chapter 28, I will briefly discuss other approaches to the theory that were gradually developed over the centuries after Newton.

11.5. FURTHER READING

The textbooks listed in section 6.9.1 may be consulted. The book by Fowles is recommended.

12

TIME AND A MATHEMATICAL GEM

s Newton put it: "*So far concerning the finding of orbits. It remains that we determine the motions of bodies in the orbits so found.*"

We know the orbits, but what about motion along them? Where on the orbit will a body be at a given time? Thus in the *Principia* we come to *Section VI: "How the Motions Are to Be Found in Given Orbits."*

The orbits to be dealt with are the conic sections produced by inverse-square-law forces.

This *Section* sees Newton at his mathematical best, producing results that we still marvel at and use today. We also have Newton at his most difficult to read, with more-intricate geometrical arguments presented in prose form and with little thought about guidance for the reader. I am sure many people find this *Section* impenetrable, and I will describe its contents largely in a modern setting. This work is too important for it to be lost in obscurities (but see the "Further Reading" section at the end of this chapter for comment on that).

12.1. THE NATURE OF THE PROBLEM

We know that, generally, a body does not move at a uniform speed along its orbit around a centripetal force center, the exception being motion in a circle. However, we can be sure that the line from the force center to the body sweeps out area at a constant rate (see figure 12.1). If a body moves from Q to P and in so doing sweeps out an area A_{QP} at the constant rate s, then the time τ taken to move from Q to P is given by $A_{QP} = \tau s$. If the total orbit encloses the area A_{tot} and the period is T, then $s = A_{tot}/T$, so

$$A_{QP} = \tau s = \frac{\tau A_{tot}}{T} \quad \text{or} \quad \tau = \frac{T A_{QP}}{A_{tot}}. \tag{12.1}$$

Thus, if a body is at point Q on its orbit and we wish to know the point P, which it will reach after a time τ, we first calculate the value of A_{QP} according to equation (12.1) and then use geometry to find the point P, which gives that area. That is conceptually simple, but it is not so easy to implement practically, as we shall see.

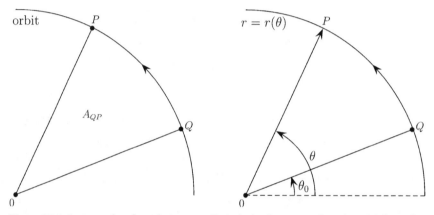

Figure 12.1. A square-law force has center O, and a body moves along its orbit from Q to P, sweeping out an area A_{QP}. When using polar coordinates (r, θ), the body at Q has the initial angle θ_0.

In the modern formulation, the equal-areas-swept-out-in-equal-times law is known as the conservation of angular momentum, as discussed in the previous chapter, section 11.3.3. If we use polar coordinates (r, θ), as in figure 12.1, we know that

$$h = r^2 \frac{d\theta}{dt}. \tag{12.2}$$

The constant h gives the area-sweeping constant s in equation (12.1) as $s = (1/2)h$ (see section 11.3.3). We can integrate equation (12.2) to relate time and orbit position if the orbit is known in the form $r = r(\theta)$. Let the body be at Q, $\theta = \theta_0$ at time $t = 0$. A time τ later, it is at P, as shown in figure 12.1. Then,

$$h\int_0^\tau dt = \int_{\theta_0}^\theta r^2(\theta)d\theta \quad \text{so that}$$

$$h\tau = \int_{\theta_0}^\theta r^2(\theta)d\theta \quad \text{or} \quad h\tau = 2A_{QP}. \tag{12.3}$$

If we carry out the θ integration, we will have an equation involving time τ and angle θ for the point P; for a given τ, we can therefore solve to find where on the orbit the body will be. Of course, this is just the same as using equation (12.1), but we are employing an algebraic description of the orbit.

12.2. MOTION ALONG A PARABOLA

This is not such an important case, but it is one where the time behavior can be found quite easily.

Proposition XXX tells us how "*to find at any assigned time the place of a body moving in a given parabolic trajectory.*" Newton's diagram is reproduced in figure 12.2. The body starts at the vertex A and moves to the point P, sweeping out the area ASP.

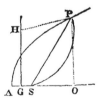

Figure 12.2. Newton's diagram for motion along a parabola from the vertex A to a point P. S is the focus, and G is midway between A and S. The line GH is perpendicular to AGS. The area swept out in the motion is ASP. The circle with center H and radius HS cuts the parabolic path at P. (From *The* Principia, trans. Andrew Motte [Amherst, NY: Prometheus Books, 1995].)

Newton explains how to find the required area and solve the time problem. Finding areas bounded by parabolas and straight lines was a problem solved by ancient geometers. In its modern form, finding the time taken to reach a point on the parabolic trajectory involves the solution of a cubic equation, and Newton effectively does the same thing using a geometrical method. Interested readers should consult the book by Chandrasekhar listed in the "Further Reading" section at the end of the chapter.

12.3. MOTION ALONG AN ELLIPSE

This is one of the great problems of celestial mechanics and applied mathematics, and it is fascinating to see Newton's approach to it.

We are now ready for *Proposition XXXI, Problem XXIII*: "*To find the place of a body moving in a given elliptic trajectory at any assigned time.*" The situation is shown in figure 12.3. The force center is at the ellipse focus F. In time τ the body moves from Q to P, sweeping out the area $A_{QP} = FQPF$. Using equation (12.1) allows us to solve the given problem.

A difficulty arises when we try to calculate the area FQP; in contrast to the parabolic case, there are no simple formulas for that area in the ellipse or for the integral giving it in a modern formulation. The problem may be tackled using the auxiliary circle, which is a circle of radius a just fitting around the ellipse, as shown in figure 12.4. Points P on the ellipse are related to those on the auxiliary circle P' in a simple way: their y values are in the ratio b/a or, using figure 12.4, $NP = (b/a)NP'$.

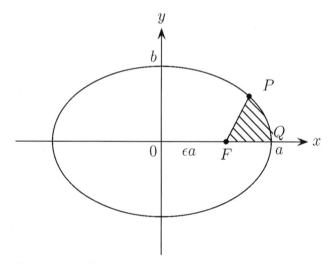

Figure 12.3. Motion on an ellipse with a major axis of length $2a$, a minor axis of length $2b$. The eccentricity is $\varepsilon = \sqrt{1 - b^2/a^2}$ and the focus F is distance εa from the ellipse center O. The shaded area $FQPF$ is swept out as the body travels from Q to P.

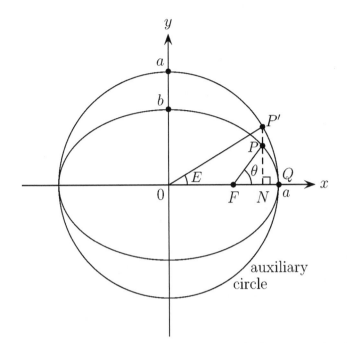

Figure 12.4. The ellipse and its auxiliary circle. The perpendicular at N cuts the ellipse at P and the circle at P'. There is a scaling property: $NP = (b/a)NP'$. Using O as the coordinate center, the point P' has polar coordinates (a,E), where E is the angle shown at O.

Areas of sectors of circles are easily obtained, and the area OQP' is simply $(1/2)a^2E$, where E is the enclosed angle at O, as shown in figure 12.4. The area of triangles is also easily calculated. Using that information and the properties of the auxiliary circle leads to

$$A_{QP} = FQP = (1/2)ab\{E - \varepsilon \sin(E)\}. \qquad (12.4)$$

(Brief details of how to prove this are given in section 12.4.1.) Substituting that into equation (12.1) and using the fact that the ellipse has a total area of $A_{tot} = \pi ab$ gives

$$\left(\frac{2\pi\tau}{T}\right) = \{E - \varepsilon \sin(E)\}. \qquad (12.5)$$

Historically, the angle E is known as the eccentric anomaly, and the angle θ at F (see figure 12.4) is the true anomaly. We also define the mean anomaly $M = (2\pi\tau/T)$. We finally arrive at the famous

$$\text{Kepler's equation: } M = E - \varepsilon\sin(E). \qquad (12.6)$$

Thus, given the time τ and the period T, we calculate M and then solve Kepler's equation to find E, which gives point P' on the auxiliary circle and, hence, the required point P on the ellipse.

However, the difficulty of calculating areas of parts of ellipses has been replaced by the need to solve Kepler's equation. That equation involves a sine, and is therefore a transcendental equation, not something we immediately recognize, as happened in the parabolic-motion case where we had to solve a relatively simple cubic equation. After the subsection (for the mathematically inclined reader) on the derivation of Kepler's equation, I will discuss approaches to equation solving and how Newton used a range of them to tackle Kepler's equation.

12.4.1. Deriving Kepler's Equation

The difficulty came in finding $A_{QP} = FQP$. Looking at figure 12.4, we see that

$$FQP = \Delta FNP + NQP.$$

It is the area NQP, with its curved boundary, that gives us trouble. However, the scaling property of the auxiliary circle tells us that the troublesome area can be related to the curved-boundary area in the circle:

$$NQP = (b/a)NQP'.$$

But NQP' is just the difference between the sector area OQP' and the triangle area $\Delta ONP'$. Finally, then we have

$$FQP = \Delta FNP + (b/a)\{OQP' - \Delta ONP'\}.$$

All of those areas can be evaluated in terms of the ellipse-length parameters a and b and the eccentric anomaly, the angle E. Equation (12.4) follows after a little algebra.

12.5. NEWTON'S APPROACHES TO KEPLER'S EQUATION

We are now ready to discuss Newton's approach to Kepler's equation. It includes some of the most remarkable mathematical results in the whole *Principia*. Unfortunately, it is also one of the more opaque parts, and you must look elsewhere (see this chapter's "Further Reading" section) if you want to battle with Newton's own details! Of course, Kepler had already had his battles with this problem long before Newton, and philosopher Alexandre Koyré gives some relevant excerpts from Kepler's work with enlightening commentaries.

The problem, then, is to solve equation (12.6) to find E for a given M. It might be useful to step back and ask what strategies we usually adopt when confronted by such an equation-solving problem. There are three common possibilities:

Approach 1: Try to find a formula. For example, many people know the formula for the roots of a quadratic, that is, for the values of x that make $x^2 + bx + c = 0$ for given constants b and c. There is also a formula for solving cubics, and that is why the parabolic-motion problem is easily solved.

Approach 2: Tackle the problem visually. For example, to find the roots of the quadratic, plot the function $y = x^2 + bx + c$ and observe where it crosses the y axis to get the x values making $y = 0$. This is a very useful method for complicated functions if a computer can be used. By drawing the graph on a larger and larger scale, we can get the axis crossing points with greater and greater accuracy.

Approach 3: Generate an approximate answer. This could involve producing a formula giving an approximation to the exact answer. Or it could be a method that produces a sequence of evermore accurate answers; such iterative methods are of great importance and allow a crude first estimate (perhaps obtained graphically) to be refined until an answer of the required accuracy is obtained.

Kepler's equation is famous, and there is an extensive literature, including one whole book about it giving many examples of all three approaches (see the "Further Reading" section at the end of this chapter). Perhaps we should not be surprised that a master mathematician like Newton pioneered those approaches and used them in the *Principia*.

12.5.1. Is There a Formula for Solving Kepler's Equation?

No doubt Newton looked for a formula, but when he could not find one, he set himself a new task: prove that no such formula exists. This question of the existence of solutions or methods is one of the great parts of mathematics, and Newton gave us one of the earliest substantial examples.

Some readers may be reminded of the problem of solving polynomial equations. By 1600, there were available the old formula for solving quadratic equations and rather similar formulas for dealing with cubics and quartics—they can be readily found today in mathematics handbooks. However, the extension to quintics and beyond turned out to be difficult. Early in the nineteenth century, mathematicians Niels Henrik Abel and Évariste Galois turned this problem around, questioned the existence of the sought-after formulas, and finally proved that they did not exist. Newton was doing a similar thing at the end of the seventeenth century.

Newton gives:

LEMMA XXVIII.

There is no oval figure whose area, cut off by right lines at pleasure, can be universally found by means of equations of any number of finite terms and dimensions.

It is necessary to read more of Newton's comments and proof to see more clearly what is involved (see figure 12.5). For example, the oval is to be algebraic, that is, it is the graph of $P(x,y)$, where P represents some sort of finite polynomial expression, such as that in equation (10.1) for the ellipse. The *Lemma* states that any area A cut off by a line $ax + by = c$, where a, b, and c are arbitrary constants, cannot be expressed in a finite polynomial form involving those constants. In particular, Newton's *Corollary* states:

> COR. *Hence the area of an ellipse, described by a radius drawn from the focus to a moving body, is not to be found from the time given by a finite equation.*

Said another way, there is no finite polynomial formula that can be used to give a solution of Kepler's equation.

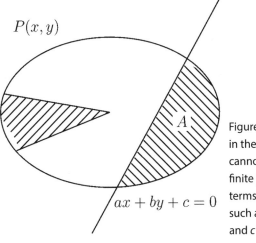

$P(x,y)$

A

$ax + by + c = 0$

Figure 12.5. The shaded areas in the oval described by $P(x,y)$ cannot be evaluated using a finite polynomial expression in terms of their defining constants, such as the line parameters a, b, and c for the area A.

It is amazing that Newton actually tackled the problem by generalizing and enlarging it from ellipses to all ovals in a certain class. Equally amazing is Newton's proof, which uses ideas about polynomials and topology that are well in advance of what might be expected in the seventeenth century. Naturally, such a stunning result raised questions about its validity, and certainly more must be added, about levels of smoothness for example, to qualify the types of ovals that may be used.

The wonder of Newton's *Lemma XXVIII* has continued to excite admiration and questioning right up to the present day. (See this chapter's "Further Reading" section for references for the proof and discussions of validity.) Here is the eminent Russian mathematician V. I. Arnol'd writing in 1990:

> In the *Principia* there are two purely mathematical pages containing an astonishingly modern topological proof of a remarkable theorem on the transcendence of Abelian integrals.[1]

The magnitude of Newton's achievement is suggested by Arnol'd's claim that "Newton's topological arguments outstripped the level of the science of his time by two hundred years."[2] Arnol'd and colleagues have gone on to generalize Newton's result, into higher dimensions, for example.

Thus, by an inspired piece of mathematics, Newton has shown that approach 1 is not useful for solving Kepler's equation.

12.5.2. The Visual, Geometric Approach

In the *Corollary*, Newton says the area of the ellipse cannot be found using a polynomial equation, and he goes on to restate that in geometric terms:

> And therefore [areas cutoff in ellipses] *cannot be determined by the description of curves geometrically rational. Those curves I call geometrically rational, all the points whereof may be determined by lengths that are definable by equations; that is by the complicated ratios of lengths. Other curves (such as spirals, quadratrixes and cycloids) I call geometrically irrational.*

In fact, it is a kind of cycloid (a curve not defined by a polynomial equation) that may be used to study Kepler's equation, as Newton himself explains. It is probable that this approach goes back to Sir Christopher Wren, but (of course) Newton gives no attribution!

A cycloid is the curve traced out by a point on the rim of a rolling circle; however, in the present case, we need a curtate cycloid, which has the tracing point displaced from the rim. An example is shown in the top diagram of figure 12.6. To present this approach to Kepler's equation in a modern form, we take the curtate cycloid with the generating circle having radius 1, and the tracing point a distance ε along a radius away from the rolling-circle center. If we set that in the x–y plane, as in the bottom diagram in figure 12.6, the curve is given in algebraic form by a pair of equations that specify a point (x,y) on the curve in terms of a parameter E:

$$x = E - \varepsilon\sin(E) \text{ and } y = 1 - \varepsilon\cos(E) \qquad (12.7)$$

The curve is generated by varying the parameter E and plotting the resulting pairs of x and y values.

Now we can see how to solve Kepler's equation (12.6). For a given M, we locate $x = M$ on the x axis, which, according to equation (12.7), just corresponds to Kepler's equation. But now the required E can be read off because we have the y value from the point on the curve corresponding to $x = M$. The path from x to y is shown in the bottom diagram in figure 12.6. If the value of y is y_M, as indicated, then we can go to equation (12.7) and get

$$y_M = 1 - \varepsilon\cos(E) \quad \text{or} \quad \cos(E) = \{1 - y_M\}/\varepsilon$$
$$E = \cos^{-1}(\{1 - y_M\}/\varepsilon).$$

Thus, the appropriate cycloid allows us to go from M to y_M and then to the required eccentric anomaly E.

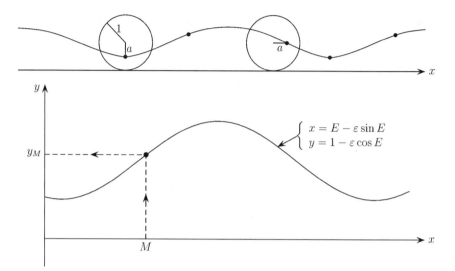

Figure 12.6. *Top*: A curtate cycloid generated by a point radial distance *a* from the center of a rolling circle. *Bottom*: Setting this cycloid in the *x–y* plane, using a unit radius circle and displacement $a = \varepsilon$, the point $x = M$ leads to the *y* value y_M.

This geometric or visual procedure, required for approach 2, gives us a way to solve Kepler's equation. But in Newton's day without computers, it would have been difficult and tedious to use, especially if considerable accuracy was required and several values for ε were to be used. As Newton himself puts it, *"But since the description of this curve is difficult, a solution by approximation will be preferable."* So, on to approach 3.

12.5.3. Approximation Methods

Newton used approximation techniques in his other writings on mathematics dealing with polynomial equations, and his name is attached to numerical-analysis methods still used today to solve nonlinear equations. (The recent book by Guicciardini, listed in the "Further Reading" section at the end of this chapter, is a good reference.) It is unfortunate then that this part of the *Principia* is very poorly expressed—at least that is my experience, and I suspect Newton's contemporaries found it virtually unreadable, as I did. Clarification came in 1882 in a paper by J. C. Adams (who we will meet again when discussing *Book III*). Here is the way Adams sees it.

To solve Kepler's equation for E when the time τ and, hence, the mean anomaly M is given, assume that we have found an approximate answer E_0 (perhaps using the graphical technique or some other estimation method). Then

$$E_0 - \varepsilon \sin(E_0) = M_0,$$

so the approximation E_0 has led to M_0 instead of M. An improved answer E_1 is then given by

$$E_1 = E_0 + \frac{M - M_0}{1 - \varepsilon \cos(E_0)}.$$

This value E_1 will lead to M_1, and a next approximation is given by

$$E_2 = E_1 + \frac{M - M_1}{1 - \varepsilon \cos(E_1)}.$$

If necessary, we can generate further approximations $E_3, E_4, \ldots$ until we have the required accuracy. We expect the sequence of answers $E_0, E_1, E_2, E_3, E_4, \ldots$ to converge to the required exact answer E, which produces the given M when substituted into Kepler's equation.

This is an application of what today is sometimes called the Newton-Raphson method:

To solve the equation $f(x) = 0$, start with an initial estimate x_0 and generate a sequence of approximations $x_1, x_2, x_3, x_4, \ldots$ using the iteration scheme

$$x_{k+1} = x_k - \frac{f(x_k)}{f'(x_k)} \qquad \text{where } f'(x_k) \text{ is } \frac{df}{dx} \text{ evaluated at } x = x_k.$$

We may find it difficult to follow how Newton carried it out, but he did use approach 3 and showed how to find solutions, noting that "*it converges so very fast, that it will be scarcely ever needful to proceed beyond the second term.*"

12.6. CONCLUSIONS

That wonderful result about equal areas being swept out in equal times (or the conservation of angular momentum, in modern terms) allows us to calculate the form of orbits and trajectories, as explained in the previous chapters. However, when we ask where a body will be at a given time, we must use it in a different way, as set out in section 12.1. The result is a scheme that is conceptually simple but that can be difficult to implement in practice. For motion governed by inverse-square-law forces, the required calculation is relatively simple for motion on a parabola, but it throws up a most challenging problem for the vitally important case of motion on an ellipse. Newton takes on the challenge and, in a virtuoso performance, demonstrates his abilities as both pure mathematician (taking on fundamental existence questions) and applied mathematician (giving practical schemes for generating solutions to any required accuracy).

It is quite startling to come across *Lemma XXVIII* when reading the *Principia*. The result is so unexpected and powerful that one can only wonder how Newton ever got to it. Kepler had beautifully set out the ellipse problem and asked for help with it; it took a Newton to see that the way forward involved an enlargement of the problem and a radically new approach to a mathematical existence proof.

12.7. FURTHER READING

See the classical-mechanics books listed in section 6.9.1 and the many books on celestial mechanics, such as:

Baittin, R. H. *An Introduction to the Mathematics and Methods of Astrodynamics*. New York: American Institute of Aeronautics and Astronautics, 1987.

Grossman, N. *The Sheer Joy of Celestial Mechanics*. Boston: Birkhäuser, 1996. [an almost whimsical book, but yes, a joy to read]

Linton, C. M. *From Eudoxus to Einstein: A History of Mathematical Astronomy.* Cambridge: Cambridge University Press, 2004.

Sterne, T. E. *An Introduction to Celestial Mechanics.* New York: Interscience Publishers, 1960.

For more about Kepler and his related work, see chapter 6 in Linton and:

Gingerich, O. *Dictionary of Scientific Biography.* Edited by Charles C. Gillespie. New York: Scribner, 1973, s.v. "Kepler."

———. *The General History of Astronomy.* Vol. 2. Edited by R. Taton and C. Wilson. Cambridge: Cambridge University Press, 1989, s.v. "Johannes Kepler."

Koyré, A. *The Astronomical Revolution.* 3rd ed. London: Routledge, 2009. [especially ch. 9, "Astronomy with the Ellipse"]

For details and explanations of Newton's mathematics as discussed in this chapter, see:

Guicciardini, N. *Isaac Newton on Mathematical Certainty and Method.* Cambridge, MA: MIT Press, 2009. [in particular, see sec. 10.2.6, ch. 11, and sec. 13.3]

For more on *Lemma XXVIII* and the areas of ovals, see:

Arnol'd, V. I. *Huygens and Barrow, Newton and Hooke.* Basel: Birkhauser Verlag, 1990.

Arnol'd, V. I., and V. A. Vasil'ev. "Newton's *Principia* Read 300 Years Later." *Current Science* 61 (1991): 89–95.

Chandrasekhar, S. *Newton's* Principia *for the Common Reader.* Oxford: Clarendon Press, 1995. [detailed discussion in sec. 37]

Pesic, P. "The Validity of Newton's Lemma 28." *Historia Mathematica* 28 (2001): 215–19.

Pourciau, Bruce. "The Integrability of Ovals: Newton's Lemma 28 and Its Counterexamples." *Archive for the History of the Exact Sciences* 55 (2001): 479–99. [highly recommended]

There is a considerable literature on Kepler's equation and methods of solution. Baittin devotes his forty-six-page chapter 5 to the subject. There is debate about the numerical analysis methods used. The book-length study by Colwell contains 432 references for further study.

Adams, J. C. "On Newton's Solution of Kepler's Problem." *Monthly Notices of the Royal Astronomical Society* 43 (1882): 43–49.

Chabert, J.-L. *A History of Algorithms.* Berlin: Springer, 1994.

Colwell, Peter. *Solving Kepler's Equation over Three Centuries.* Richmond, VA: Willmann-Bell, 1993.

Dutka, J. "A Note on Kepler's Equation." *Archive for the History of the Exact Sciences* 51 (1997): 59–65.

Goldstine, H. H. *A History of Numerical Analysis.* New York: Springer-Verlag, 1997.

Kollerstrom, N. "Thomas Simpson and 'Newton's Method of Approximation': An Enduring Myth." *British Journal for the History of Science* 25 (1992): 347–54.

13

COMPLETING THE
SINGLE-BODY FORMALISM

We have now seen how Newton established the conceptual basis of classical mechanics, set out the laws of motion, showed how to deduce general properties (such as the conservation of both momentum and angular momentum), and explained the link between forces, inertial motion, and a body moving on an orbit. He gave a general theory for finding the centripetal or central force responsible for any particular given orbit or trajectory (his direct problem). The details were given for inverse-square-law forces. What remains to be done? Newton has yet to give a general formalism for deducing the motion that results when a body is launched onto a trajectory governed by an arbitrary centripetal force. **That is Newton's inverse problem, but today it is the main thrust in the presentation of classical mechanics.** He fills in this gap in *Sections VII* and *VIII*, which I discuss in this chapter. **This work must rank as one of the great pinnacles in the *Principia*.**

A good, very detailed explanation of Newton's work in these sections is given by Chandrasekhar (see the "Further Reading" section at the end of this chapter). I will give enough mathematical details to link Newton's work with the modern formalism, and I urge all readers to follow the conceptual flow, even if they are not overfamiliar with the precise manipulations involved. As is my intention with the whole book, the main goal is to appreciate the overall scheme.

13.1. MOTION IN ONE DIMENSION

A modern textbook will use the one-dimensional case as the easiest vehicle for demonstrating how the theory works. Newton has not followed that strategy in the *Principia* until now, when we find *Section VII*: "*Concerning the Rectilinear Ascent and Descent of Bodies.*"

Propositions *XXXII* to *XXXVIII* solve problems of bodies moving, falling, or rising along a line and with various forces acting on them. As we now expect from Newton, a clever approach is involved: take the motions already described (like motion along an ellipse) and consider the limiting case when the orbit shrinks to a line (the width of the ellipse minor axis shrinks to zero). Using the array of results given in earlier *Propositions* and their *Corollaries* allows Newton to solve several specific cases of motion in one dimension.

However, things change when we reach *Proposition XXXIX*, which tells us how to deal with motion in one dimension in general. Before coming to that, readers may like to stop and check what is meant by a general solution and what will be involved.

13.1.1. What Is Required?

For motion along the x axis, say, we will need to use the second law:

$$m\frac{d^2x}{dt^2} = F(x). \tag{13.1}$$

Can we give advice about how to find the resulting motion $x(t)$ other than "solve the differential equation when given the force F"? It may be useful to step back and ask what is actually involved.

Given a function $x = x(t)$, a chain of derivatives may be created

$$x = x(t) \quad \rightarrow \quad \text{differentiate to get } \frac{dx}{dt} \quad \rightarrow \quad \text{differentiate to get } \frac{d^2x}{dt^2}.$$

The chain is reversed when we use integration to reduce the order of the derivative:

given $\dfrac{d^2x}{dt^2}$ → integrate to get $\dfrac{dx}{dt} = \displaystyle\int \dfrac{d^2x}{dt^2}\,dt$ → integrate to get $x(t) = \displaystyle\int \dfrac{dx}{dt}\,dt.$

Thus, **to solve the second-order differential equation (13.1), two integrations will be involved.** For example, for the simplest case of a constant force (Galileo's problem), we find

given $\dfrac{d^2x}{dt^2} = k$ → integrate to get $\dfrac{dx}{dt} = kt + A$ → integrate to get $x(t) = \left(\tfrac{1}{2}\right)kt^2 + At + B.$

The "constants of integration," A and B, are used to fit initial conditions, as explained in chapter 11. In this case of a constant force, the integrations are almost trivially simple, but that will not be the case in general. Is there any help that can be given? Is there a formula that says, "Plug your given $F(x)$ in here, and these integrals will lead to the required answer"?

13.1.2. Newton's Response

PROPOSITION XXXIX. PROBLEM XXVII.

Supposing a centripetal force of any kind, and granting the quadratures of curvilinear figures; it is required to find the velocity of a body, ascending or descending in a right line, in the several places through which it passes, as also the time in which it will arrive at any place; and vice versa.

Note that the force is now "*of any kind,*" and "*granting the quadratures of curvilinear figures*" means assuming we can do any required integrations. Newton's figure is reproduced in the left diagram of figure 13.1. The body starts at A and moves toward the force center C; we wish to know the velocity and time associated with some point E. You can see that Newton will be using the small (infinitesimal) motion from D to E in his derivation. I will briefly go through Newton's instructions here and then put them into a modern (and, I suspect, more understandable) form later.

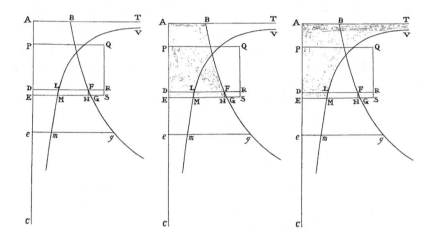

Figure 13.1. *Left*: Newton's figure for motion from *A* toward the force center at *C*. *Middle*: The shaded area is proportional to the square of the velocity at *E*. *Right*: The shaded area leads to the time taken to reach *E*. (Originally from *The* Principia, trans. Andrew Motte [Amherst, NY: Prometheus Books, 1995]; modified by Annabelle Boag.)

Newton tells us to construct the curve BFGg so that the distance of each point on that curve from the line of motion APDEC is equal to the force acting; for example, AB will be equal to the magnitude of the force acting at A, DF will be the magnitude of the force acting at D. Then:

> *the velocity of the body in any place* E *will be as a right line whose square is equal to the curvilinear area* ABGE. *Q.E.I.*

Thus we can get the velocity by knowing the area shown shaded in the middle diagram in figure 12.1, and that is essentially the first integral to be done. Now we can construct the curve VLMm: for the point M on that curve, the length EM is inversely proportional to the square root of the area just considered, which is ABGE, in this case. Then:

> *the time in which the body in falling describes the line* AE *will be as the curvilinear area* ABTVME. *Q.E.I.*

The required shaded area is shown in the right diagram in figure 13.1, and that is the second integral in the solving process.

13.1.3. In Modern Terms

A little calculus converts equation (13.1) into

$$m\frac{dx}{dt}\frac{d}{dx}\left(\frac{dx}{dt}\right) = F(x) \qquad \text{or} \qquad \frac{m}{2}\frac{d}{dx}\left(\frac{dx}{dt}\right)^2 = F(x).$$

Now one integration is easily done to get

$$\frac{m}{2}\left(\frac{dx}{dt}\right)^2 = \int F(x)\,dx \qquad \text{or} \qquad \frac{dx}{dt} = \sqrt{(2/m)\int F(x)\,dx}. \quad (13.2)$$

This is exactly Newton's prescription for finding the velocity, expressed here in terms of equations, whereas Newton uses proportions. Since the right-hand side of equation (13.2) is a function of x alone, the equation is easily integrated to get

$$\int \frac{dx}{\sqrt{(2/m)\int F(x)\,dx}} = \int dt = t. \quad (13.3)$$

This is Newton's second integral. When the two integrals are evaluated, there will be the usual constants of integration for fitting the initial conditions: the value of x and the initial speed at Newton's starting point A in the left diagram in figure 13.1.

Newton has clearly understood the two-step integration process described in section 13.1.1, and he has given the required integrals expressed as areas under curves rather than in analytic form, as in equations (13.2) and (13.3).

13.2. SOME GENERAL CONSIDERATIONS

Some readers may recognize that Newton is doing indirectly (as we illustrated in section 13.1.3) what today we do more directly using energy concepts. Today we talk about

$$\text{the kinetic energy, } K \; = (1/2)mv^2$$

$$= (1/2)m\left(\frac{dx}{dt}\right)^2 \quad \text{in one dimension,} \tag{13.4}$$

$$\text{the potential energy, } V = -\int F(x)dx,$$

$$\text{so } F(x) = -\frac{dV}{dx} \quad \text{[and in three dimensions } \mathbf{F} = -\nabla V(\mathbf{r})\text{].} \tag{13.5}$$

Then the conservation of energy, $K + V = E$, a constant, leads directly to equation (13.2) and Newton's equivalent, geometrically expressed result.

However, it is important to note that Newton did not introduce the idea of potential energy or the conservation of energy in the *Principia*. Those ideas were developed much later, as I will outline in chapter 28. Nevertheless, as we have seen in the equations above, he was manipulating terms like K and V in his own version of a type of energy argument. In section 13.3, we will see how he makes that explicit for the general centripetal case.

13.2.1. A Historical Aside

Some ideas about energy were lurking in the early work of people like Galileo and Huygens. Their main result was very clearly set out by Huygens (see this chapter's "Further Reading" section):

Huygens's Proposition VIII

If, from the same height, a body descends by a continuous motion through any number of contiguous planes having any inclinations whatsoever, it will always acquire, at the end, the same velocity; namely, a velocity equal to that which would be acquired by falling perpendicularly from the same height.

Huygens's Proposition IX

If, after falling, a body converts its motion upward, it will rise to the same height from which it came, no matter how many contiguous plane surfaces it may have crossed, and no matter what their inclinations are.[1]

We would say that the initial potential energy is converted into the same amount of kinetic energy no matter what path is taken to the new lower level. Equally, that kinetic energy is converted back into potential energy as the particle rises, again no matter which path is taken.

This idea of path independence is the key to the definition of work done and conservative forces, as well as the definition of a potential from which forces may be calculated by differentiation, as in equation (13.5). The discovery of that idea and its use by Galileo, Huygens, and Newton was vital for the progress of classical mechanics, even though the step to potential theory was not explicitly taken and a statement of the conservation of energy was still some way off (see chapter 28).

In fact, even the notion of kinetic energy was not well established in Newton's day. Leibnitz championed the importance of *vis viva*, or "living force" (mv^2), and its conservation in collisions. Of course, we now know that both linear momentum and kinetic energy are conserved in elastic collisions. Leibnitz also introduced *vis mortua*, or "dead force," which has connections with what today we would call potential energy, but he did not have the concept of general conservation laws.

13.3. GENERALIZING THE PATH-FREE RESULT

Newton is now ready to move on to the more complex case in *Section VIII*: "*Of the Invention of Orbits Wherein Bodies Will Revolve, Being Acted upon by Any Sort of Centripetal Force.*"

There are three *Propositions* in this section, and the first extends the Galileo-Huygens result given above.

PROPOSITION XL. THEOREM XIII.

If a body, acted upon by any centripetal force, is any how moved, and another body ascends or descends in a right line, and their velocities be equal in any one case of equal altitudes, their velocities will be also equal at all altitudes.

Notice that we now have any "*centripetal force*," not just the gravitational force of Galileo and Huygens, and we have a clear statement linking position and velocities. Newton's diagram is reproduced in figure 13.2. Although it is not emphasized, the motion takes place in a plane—something we know right back from *Proposition I.*

Essentially, Newton is saying that if a body moves from a given point to another point nearer to the force center, its new velocity does not depend on the path taken to that second point. The proof involves the motion from the circle *ID* to *KNE*, and Newton uses the fact that the influence of the force is only in the direction toward the force center. (Again we note that Newton has the key concept—path independence—that leads on to work done and potential energy, although he himself does not explicitly make that step.)

The same result holds for motion away from the force center:

By the same reasoning, bodies of equal velocities and equal distances from the centre will be equally retarded in their ascent to equal distances.

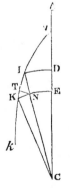

Figure 13.2. Newton's diagram showing a body starting at *A* and moving toward a force center *C*. *ID* and *KNE* are parts of circles centered on *C*. The body may move from *V* to *E* along the line *AVDEC*, or to *K* along the curve *VITK* so that it is still the same distance from the force center *C*. *Proposition XL* says that the velocities at *E* and *K* are equal. (From *The* Principia, trans. Andrew Motte [Amherst, NY: Prometheus Books, 1995].)

13.3.1. Application to Constrained Motion

Newton now tells us that the theory holds in a more complicated situation, one we consider in greater detail in chapter 15.

When a body is constrained, as by the string in a pendulum or by a smooth surface on which it slides, the forces responsible for the constraint (the tension in the string or the reaction force of the surface) do no work on the body and so do not affect its speed. Newton beautifully sets out this in a *Corollary* that is worth quoting in full:

> COR. 1. *Therefore if a body oscillates by hanging to a string, or by any polished and perfectly smooth impediment is forced to move in a curved line; and another body ascends or descends in a right line, and their velocities be equal at any one equal altitude, their velocities will be also equal at all other altitudes. For by the string of the pendulous body, or by the impediment of a vessel perfectly smooth, the same thing will be effected as by the transverse force NT [see figure 13.2]. The body is neither accelerated nor retarded by it, but only obliged to leave its rectilinear course.*

(Note that Newton should probably say "speed" rather than "velocity," as velocity usually implies a direction as well as a speed.) In essence, this is very close to the conservation-of-energy law that we use to solve such problems, but the potential energy subtly enters via the speed, known through Galileo's work for simple vertical motion through a given height.

13.3.2 A Mathematical Example

As an example, Newton takes a power-law attractive force, so that in our terms $F(r) = -kr^{n-1}$, where k is the force-strength constant. Again it might be apt to quote Newton's own words:

> COR. 2. *Suppose the quantity P to be the greatest distance from the centre to which a body can ascend, whether it be oscillating, or revolving in a trajectory [so the speed at P will be zero]. . . . Let the quantity A be the distance of the body from the centre in any other point of the orbit; and let the centrip-*

etal force be always as the power A^{n-1}, *of the quantity* A, *the index of which power* n−1 *is any number* n *diminished by unity. Then the velocity in every altitude* A *will be as* $\sqrt{(P^n - A^n)}$, *and therefore will be given. For by Prop. XXXIX, the velocity of a body ascending or descending in a right [straight] line is in that very ratio.*

In modern terms, we say that

the force $F(r) = -kr^{n-1}$ corresponds to the potential $V(r) = (k/n)r^n$,

and the conservation of energy for the points $r = A$ and $r = P$ leads to

$$(1/2)mv_A^2 + (k/n)A^n = (1/2)mv_P^2 + (k/n)P^n = (k/n)P^n.$$

I have set v_P to zero, as required, and then we can solve to get the speed at A as

$$v_A = \left(\sqrt{2k/mn}\right)\sqrt{(P^n - A^n)},$$

which is just Newton's result. As we know from earlier discussions, Newton is actually doing something very similar when he uses his *Proposition XXXIX*.

13.4. THE GENERAL FORCE-TO-ORBIT RESULT

Everything is now in place for Newton to tell us how to find the orbit corresponding to any given central force. Here is one of the *Principia's* most famous *Propositions*:

PROPOSITION XLI. PROBLEM XXVIII.

Supposing a centripetal force of any kind, and granting the quadratures of curvilinear figures; it is required to find as well the trajectories in which bodies will move, as the times of their motion in the trajectories found.

Newton gives a diagram (here, figure 13.3) and begins:

Let any centripetal force tend to the centre C, and let it be required to find the curve VIKk.

I will make no attempt to explain Newton's working, but refer the interested reader to Chandrasekhar's book. However, I will point out that the circles drawn with center C to give the points Y and X, and points K and N, indicate that Newton is, in effect, using polar coordinates. The change from I to K involves what we would call infinitesimal radial (*dr*) and infinitesimal angular (*dθ*) changes. The results involve the various areas arising in the figure—you might recognize the area VIC as the area swept out as the body moves from V to I and, hence, proportional to the time, as Newton states.

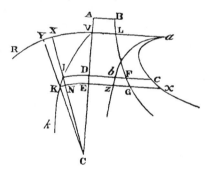

Figure 13.3. Newton's diagram to support *Proposition XLI*. (From *The Principia*, trans. Andrew Motte [Amherst, NY: Prometheus Books, 1995].)

In modern terms, using the theory presented in chapter 11, we can derive the following equations:

$$t = \int \frac{dr}{\sqrt{(2/m)\int F(r)dr - (h^2/r^2)}}$$

$$\theta = \int \frac{h\,dr}{r^2\sqrt{(2/m)\int F(r)dr - (h^2/r^2)}}.$$

(13.6)

Chandrasekhar's book should be consulted to see how that is equivalent to Newton's work. Thus, as long as we can do the integrations (*"granting the quadratures of curvilinear figures"*), we can find the orbit equation and the time taken to move along it by substituting the given force $F(r)$ into those formulas.

In *Corollaries 1* and *2*, Newton explains how we can find the apsides of the orbits, that is, the places where the body is nearest or farthest away from the force center. I will return to this in the next chapter.

13.4.1. The Famous (or Infamous) *Corollary 3*

Of course, we expect Newton to give an example of the use of such an important result, and he does so. However, it is not the one many people expect to read. Recall that in chapter 10 we saw how Newton caused great controversy by not giving an explicit proof of the fact that the trajectories produced by an inverse-square-law force are just the conic sections. Now, this *Corollary* seems to provide the perfect opportunity for him to supply the required proof; simply set $F(r) = -k/r^2$ and use the defined procedure to spit out the conic sections. But he does not do that! Some people have claimed that Newton could not do the required integrations, but that is not correct, as he tabulated such integrals elsewhere. Or, he may have genuinely thought his working (as described in chapter 10) was sufficient, so why labor the point?

Newton chose the inverse-cube-law force, $F(r) = -k/r^3$, as his example. Chandrasekhar gives the details of Newton's work, and the old book by Horace Lamb gives a complete treatment of the inverse-cube-law case. Newton ended his *Corollary* with words that must have exasperated any reader:

> All these things follow from the foregoing Proposition, by the quadrature of a certain curve, the invention of which, as being easy enough, for brevity's sake I omit.

Little wonder that Newton managed to aggravate so many people! One person who asked Newton for help was David Gregory, and in 1694, Newton wrote to him explaining how it works out. Amazingly the letter survives, and you can see (in figure 13.4) that Newton is using a form of calculus that is quite familiar to us today to carry out the integrations. For more on the whole story, I refer you to the "Further Reading" section at the end of this chapter.

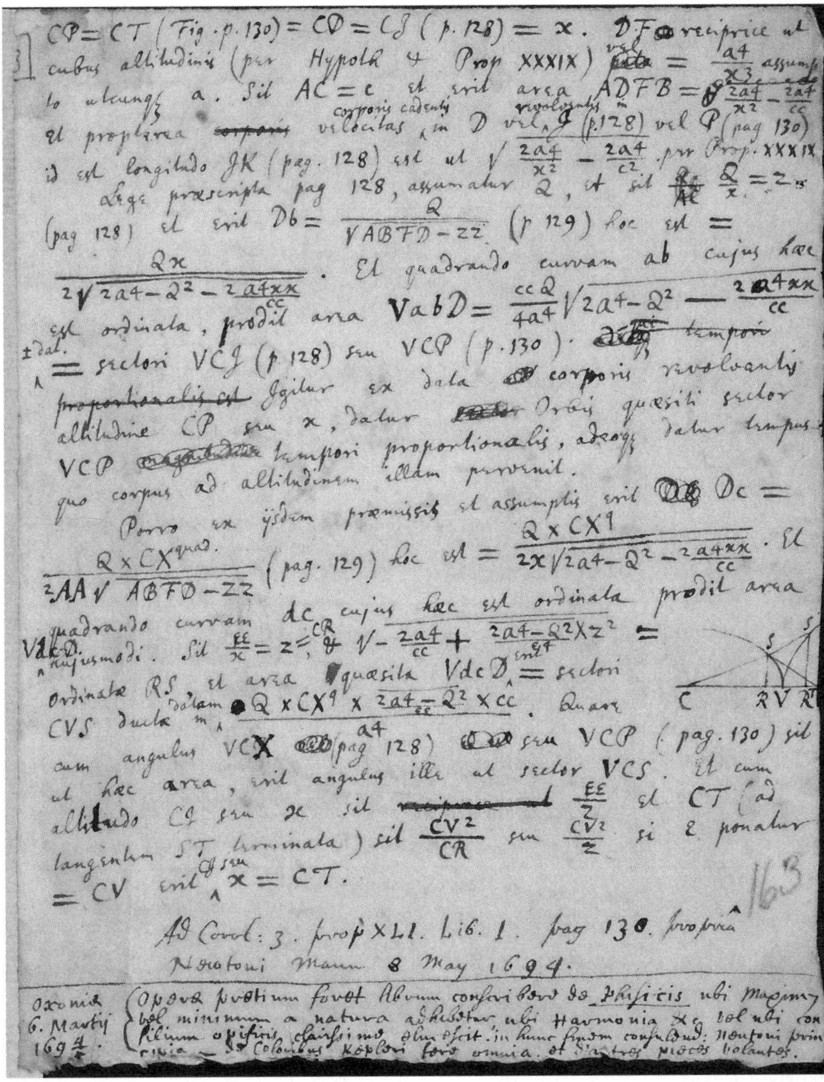

Figure 13.4. Part of Newton's letter written to David Gregory in 1694 explaining how to deal with the inverse-cube case used in *Corollary 3*. (Image reproduced by permission of the Royal Society.)

The inverse-power-law forces $F(r) = -k/r^n$ for which the integral in equation (13.6) can be evaluated in terms of readily available functions (such as elliptic functions) have been identified; details are in the classic texts by Goldstein and Whittaker.

13.5. INITIAL CONDITIONS

Section VIII concludes with

PROPOSITION XLII. PROBLEM XXIX.

The law of centripetal force being given, it is required to find the motion of a body setting out from a given place, with a given velocity, in the direction of a given right line.

Newton is now saying that, having found the possible paths for a body with the "*law of centripetal force being given,*" this is how to use the initial conditions (that is, the launching position and velocity) to pick out the particular path taken. Of course—as always—his result is couched in geometric terms.

In modern terms, we use the initial conditions to fit the integration constants, as explained in section 13.1.1. Because we began with two second-order differential equations, there will be four constants to cover all the initial conditions. (That could be the starting values of the polar coordinates r and θ and the two components of the initial velocity vector, perhaps the speeds in the radial and angular directions.) Using equations (13.6), there are three integrations to be carried out, each one with an associated constant of integration. The angular-momentum constant h is also involved, making the fourth constant for dealing with those initial conditions.

It might seem that this is a rather mundane final step in the whole process, but Nobel laureate Eugene Wigner has a different opinion: "The surprising discovery of Newton's age is just the clear separation of laws of nature on the one hand and initial conditions on the other."[2]

He explains the importance of this discovery as follows:

The world is very complicated and it is clearly impossible for the human mind to understand it completely. Man has therefore devised an artifice which permits the complicated nature of the world to be blamed on something which is called accidental and thus permits him to abstract a domain in which simple laws can be found. The complications are called initial conditions; the domain of regularities, laws of nature. Unnatural as such a division of the world's structure may appear from a very detached point of view, and probable though it is that the possibility of such a division has its own limits, the underlying abstraction is probably one of the most fruitful ones the human mind has made. It has made the natural sciences possible.[3]

Newton certainly understood and used this division. For example, he discovered that an inverse-square-law force gives rise to paths that may be ellipses, parabolas, or hyperbolas, and, in *Proposition XVII*, he explained how varying the initial conditions could lead to a body following each of them. Whether he realized that such a division "*has made the natural sciences possible*" seems unlikely, but from what we know of the man, he would have happily taken the attached credit and kudos!

13.6. DISCUSSION

Section VIII of the *Principia* is surely one of the most momentous in the whole book. Newton has reached a pinnacle: he has drawn together all previous results to present a complete prescription for the solution of the problem of a body moving under the influence of an arbitrary centripetal force. *Proposition XLI* surely represents one of science's greatest achievements. *Proposition XLII* then tells us how to use that result in any particular circumstance.

One can only wonder at the leap from the working of Galileo, Descartes, and even Huygens to the mathematically advanced and generalized theory described in the *Principia*.

Newton does use what today we call the conservation of angular momentum, but he does not have the concept of mechanical energy and its conservation, which is a key element in the modern approach and solution of dynamical problems. However, in the fully general *Proposition XL*, he does exhibit the vital property that lies behind the energy method, and he masterfully uses that property to produce his route to that general solution of dynamical problems.

What Newton achieves in *Section VIII* is equivalent to our modern formulation of classical mechanics for the one-body case with central (centripetal) forces. Of course, Newton's mathematical language is completely different and bewildering; we are reminded of Whewell's analogy to some ancient "gigantic implement of war" that "makes us wonder what manner of man he was who could wield as a weapon what we can hardly lift as a burden" (see section 7.6).[4] This may be a good time to ask why Newton did not help readers a little more and include more of the mathematical notation and formalism that was being developed at that time. It is fascinating to read Newton's own view of the situation written in the late 1710s:

> *To the mathematicians of the present century, however, versed almost wholly in algebra as they are, this* [the *Principia's*] *synthetic style of writing is less pleasing, whether because it may seem too prolix and too akin to the method of the ancients, or because it is less revealing of the manner of discovery. And certainly I could have written analytically what I had found out analytically with less effort than it took me to compose it. I was writing for Philosophers steeped in the elements of geometry, and putting down geometrically demonstrated bases for physical science. And the geometrical findings which did not regard astronomy and physics I either completely passed by or merely touched lightly upon.*[5]

Some of this must be read in the context of the controversies discussed in section 7.5. Whether or not Newton "*had found out analytically*" as much as he claimed is possibly still debatable, but he is certainly correct in saying that we later readers would have been happier with less geometry and more analysis!

Section VIII also contains the celebrated *Corollary 3*, with its mysterious choice of inverse-cube-force law as an example. In some ways, it is sad to find that the magnificence of *Proposition XLI* is almost ignored, or perhaps nowadays taken for granted, but it is also testament to the enduring curiosity about Newton the man and the controversies in which he became embroiled.

13.7. FURTHER READING

As always, the reader may consult the books recommended in section 6.9.1. The book by Chandrasekhar is very useful for unpicking Newton's arguments in *Section VIII*. The books by Goldstein (particularly sec. 3.5) and Whittaker (sec. 48) give details of the integrable-power-law cases.

Huygens's propositions are given in:

Huygens, Christiaan. *The Pendulum Clock; or, Demonstrations concerning the Motion of Pendula as Applied to Clocks*. Translated from the original 1673 *Horologium Oscillatorium* by Richard J. Blackwell. Ames: Iowa State University Press, 1986. [see part 2, in particular]

For details about work done by forces and path independence, see any of the dynamics texts, such as Whittaker (specifically, sec. 22) and Goldstein (ch. 1), and, for deeper discussion and historical details, see:

Lindsay, R. B., and H. Margenau. *Foundations of Physics*. New York: Dover, 1957.

For *vis viva* and Leibnitz's contributions, see:

McDonough, Jeff. *Leibnitz's Philosophy of Physics*. In *Stanford Encyclopedia of Philosophy*. Stanford University Metaphysics Research Lab. 2007. http://plato.stanford.edu/entries/Leibnitz-physics/.
Smith, George E. "The Vis Viva Dispute: A Controversy at the Dawn of Dynamics." *Physics Today* 59 (October 2006): 31–36.

For greater discussion of *Corollary 3*, see Guicciardini as cited in section 6.9.1; also see:

Brackenridge, J. B. "Newton's Easy Quadratures 'Omitted for the Sake of Brevity.'" *Archive for the History of the Exact Sciences* 57 (2003): 313–36.

Erlichson, H. "The Visualization of Quadratures in the Mystery of Corollary 3 to Proposition 41 of Newton's *Principia*." *Historia Mathematica* 21 (1994): 148–61.

Guicciardini, N. *Isaac Newton on Mathematical Certainty and Method.* Cambridge MA: MIT Press, 2011. [see sec. 12.2.1]

Lamb, Horace. *Dynamics.* 2nd ed. Cambridge: Cambridge University Press, 1945. [sec. 91 gives a full analysis of motion under the influence of an inverse-cube force]

14

ROTATING ORBITS

With a theory in place for finding orbits corresponding to any given centripetal force, it seems natural to ask whether there are any general properties of such orbits that may now be deduced. For the inverse-square-law case, the orbit is an ellipse and the body continually traces out the same, fixed path with exactly the same apsides (points where the distance from the force center is greatest or least). Knowing about such "*motions of bodies in immovable orbits*," Newton says, "*it remains now to add something concerning their motions in orbits which revolve around the centres of force.*" Thus we get to *Section IX*: "*The* Motion of Bodies *in Movable Orbits; and the Motion of the Apsides.*"

This might sound like the theorist at play, and in some ways it is, but there is more to it than that. To appreciate what Newton is doing, we need to go back to his *Preface* and recall the following statement:

I offer this work as the mathematical principles of philosophy, for all the difficulty of philosophy [what we today might call natural philosophy or science] *seems to consist in this—from the phenomena of motions to investigate the forces of nature, and then from these forces to demonstrate the other phenomena; and to this end the general propositions in the first and second books are directed.*

Newton is seeking results that will help with that task of "*from the phenomena of motions to investigate the forces of nature.*" In particular, as we shall see when we get to *Book III*, the results Newton presents in this chapter play a vital role when he applies his theory to gravity, the solar system, and the motion of the Moon.

Section IX also shows us some wonderfully inventive mathematical arguments and emphasizes yet again the sophistication of Newton's work. He presents general results, gives useful examples, and comes close to giving us one of the classic theorems in dynamics—one not proven until almost two hundred years after the *Principia* first appeared.

14.1. BOUNDED ORBITS

Much of our thinking is colored by the inverse-square-law case: we have unbounded parabolic and hyperbolic motion, as well as bounded motion on an ellipse, a special case of which is the circle. When discussing his definitions, Newton recognized that a body *"might never fall to earth but go forward into the celestial spaces, and proceed in its motion in infinitum,"* if it has a suitable velocity, which we call the escape velocity (see section 5.4.1). Thus the boundary between the two types of motion is clear, but what can we say about the nature of bounded motion?

At the apsides, where r takes on its maximum and minimum values, dr/dt will be zero. In the modern formulation, this can be investigated using the energy equation, equation (11.26), which at the apsides becomes

$$\frac{1}{2}m\frac{h^2}{r^2} + V(r) - E = 0. \tag{14.1}$$

For the types of potentials $V(r)$ of interest (for example, see section 13.3.2), this equation has two solutions $r = r_{min}$ and $r = r_{max}$. We may conclude that the bounded motion in the r–θ plane is confined between two circles with radii r_{min} and r_{max}, as shown in figure 14.1.

For motion on an ellipse, the apsides are fixed and the body repeatedly touches the limiting circles at the same points, as shown in figure 14.1(a). However, for some types of motion, the body's trajectory may wander around in the space between the two circles, as shown in figure 14.1(b).

Thus the apsides may move around, they may rotate around the limiting circles, and that is basically what Newton is concerned with in *Section IX*. When do we get rotation? And how fast will it be?

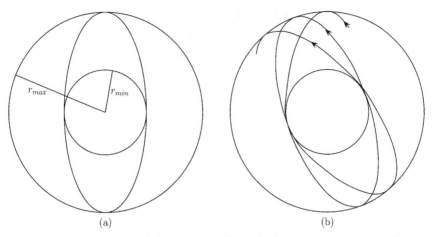

Figure 14.1. The diagram in (a) shows a fixed elliptical orbit as given by an attractive force proportional to distance (see section 11.2.2); the diagram in (b) shows a rotating orbit from a different type of force.

14.2. ROTATING ORBITS

Newton begins his investigation by introducing an ingenious problem in two propositions:

PROPOSITION XLIII. PROBLEM XXX

It is required to make a body move in a trajectory that revolves about the centre of force in the same manner as another body in the same trajectory at rest.

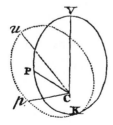

Figure 14.2. Newton's diagram for *Proposition XLIII*. The center of force is at *C*, and the orbit *VPK* is fixed while the dotted orbit is rotating. (From *The* Principia, trans. Andrew Motte [Amherst, NY: Prometheus Books, 1995].)

Both bodies start at V, and one follows the fixed orbit VPK. The second body follows the same orbit, but that orbit now has a rotation around the center of force C. Thus the initial orbit point V has moved to u, and when the first body is at P on the fixed orbit, the second is at p on the rotating orbit (see figure 14.2).

Newton shows that, for the second motion, we still have equal area swept out in equal time by the line Cp. This tells us (according to *Proposition II*) that there is a centripetal force that can produce the second motion on the rotating orbit. However, we know how to find the force when the orbit is given (as reviewed in chapter 9), and that leads Newton to the following:

PROPOSITION XLIV. THEOREM XIV

The difference of the forces, by which two bodies may be made to move equally, one in a quiescent [fixed], the other in the same orbit revolving, varies inversely as the cube of their common altitude.

This result may be appreciated more easily if I translate it into modern terms. Suppose the first body under the influence of a centripetal (central) force has a fixed orbit given by $r_1 = g_1(\theta_1)$, and the second body on the rotating orbit has $r_2 = g_2(\theta_2)$. When the first body is at θ_1, the second body is at $\theta_2 = \lambda\theta_1$, where λ characterizes the rotation rate for the orbit. Because the r values are the same, we must have

$$g_2(\theta_2) = g_2(\lambda\theta_1) = g_1(\theta_1).$$

The angular momentum or equal-areas-law constants are also linked by λ, the rotation rate, $h_2 = \lambda h$. There is orbit rotation when λ differs from 1. Now we can show that, if the first body is moving under the influence of an attractive force $F(r) = -f(r)$, then the second body is moving on an orbit corresponding to an attractive force

$$f_2(r) = f(r) + \frac{mh^2(\lambda^2 - 1)}{r^3},$$

$f(r)$ gives a fixed orbit; the parameter λ gives the orbit rotation.

$$(14.2)$$

That result follows by using the theory in chapter 9 (refer to the books by Lamb and Whittaker for details). Thus the second body moves under the influence of a force that is the original attractive force with an added component varying as an **inverse cube** of the distance from the force center. Its apsides move around according to the rotation parameter λ; we can go from the force law, equation (14.2), to the apsides rotation.

What a brilliant yet simple result is encapsulated in equation (14.2)! It immediately greatly enlarges the set of problems we can study. As Newton puts it, *"and thence new fixed orbits may be found in which bodies may revolve with new centripetal forces."*

In *Corollary V*, Newton (showing off!?) demonstrates how this theory can be used to solve again the problem of motion under an inverse-cube force: set $f(r) = 0$, and the original, now force-free motion, is in a straight line; the rotated form of that will give the spirals as discussed earlier in the *Principia*. How ingenious is that?

14.3. MOTION OF THE APSIDES

Now Newton is ready to do something really clever. In effect, he wants to reverse the argument spelled out above. If a force gives rise to an orbit that gradually rotates so that its apsides move around, how can we relate that movement to the force-law details? Newton's strategy is to begin with a general form containing adjustable parameters for that force, then use the above theory to find formulas for the rotation; as those formulas must involve the force parameters, he will be able to deduce their relevant values for any given rotation. In essence: we know the force-to-rotation link, now we want the rotation-to-force step.

In this case, the fixed orbit will be generated by an attractive inverse-square-law force, so $f(r)$ in equation (14.2) will be k_2/r^2, where k_2 is the force-strength constant and the orbit will be as in the left diagram in figure 14.3. Newton is interested in deviations from the inverse-square law and the rotations they produce (see the right diagram in figure 14.3).

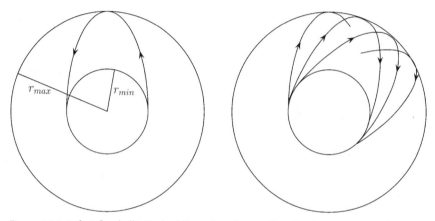

Figure 14.3. *Left*: A fixed elliptical orbit as given by an attractive inverse-square-law force (see section 11.3.5). *Right*: A rotating orbit when the force varies a little from that giving the fixed elliptical orbit seen in the left diagram.

Newton works as follows: Suppose the given force is $f_G(r)$, then we try to approximate it so that we can use equation (14.2). This is possible if we consider nearly circular orbits, which is actually the case for the solar system planetary orbits. (In figure 14.3, for illustrative purposes, the orbits are far from circular.) Then we can put $r = R + x$, where R is the radius of the approximating circle and x measures the small variations around that circle. Thus, we write

$$\text{match a given } f_G(r) \quad \text{to} \quad \frac{k_2}{r^2} + \frac{k_3}{r^3},$$
$$\text{find the approximation} \quad f_G(R+x) = \frac{k_2}{(R+x)^2} + \frac{k_3}{(R+x)^3} \quad \text{for small } x. \tag{14.3}$$

If we can do that, then by comparing with equation (14.2), we can see how k_3 tells us what λ and, hence, the orbit rotation must be. We thus have the required rotation-force link.

In the next subsection, I will give a few mathematical details. They reveal another branch of mathematics that Newton uses in the *Principia*, but less mathematically inclined readers may gloss over this until the results in section 14.3.2.

14.3.1. Mathematical Details

For the interested reader, I present brief details in modern form for the power-law example where $f_G(r) = k/r^p$ is to be used in equation (14.3). According to equation (14.3), we then require

$$\frac{k}{r^p} \approx \frac{k_2}{r^2} + \frac{k_3}{r^3} \quad \text{or} \quad \frac{k}{r^{p-3}} \approx k_2 r + k_3. \tag{14.4}$$

For orbits that are nearly a circle of radius R, we put $r = R + x$, where x measures the small deviations from the circle. This is where Newton uses his algebra to write

$$\frac{1}{r^{p-3}} = \frac{1}{(R+x)^{p-3}} \quad \frac{1}{R^{p-3}}\left(1+\frac{x}{R}\right)^{3-p} = \frac{1}{R^{p-3}}\left(1+(3-p)\frac{x}{R}+\ldots\right).$$

Ignoring the higher-order terms in x/R and substituting into equation (14.4) gives

$$\frac{k}{R^{p-3}}\left(1+(3-p)\frac{x}{R}\right) = k_2(R+x) + k_3.$$

We now equate the powers of x (or, as Newton says: "*collating together the homologous terms*") to get

$$\frac{k}{R^{p-3}} = k_2 R + k_3 \quad \text{and} \quad \frac{k(3-p)}{R^{p-2}} = k_2.$$

These two equations may be solved to get the force strengths k_2 and k_3 in terms of the original given force parameters k and p. The orbit rotation is then calculated as explained above.

14.3.2. Results

Let α be the angle through which the body moves as it goes from its upper apse (farthest from the force center) to its lower apse (nearest to the force center). For the closed inverse-square-law elliptical orbit $\alpha = 180°$ (or π radians), see the diagram on the left in figure 14.3. Newton replaces the inverse-square-law exponent 2 by 3−n (so n = 1 corresponds exactly to the inverse-square law). Using the theory outlined above, he then finds that the line-of-apsides rotation angle is fixed according to

$$\text{if } f_G(r) = \frac{\text{constant}}{r^{3-n}}, \quad \text{then the rotation angle } \alpha = \frac{180°}{\sqrt{n}}. \quad (14.5)$$

Thus, if an observed rotation is not 180°, we can fit the answer to find the force parameter n.

Newton mentions some special cases. If n = 4, we get the force directly proportional to the distance r, and $\alpha = 180°/\sqrt{4} = 90°$, which is just what we observe in the diagram on the left in figure 14.1 (which is a good check on the method). If the force is inversely proportional to the distance, we take n = 2 and then $\alpha = 180°/\sqrt{2} = 127° \ 16' \ 45''$.

An extreme case is given by n = 1/4 when the force varies inversely like $r^{11/4}$ and $\alpha = 360°$: "*therefore the body parting from the upper apse, and from thence continually descending, will arrive at the lower apse when it has completed one entire revolution; and thence ascending perpetually, when it has completed another entire revolution, it will arrive at the upper apse; and so alternately forever.*"

Newton gives another example for a more elaborate $f_G(r)$, perhaps to slip in more of his mathematical "*method of converging series.*" Here is another case of a type of mathematics other than geometry occurring in the *Principia*. He then comes to his major point, summarized in a *Corollary*:

COR. 1. *Hence if the centripetal force be as any power of the altitude, that power may be found from the motion of the apsides; and so contrariwise* [conversely]. *That is, if the whole angular motion, with which the body returns to the same apse, be to the angular motion of one revolution, or*

360°, as any number m *as to another number* n, *and the altitude be called* A, *the force will be as the power* $A^{nn/mm-3}$ *of the altitude* A; *the index of which power is* nn/mm − 3.

In our notation,

if the rotation $2\alpha = \left(\dfrac{m}{n}\right)360°$, the responsible force is $f_G(r) = \dfrac{\text{constant}}{r^s}$,

$$\text{where } s = 3 - \left(\frac{n^2}{m^2}\right). \quad (14.6)$$

14.3.3. Stability

If you look at topics like this in a modern textbook, you will most likely be led to a section on the stability of orbits. Any attractive force can give rise to a circular orbit (just balance centrifugal acceleration and force to find the suitable radius and speed). However, if the body is given a slight push, it may settle down into a nearby similar orbit, or it may radically change its motion and totally move away from the orbiting situation. This is the stability question, and it is important for things like satellite positioning. If we assume a power-law force, the standard analysis shows that the orbit is stable as long as the exponent is greater than −3. Thus the important square-law force (exponent −2) has stable circular (and elliptical) orbits, but the inverse-cube law does not. Newton basically deduces this property, although he never uses the word *stability* when he analyzes equation (14.6); as the cubic case is approached, he shows that the body may spiral into the center of force or away from it. As Newton explains, orbits corresponding to force-law exponents less than −3 will be unstable.

14.3.4. Motion of the Moon

As we shall see later, the Moon has a complicated orbital motion that is not at all easy to come to grips with. It is orbiting the Earth, but the Sun provides an extra disturbing force. Newton struggled greatly with the theory of the Moon's motion This is the first mention of it in the *Principia*, and Newton uses his theory assuming a disturbing force to first estimate that the Moon's "*upper apse in each revolution will go forward 1° 31' 28″.*" An ominous admission follows: "*The apse of the moon is about twice as swift.*"

14.4. FUTURE DEVELOPMENTS

The topics in *Section IX*—orbit properties, rotations, and stability—have continued to stimulate research and applications ever since Newton introduced them. Two examples are particularly interesting.

14.4.1. Closed Orbits and Bertrand's Theorem

Newton knew that the orbit is a closed ellipse in two important cases: the force is directly proportional to the distance from the force center (the linear force), or it is inversely proportional to the square of the distance. These are the cases for which I gave the full modern analysis in sections 11.2.2 and 11.3.5. He has now demonstrated how changes or additions to those forces cause the orbits to rotate and so no longer close back on the starting point. The question arises whether the linear-law force and inverse-square-law force are unique in that way. Newton does not appear to have explicitly asked that question—or, if he did, he could not provide an answer and so did not mention it in the *Principia*.

The above analysis still allows orbits to close back exactly on themselves after a certain number of loops (basically because, according to equation [14.5], for some forces, the rotation is a rational multiple of 2π, and so it can join up exactly after the requisite number of loops). We must remember that this whole analysis is based on near-circular orbits and approximations made in the theory on that basis. If the analysis is carefully taken further, we arrive at a theorem first proved by J. Bertrand in 1873:

The only central (centripetal) forces that result in closed orbits for all bound particles are the inverse-square-law forces and the linear-law force. If we also impose the physical requirement that the force tends to zero as the distance becomes very large (approaches infinity), then only the inverse-square-law force can be used to get all closed, bounded orbits.

See the "Further Reading" section at the end of this chapter for follow-up references. I believe the subtle, careful analysis would have greatly appealed to Newton, as he was obviously keen to show his skills in that area in this section of the *Principia*. I guess he would have also greatly enjoyed hearing about the following work.

14.4.2. Einstein and Rotating Orbits

In *Book III*, we will find Newton using the results in this chapter as he examines the orbits of the planets. By the end of the nineteenth century, most of the details of the solar system were reasonably well understood except for an anomalous rotation of the orbit of the planet Mercury. This famous small rotation (just forty-three seconds of arc per century) was finally explained by Einstein's general theory of relativity. To a first approximation, we can still use Newtonian mechanics, but the gravitational force on a planet must have a correction term added in:

$$\text{gravitational force} = \frac{GmM}{r^2} + \left(\frac{3GmMh^2}{c^2}\right)\frac{1}{r^4}. \qquad (14.7)$$

The second term depends inversely on r^4, rather than r^2 as in the standard force formula, but because c is the speed of light, it represents a very small relativistic correction term (see, for example, the book by Thornton and Marion for further details).

Newton's approach lets us find the orbit rotation caused by that correction term. First, use the near-circular orbit approximation, as in section 14.3.1, to write the force correction term as a sum of an inverse-square-law force and an inverse-cube-law force; the first of those terms gives a very small (and ignorable) change to the main inverse-square-law force in equation (14.7). Second, use that inverse-cube-law force, as in equation (14.2), so that the rotation constant λ is determined. Finally, calculate the orbit rotation generated by the relativistic correction to get

$$\text{orbit rotation} \quad = \quad 6\pi \left(\frac{GM}{ch} \right)^2 .$$

Substituting the right data for Mercury gives the missing forty-three seconds of rotation.

Surely Newton would have been delighted by this extension of his theory and the paying of homage to his mathematical techniques.

14.4. CONCLUSION

Section IX of the *Principia* does not receive a great deal of attention. That is a pity because it contains a brilliant and subtle analysis of orbit properties, and it is a wonderful example of the theoretical physicist at work. Today we take the inverse-square law for gravity so much for granted that we forget how important it was for Newton to discover methods for fixing on that law, rather than on one with a different distance dependence.

This section is a good example of Newton's command of mathematical techniques and how to apply them. His ingenious approach of studying near-circular orbits has remained of significance ever since.

14.5. FURTHER READING

The textbooks listed in section 6.9.1 should be consulted. I particularly recommend the book by Fowles, as well as the book by Thornton and Marion. Newton's orbit-rotation theorem is covered by Whittaker and thoroughly by:

Lamb, Horace. *Dynamics*. Cambridge: Cambridge University Press, 1945. First published 1914. [sections 89–90, especially example 2 in sec. 90; in the example at the end of sec. 91, Lamb explicitly shows how a rotating ellipse orbit corresponds to a combination of inverse-square and inverse-cube forces]

See also:

Valluri, S. R., C. Wilson, and W. Harper. "Newton's Apsidal Precession Theorem and Eccentric Orbits." *Journal for the History of Astronomy* 28 (1997): 13–27.

Waff, C. B. "Isaac Newton, the Motion of the Lunar Apogee, and the Establishment of the Inverse Square Law." *Vistas in Astronomy* 20 (1976): 99–103.

For more on Bertrand's theorem, see Goldstein (see specifically sec. 3.6) and:

Arnol'd, V. I. *Mathematical Methods of Classical Mechanics.* 2nd ed. New York: Springer Verlag, 1989. [sec. 8D gives a proof via a series of problems]

Grandati, Y., and A. Bernard. "Inverse Problems and Bertrand's Theorem." *American Journal of Physics* 76 (2008): 782–87.

Grossman, N. *The Sheer Joy of Celestial Mechanics.* Boston: Birkhäuser, 1996. [see ch. 5]

For more on the rotation of Mercury's orbit, see Thornton and Marion (see specifically sec. 8.9), Fowles (specifically sec. 6.14), and relativity texts such as:

Lambourne, R. J. A. *Relativity, Gravitation and Cosmology.* Cambridge: Cambridge University Press, 2010.

Weinberg, Steven. *Gravitation and Cosmology.* New York: John Wiley, 1972.

Will, Clifford. *Was Einstein Right?* 2nd ed. New York: Basic Books, 1993.

PART 4

ON TO MORE COMPLEX SITUATIONS

Newton has given the complete solution for the one-body problem—the motion of a single particle acted upon by a given centripetal (central) force. But the real world is not like that, and now Newton must show how to deal with constraints on motion, interacting particles, and composite bodies in order to produce a physically applicable theory.

Reading plans: If you want to see all the theory that Newton developed, and in particular some of his most impressive, sophisticated results, then do not miss chapters 15 to 17. Whether or not you read chapters 15 to 17, you should read chapter 18 to get a summary of everything that has been achieved in Book I.

15

CONSTRAINED MOTION

In part 3 we have seen how, in *Sections II* to *IX* of the *Principia*, Newton gave a complete theory for the motion of a single body moving under the influence of a centripetal force. His next job is to elaborate that theory so that it will better match what we find in the real world.

Newton's first step is to deal with constrained motion, that is, bodies held on strings (pendulums) or sliding on rigid surfaces. This is not a step on the way to the grand prize—a theory of the solar system—but it does cover those experimental arrangements first used to explore the principles of mechanics. It was with pendulums and bodies moving on inclined planes that people like Galileo, Riccioli, Mersenne, and Huygens investigated dynamical problems; Newton followed suit, as we saw in chapter 6, for example. Newton sets out his intentions at the end of *Section IX*:

> So much for the motions of bodies in orbits whose planes pass through the centre of force. It now remains to determine those motions in eccentrical planes. For those authors who treat of the motion of heavy bodies used to consider the ascent and descent of such bodies, not only in a perpendicular direction, but at all degrees of obliquity upon given planes [the inclined-plane problem]; and for the same reason we are to consider in this place the motion of bodies tending to centres by means of any forces whatsoever, when those bodies move in eccentrical planes.

Newton is generalizing the Galilean inclined-plane situation so that any force may be acting, not just an assumed uniform gravity. Newton goes on to specify other theoretical assumptions:

These planes are supposed to be perfectly smooth and polished, so as not to retard the motion of the bodies in the least [modern frictionless motion]. *Moreover, in these demonstrations, instead of the planes on which those bodies roll or slide, and which are therefore tangent planes to the bodies, I shall use planes parallel to them, in which the centres of the bodies move, and by that motion describe orbits* [meaning the point of contact is on the constraining plane and the center of a body moves in a plane parallel to that]. *And by that same method I afterwards determine the motions of bodies performed in curved superficies* [surfaces].

Notice that Newton mentions "*bodies roll or slide,*" but Galileo was not clear on that point (see the "Further Reading" section at the end of this chapter); when discussing *Book III,* I will comment on why Newton too did not properly deal with rolling bodies.

So we are to expect a general theory for bodies moving on smooth (not necessarily plane) surfaces and acted on by any type of centripetal force with a force center not in the surface of motion. Of course, a body suspended on the end of a string can be thought of as moving on a smooth surface given by the extent and angle of the string. Newton is saying, "I can do it all in complete generality!" Thus we are led to *Section X: "Of the Motion of Bodies in Given Surfaces; and the Oscillating Motion of Pendulous Bodies."*

Newton begins with some basic dynamics.

15.1. FORCES AND MOTION ON SURFACES

Newton again considers motion under the influence of a centripetal force, but now that motion is confined to a surface not containing the center of force. The surface thus impedes the full effect of the force upon the particle. He poses the problem in:

PROPOSITION XLVI. PROBLEM XXXII.

Any kind of centripetal force being supposed, and the centre of force, and any plane whatsoever in which the body revolves, being given, and the quadratures of curvilinear figures being allowed [we can do any necessary integrations];

it is required to determine the motion of a body going off from a given place with a given velocity, in the direction of a given right line in that place.

The full action of the force is not effective; we need the component acting parallel to the surface, as this is the force that actually moves the body on that surface. Once we have specified the force component, we can use all the machinery built up in earlier sections to determine the trajectory. In this case, Newton's description is quite easy to follow, so this is a good chance to read the words of the Master—and even understand them! He begins with a figure and then talks us through the argument.

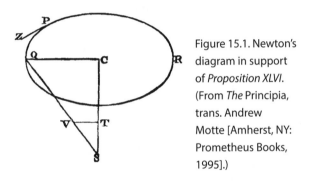

Figure 15.1. Newton's diagram in support of *Proposition XLVI*. (From *The* Principia, trans. Andrew Motte [Amherst, NY: Prometheus Books, 1995].)

Describing the physical situation: "*Let S be the centre of force, SC the least distance of that centre from the given plane, P a body issuing from the place P in the direction of the right [straight] line PZ, Q the same body revolving in its trajectory, and PQR the trajectory itself which is required to be found, described in that given plane.*"

Now to find the effective force acting at Q by resolving: "*Join CQ, QS, and if in QS we take SV proportional to the centripetal force with which the body is attracted towards the centre S, and draw VT parallel to CQ, and meeting SC in T; then will the force SV be resolved into two (by Cor. 2 of the Laws of Motion), the force ST and the force TV; of which ST attracting the body in the direction of a line perpendicular to that plane, does not at all change its motion in that plane.*"

Thus only the component TV affects the motion, so we must use that:

"*But the action of the other force TV, coinciding with the position of the plane itself, attracts the body directly towards the given point C in that plane; and therefore causes the body to move in the plane in the same manner as if the force ST were taken away, and the body were to revolve in free space about the centre C by means of the force TV alone.*"

We have obtained an equivalent centripetal-force motion and we know how to solve for that: "*But there being given the centripetal force TV with which the body Q revolves in free space about the centre C, there is given (by Prop. XLII* [the method for finding orbits for any given force] *the curve PQR which the body describes; the place Q, in which the body will be found at any given time; and, lastly the velocity of the body in that place Q. And conversely. Q.E.I.*"

That is a beautiful example of the power of the theoretical structure Newton has been slowly building. So that we can test it all out, he gives an example:

PROPOSITION XLVII. THEOREM XV.

Supposing the centripetal force to be proportional to the distance of the body from the centre; all bodies revolving in any planes whatsoever will describe ellipses, and complete their revolutions in equal times; and those which move in right lines, running backwards and forwards alternately, will complete their several periods of going and returning in the same times.

This is the old example of the linear force or $\mathbf{F}(r) = -k\mathbf{r}$, which Newton has considered several times before and for which I have given the modern discussion in section 11.2.2.

In this case, the resolved component VT in the plane of motion (PQR in Newton's diagram) is also a linear force proportional to the distance QC from the effective force center C. Thus the motion is just that for this linear force and we can refer to the previous results for details.

In a short *Scholium*, Newton (somewhat cryptically) explains that the same ideas can be used when the constraining plane becomes a curved surface.

Note that the key step was resolving the force to find the component controlling the body's motion. For a pendulum, the force component along the line of the string only concerns the tension in the string, which is providing the constraint on motion, and we need the component perpendicular to that to evaluate the motion of the swinging body.

15.2. THE PENDULUM

The pendulum played a vital role in the development of science, and it features extensively in the writings of Galileo. He believed that the pendulum was isochronous (period independent of the amplitude of the swing), but this was soon corrected by Mersenne and Huygens. Huygens made the revolutionary step of asking how the pendulum must swing so that it actually is isochronous. His answer (see figure 15.2) was that the pendulum bob must follow the arc of a cycloid rather than that of the circle, as in the simple pendulum. (I have previously introduced the cycloid in section 12.5.2.) He brilliantly also showed that if the string was constrained in its swing by suitable cycloid "cheeks," as shown in figure 15.2, then the bob travels along a cycloid as required.

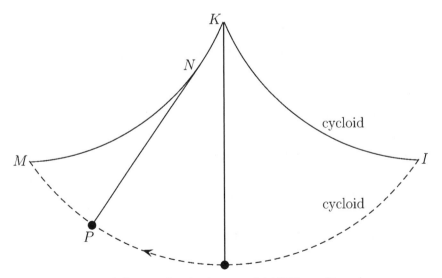

Figure 15.2. Huygens's diagram showing how a cycloid *MKI* is used to make a pendulum bob *P* swing along a cycloid. The string wraps around the top cycloid and, in the situation as shown, it is leaving the cycloid *KM* branch as a tangent at *N*.

In another significant advance, Huygens solved the old problem of measuring the rate of fall under gravity. He showed that if a pendulum of length L has a period of 1 second, then the distance D fallen under gravity in 1 second will be $2\pi^2 L$. Thus the whole experiment is reduced to constructing a "seconds" pendulum and measuring its length. Huygens got the answer as 15 Rhenish feet 7.5 inches, which corresponds closely to our value of 981 cm × s^{-2} for the gravitational constant g. This result will be of importance when we discuss *Book III*.

To see the basis for Huygens's result in modern terms, note that for a small-amplitude simple pendulum,

$$\text{period } T = 2\pi \sqrt{\frac{L}{g}}, \text{ and the distance fallen in that time is } D = \frac{1}{2}gT^2 = 2\pi^2 L.$$

Putting in the L corresponding to T = 1 second then gives the required fall distance D.

So, much was known about the pendulum when Newton came on the scene. Now to some of Newton's results in *Section X* (and we will find more when we come to both *Books II* and *III*).

15.3. NEWTON'S RESULTS FOR THE PENDULUM

Newton begins by defining the cycloids obtained when a circle revolves around the circumference of a larger circle, which could be a great circle on a sphere. The original or ordinary cycloid is generated by a circle rolling on a plane surface. A point on the edge of the revolving circle traces out a cycloid, as shown in Newton's diagram in figure 15.3. He gives some properties of those cycloids and also repeats Huygens's method for causing a body to oscillate along the path of a cycloid, as in figure 15.2. (See the paper by Gauld, listed in the "Further Reading" section at the end of this chapter, for a discussion.)

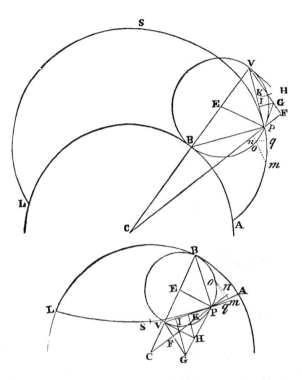

Figure 15.3. Newton's diagram showing how cycloids are generated by a point on a circle that is revolving outside (*top*) or inside (*bottom*) a larger circle. (From *The Principia*, trans. Andrew Motte [Amherst, NY: Prometheus Books, 1995].)

Next comes a remarkable result:

PROPOSITION LI. THEOREM XVIII.

If a centripetal force tending to all sides to the centre C of a globe, be in all places as the distance of the place from the centre; and, by this force alone acting on it, the body T oscillate (in the manner above described) in the perimeter of the cycloid QRS: I say, that all the oscillations, howsoever unequal in themselves, will be performed in equal times.

We are back to the linear force again, but this time the force center C is at the center of a globe, and forming a cycloid on the inside of that globe allows the generation of isochronous pendulum oscillations (see figure 15.4). Newton gives various results for this pendulum motion.

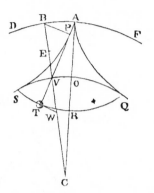

Figure 15.4. Newton's diagram showing pendulum motion along the cycloid *STRQ* formed on the inside of the circle or globe *SVQ* with center *C*, which is also the center for a linear centripetal force acting on the pendulum bob. (From *The* Principia, trans. Andrew Motte [Amherst, NY: Prometheus Books, 1995].)

Newton then deals with a limiting case: "*if the diameter of the globe be infinitely increased, its spherical surface will be changed into a plane, and the centripetal force will act uniformly in the direction of lines perpendicular to that plane, and this cycloid of ours will become the same with the common cycloid.*" Thus he makes contact with results given by Sir Christopher Wren and Christiaan Huygens for the "*common cycloid.*"

It may seem that Newton has gone through some flashy exercise with his elaborate cycloids as in figure 15.3. However, it turns out (as I will discuss in chapter 17 and when we examine *Book III*) that the linear-force case does have a practical importance:

> *The Propositions here demonstrated are adapted to the true constitution of the earth, . . . and pendulums in mines and deep caverns of the earth, must oscillate in cycloids within the globe, that those oscillations may be performed in equal times. For gravity (as will be shown in the third book) decreases in its progress from the surface of the earth . . . downwards in a simple ratio of these distances* [that is, like a linear force].

Thus the linear-force case, as in *Proposition LI*, turns out to be important when we analyze gravitational forces inside the Earth.

15.4. COMPLETING THE STORY

To round off this section, Newton, as usual, demonstrates his mathematical skills by going to some general cases. In *Proposition LIII*, he considers the inverse problem: Given the curve along which a body performs oscillations, what force must be acting if they are to be isochronous? He briefly comments on the design of clocks.

In *Propositions LIV* to *LVI*, Newton gives the general theory for calculating motion on surfaces, partly by using his general formalism in *Section VIII*. He finds some equal-areas- swept-out-in-equal-times results, which inspired a recent paper by James Casey (see the "Further Reading" section at the end of this chapter).

15.5. CONCLUSION

In *Section X*, Newton has extended the work of Galileo, Huygens, and others on the constrained motion of bodies along curves and surfaces. The major change is that now the theory is developed using forces and the laws of motion. The key step is to resolve the forces into components, one of which has no effect because of the constraint, while the other controls a body's motion along the constraining surface. In this way, Newton can consider a variety of surfaces (beyond Galileo's inclined planes) and explore the effects of forces other than an assumed uniform gravitational force.

15.6. FURTHER READING

The motion of a pendulum is a standard topic in books on classical mechanics, such as those listed in section 6.9.1. Sommerfeld, in his section 17, gives a lovely analysis of the cycloidal pendulum. A large collection of papers covering many aspects of the pendulum and its history is in:

Matthews, M. R., C. F. Gauld and A. Stinner, eds. *The Pendulum: Scientific, Historical, Philosophical and Educational Perspectives.* New York: Springer, 2005. In that book, see especially: Gauld, C. *The Treatment of Cycloidal Pendulum Motion in Newton's* Principia.

The work of Huygens is covered in:

Huygens, Christiaan. *The Pendulum Clock; or, Demonstrations concerning the Motion of Pendula as Applied to Clocks.* Translated from the original 1673 *Horologium Oscillatorium* by Richard J. Blackwell. Ames: Iowa State University Press, 1986.

Yoder, J. G. *Unrolling Time: Christiaan Huygens and the Mathematization of Nature.* Cambridge: Cambridge University Press, 1988. [this takes the reader through many examples of Huygens's working]

A highly recommended article on the pendulum and its early use in measurements is:

Koyré, Alexandre. "An Experiment in Measurement." *Proceedings of the American Philosophical Society* 97 (1953): 222–37. Reprinted in *Metaphysics and Measurement.* London: Chapman and Hall, 1968.

The literature on Galileo is vast; here are two relevant papers:

Ariotti, P. "Galileo on the Isochrony of the Pendulum." *Isis* 59 (1968): 414–26.

Crawford, F. S. "Rolling and Slipping down Galileo's Inclined Plane: Rhythms of the Spheres." *American Journal of Physics* 64 (1996): 541–46.

A recent paper discussing swept-out areas for particles moving on curved surfaces is:

Casey, James. "Areal Velocity and Angular Momentum for Non-planar Problems in Particle Mechanics." *American Journal of Physics* 75 (2007): 677–85.

MANY BODIES
TRIUMPHS AND CHALLENGES

This chapter covers one of the monumental, vital steps in mechanics, though that does make for a longer chapter. So far, Newton has told us how to solve the problem of a body moving under the influence of a centripetal force—the "single-body problem." Now he explains how to deal with two (and more) interacting bodies, which could be the Earth and the Sun, for example. We know, from right back when the laws of motion were introduced, that the motion of the center of mass of a system of bodies is easily determined (see sections 6.3 and 6.6.2). Newton now tells us how to calculate the motion of the bodies relative to that center of mass. Further, for two interacting bodies, that calculation only requires the solution of an equivalent single-body problem. Thus the complete solution for the problem of two interacting bodies is known. Amazing!

Sections 16.1 and 16.2 are of particular importance. The other sections give more details, present the modern formulation, and discuss the extension to many-body systems, especially the famous three-body problem.

16.1. INTRODUCTION

This chapter deals with the *Principia* Section XI: "*Of the Motion of Bodies Tending to Each Other with Centripetal Forces.*"

This is one of the most significant sections in the whole book, as

Newton simply but brilliantly shows us how his theory may be extended to make it more applicable to real-life problems. There are struggles too; typically it is Newton who discovers possibly the most famous of all difficult problems in dynamics! Newton begins this section with a clear and careful introduction:

> I have hitherto been treating of the attraction of bodies towards an immovable centre; though very probably there is no such thing existent in nature. For attractions are made towards bodies, and the actions of the bodies attracted and attracting are always reciprocal and equal, by Law III; so that if there are two bodies, neither the attracted nor the attracting body is truly at rest, but both (by Cor. 4 of the Laws of Motion), being as it were mutually attracted, revolve about a common centre of gravity [or center of mass]. And if there be more bodies, which either are attracted by one single one, which is attracted by them again, or which all of them attract each other mutually, these bodies will be so moved among themselves, as that their common centre of gravity will either be at rest, or move uniformly forwards in a right line. I shall therefore at present go on to treat of the motion of bodies mutually attracting each other; considering the centripetal forces as attractions.

The introduction continues with comments that take us back to Newton's wish to set up a mathematical formalism, which is later to be applied to physical situations, rather than to become embroiled in debate about the mechanisms responsible for the forces he uses (for example, see section 5.4.2):

> though perhaps in a physical strictness they may more truly be called impulses. But these Propositions are to be considered as purely mathematical; and therefore, laying aside all physical considerations, I make use of a familiar way of speaking, to make myself the more easily understood by a mathematical reader.

Newton begins with the simplest case, two interacting bodies, the **two-body problem.**

16.2. MOTION AROUND THE CENTER OF MASS

In his introduction, Newton tells us that the bodies must move around the center of mass, which will be at rest or moving uniformly in a straight line if there are no external forces acting on the bodies (see section 6.6.2). For simplicity, we can imagine using a reference frame in which the center of mass is at rest. He formalizes that result and explains what it implies for orbits in:

PROPOSITION LVII. THEOREM XX.

Two bodies attracting each other mutually describe similar figures about their common centre of gravity [or center of mass], *and about each other mutually.*

Recall that if two masses m_1 and m_2 are a distance L apart, the center of mass is that point directly between them, which is a distance (m_2/M) L from mass 1 and a distance $(m_1/M)L$ from mass 2, where M is the total mass $m_1 + m_2$ (see figure 16.1[a]). (When no external forces are present or in a uniform gravitational field, the centers of gravity and mass are identical.)

Newton is saying that the interacting bodies revolve around the center of mass, as shown for assumed circular motion in figure 16.1(b) (for $m_1 > m_2$) and 16.1(c) (for $m_1 = m_2$). Also, viewed from one mass, the second mass appears to travel around it in a circle of radius L, as shown in figure 16.1(d).

For the circular-motion case, both bodies are at constant distances from the center of mass at all times. This is not the case for motion on ellipses, as shown in figures 16.1(e) and 16.1(f). In this case, L varies but, of course, the bodies remain at proportionally the same distances from the center of mass; if the bodies at some time are a distance L' apart, then body 1 is a distance $(m_2/M)L'$ from the center of mass, with a similar distance $(m_1/M)L'$ for body 2.

After a little thought, this probably seems obvious to you; I will give the modern algebraic form in section 16.3. For those who would like to read the original, here is Newton's somewhat wordy explanation:

> For the distance of bodies from their common centre of gravity are inversely as the bodies; and therefore in a given ratio to each other; and thence, by composition of ratios, in a given ratio to the whole distance between the bodies. Now these distances revolve about their common term [extremity] with an equable angular motion, because lying in the same right line they never change their inclination to each other mutually. But right lines that are in a given ratio to each other, and revolve about their terms [extremities] with an equable angular motion, describe upon planes, which either rest with those terms, or move with any motion not angular, figures entirely similar round those terms. Therefore the figures described by the revolution of these distances are similar. Q.E.D.

The obvious question is what are those "*similar figures about their common centre of gravity*"? Furthermore, how do we find them using the force describing the interaction of the bodies? I will deal with this in section 16.3, but first, a few comments about the importance of *Proposition LVII*.

16.2.1. Implications Old and New

Proposition LVII seems like an almost trivially simple result, but its consequences are really quite profound. It tells us that when two bodies interact, they will **both** move; only in the unphysical case, when one is infinitely larger than the other, will we find it not moving. Furthermore, it tells us something quite general about how they move, as shown in figure 16.1.

The Earth and the Moon may be taken as an interacting pair, and the motion of their center of mass may be studied. Of course, the effect of the Sun is important, and I will come to that later in this chapter. When we come to study Newton's treatment of the solar system in *Book III*, we shall find arguments based on necessary motions about the center of mass of great importance for settling old controversies about just what points might be at rest and considered a center for the whole system.

Two examples will illustrate the importance of *Proposition LVII* in astrophysics. Sirius is the brightest star in the sky and, as such, has garnered the attention of astronomers through the ages. Studying its motion in 1834, F. W. Bessel observed oscillations and eventually concluded that it was part of a binary system with an orbital period of about fifty years. It was not until 1862 that Alvan Clark observed the faint "companion" that we now know to be a white dwarf and that causes Sirius to oscillate. Figure 16.2 shows the orbits of Sirius and its companion with more recently dated positions.

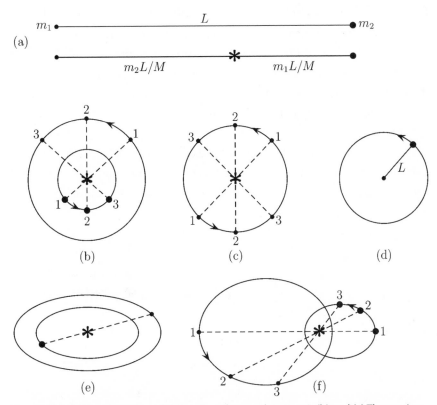

Figure 16.1. (a) The definition of the center of mass, shown as *. (b) and (c) The motion of two masses in circles around their center of mass, masses assumed equal in (c). (d) The circular motion of body 2 as seen from body 1. (e) and (f) The motion of two bodies on ellipses around their center of mass.

In 1600, just forty-two years before Newton was born, the monk and philosopher Giordano Bruno was burned at the stake for suggesting (among other things) that there could be other "earths" orbiting other "suns." There has been speculation about "other worlds" orbiting their own suns for centuries; in 1698, Huygens published *Cosmotheoros; or, Conjectures concerning the Planetary Worlds and Their Inhabitants*. But it was not until 1995 that, with great excitement, astronomers Michel Mayor and Pierre-Yves Queloz finally announced that a planet around another star had been discovered. Now, many hundreds of "exoplanets" have been found. How can it be done? The clue comes from an application of *Proposition LVII*: if a star is known to "wobble around," then it may have a companion planet heavy enough to give it an extra motion, as explained in figure 16.1. Mayor and Queloz discovered that the star 51 Peg has a companion that causes it to wobble around with a period of about four days. Because 51 Peg will sometimes be moving toward us and sometimes away, its emitted light suffers different Doppler shifts, and the subsequent wavelength variations may be used to deduce orbital characteristics (see the "Further Reading" section at the end of this chapter). One of the most exciting and significant events of twentieth-century astronomy depended on an application of Newton's *Proposition LVII* some three hundred years after he first wrote it down!

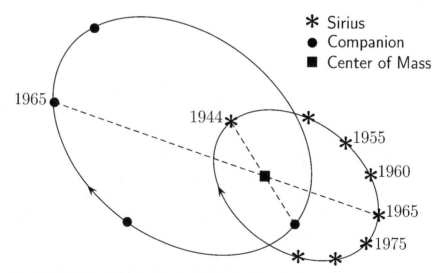

Figure 16.2. Orbits of the star Sirius and its companion, relative to their center of mass. (Drawn from data found in *Essentials of Astronomy*, by L. Motz and A. Duveen [London: Blackie and Sons, 1977].)

16.3. THE WONDERFUL STEP: FINDING ORBIT DETAILS

In the next part of *Section XI*, in one of the greatest triumphs of classical mechanics, Newton tells us how to calculate those "*similar figures*" or orbits, as shown in figure 16.1. We know how the center of mass behaves, and if we can find those orbits around it, we will have completely solved the two-body problem; that is, we will know how to calculate the details of the motion of two interacting bodies.

Newton shows that the required orbits may be found using a particular **single-body** problem, the problem of **one body moving under the influence of a centripetal force.** Since the first sections of the *Principia* have explained how to do just that, everything will be in place for solving the two-body problem.

The single-body problem to be solved uses the given force of interaction to solve for the size and orientation of the vector joining the two bodies. That will allow us to write down the positions of the bodies relative to the center of mass. We can also calculate the period of this orbital motion and see how to correct what seems like a more obvious result.

I think this may be easier to appreciate as it is set out in the modern formulation, and I will do that in section 16.4.2. I will summarize the whole result in section 16.4.3. But first, here is the way Newton does things, and I leave it to you, if you so wish, to follow his steps and to consult the *Principia* for details of his proofs.

16.3.1. Newton's Propositions Solving the Two-Body Problem

PROPOSITION LVIII. THEOREM XXI.

If two bodies attract each other mutually with forces of any kind, and in the mean time revolve about the common centre of gravity [mass]; I say, that, by the same forces, there may be described round either body unmoved a figure similar and equal to the figures which the bodies so moving describe around each other mutually.

Newton stresses again that the center of gravity (mass) may be at rest or moving uniformly in a straight line, which is the way the plane containing the bodies must move. In the *Corollaries*, he points out that the figures will be ellipses for the cases of linear-law forces and inverse-square-law forces. He also notes that an equal-areas-swept-out-in-equal-times law will hold in all the cases:

> COR. 3. *Any two bodies revolving round their common centre of gravity describe areas proportional to the times, by radii drawn both to that centre and to each other mutually.*

Now he moves on to the period of revolution:

PROPOSITION LIX. THEOREM XXII

The periodic times of two bodies S and P revolving round their common centre of gravity C, is to the periodic time of one of those bodies P revolving round the other S remaining unmoved, and describing a figure similar and equal to those which the bodies describe about each other mutually, as $\sqrt{S}$ is to $\sqrt{S+P}$).

What Newton is saying here is that, if you were to take one body as fixed and solve for the motion of the second one around it (hence, a single-body problem), you will get the period wrong and must correct it as stated. Said another way, if you do not use the proper center-of-mass argument (but assume instead that one body is fixed), you will make an error in finding the period. Now comes a result about the various orbit sizes:

PROPOSITION LX. THEOREM XXIII

If two bodies S and P, attracting each other with forces inversely proportional to the square of their distance, revolve about their common centre of gravity; I say, that the principal axis of the ellipse which either of the bodies, as P, describes by this motion about the other S, will be to the principle axis of the ellipse, which the same body P may describe in the same periodic time about the other body S quiescent [fixed], as the sum of the two bodies S + P to the first of two mean proportionals between that sum and the other body S.

Finally, the motion of the bodies can be thought of as generated by forces centered on the center of gravity:

PROPOSITION LXI. THEOREM XXIII.

If two bodies attracting each other with any kind of forces, and not otherwise agitated or obstructed, are moved in any manner whatsoever, those motions will be that same as if they did not at all attract each other mutually, but were both attracted with the same forces by a third body placed in their common centre of gravity; and the law of the attracting forces will be the same in respect of the distance of the bodies from the common centre, as in respect of the distance between the two bodies.

Newton gives the example of two bodies at released at rest, saying that they will gradually move toward each other along the line through their (at rest) shared center of mass. He also points out that the general one-body formalism presented earlier can be used generally to solve the two-body problem for any given initial conditions.

16.4. THE TWO-BODY PROBLEM: MODERN FORMULATION

A key to understanding the theory is the specification of the positions of the two bodies in space, and then classical mechanics will tell us how those positions vary as forces act.

16.4.1. Defining Position Vectors

For a given origin O, we specify the position vectors $\mathbf{r}_1$ and $\mathbf{r}_2$ as shown in figure 16.3(a). The vector $\mathbf{r}_{12}$ joins the two bodies and tells us how to move from body 2 to body 1:

$$\mathbf{r}_{12} = \mathbf{r}_1 - \mathbf{r}_2 \text{ so } \mathbf{r}_1 = \mathbf{r}_2 + \mathbf{r}_{12}. \tag{16.1}$$

The dynamical problem is to find $\mathbf{r}_1(t)$ and $\mathbf{r}_2(t)$ for all times t for given initial conditions.

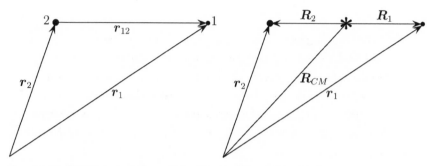

Figure 16.3. Different ways of specifying the positions of two bodies.

The situation may also be described using other vectors related to the center of mass. I will use capital letters for these vectors to emphasize that they are a different and alternative specification, and they are shown in the right diagram of figure 16.3. The vector from the origin to the center of mass, indicated by *, is

$$\mathbf{R}_{cm} = \left(\frac{m_1}{m_1 + m_2}\right)\mathbf{r}_1 + \left(\frac{m_2}{m_1 + m_2}\right)\mathbf{r}_2 = \left(\frac{m_1}{M}\right)\mathbf{r}_1 + \left(\frac{m_2}{M}\right)\mathbf{r}_2, \quad (16.2)$$

where m_1 and m_2 are the masses of the bodies and M is the total mass $m_1 + m_2$. The right diagram in figure 16.3 shows the position of the bodies relative to the center of mass, as given by the vectors $\mathbf{R}_1$ and $\mathbf{R}_2$.

Thus we may choose to work with either set of vectors ($\mathbf{r}_1$, $\mathbf{r}_2$ and $\mathbf{r}_{12}$, or $\mathbf{R}_{cm}$, $\mathbf{R}_1$, and $\mathbf{R}_2$) and move between them using equation (16.2) and

$$\mathbf{R}_1 = \left(\frac{m_2}{M}\right)\mathbf{r}_{12} \quad \text{and} \quad \mathbf{R}_2 = -\left(\frac{m_1}{M}\right)\mathbf{r}_{12}, \quad (16.3\text{a})$$

$$\mathbf{r}_1 = \mathbf{R}_{cm} + \mathbf{R}_1 \quad \text{and} \quad \mathbf{r}_2 = \mathbf{R}_{cm} + \mathbf{R}_2. \quad (16.3\text{b})$$

16.4.2. Dynamics

If we assume there are no external forces, so we are dealing with an isolated two-body system, the equations of motion are

$$m_1 \frac{d^2 \mathbf{r}_1}{dt^2} = \mathbf{F}_{12} \quad \text{and} \quad m_2 \frac{d^2 \mathbf{r}_2}{dt^2} = \mathbf{F}_{21}, \tag{16.4}$$

where $\mathbf{F}_{12}$ is the force exerted by particle 2 on particle 1, and $\mathbf{F}_{21}$ is the force exerted by particle 1 on particle 2 (see figure 6.4). Now, by Newton's third law, those two forces are equal in magnitude but opposite in direction, so $\mathbf{F}_{21} = -\mathbf{F}_{12}$ and equations (16.4) can be written as

$$m_1 \frac{d^2 \mathbf{r}_1}{dt^2} = \mathbf{F}_{12} \quad \text{and} \quad m_2 \frac{d^2 \mathbf{r}_2}{dt^2} = -\mathbf{F}_{12}. \tag{16.4a}$$

As explained in section 6.6.2, adding the equations in (16.4) leads to the center-of-mass theorem and the result that

$$\mathbf{R}_{cm}(t) = \mathbf{v}_0 t + \mathbf{R}_{cm0} \tag{16.5}$$

where $\mathbf{v}_0$ is a constant velocity and $\mathbf{R}_{cm0}$ is the initial position of the center of mass. If the center of mass is initially at rest, $\mathbf{v}_0 = 0$, then it remains at rest for all future times. Thus $\mathbf{R}_{cm}$ is completely determined.

There is another way to combine the equations in (16.4a). Multiply the first one by m_2 and from that subtract the second equation after multiplying it by m_1 to get

$$m_1 m_2 \frac{d^2 \mathbf{r}_1}{dt^2} - m_1 m_2 \frac{d^2 \mathbf{r}_2}{dt^2} = m_2 \mathbf{F}_{12} - (-m_1 \mathbf{F}_{12}).$$

That can be tidied up to reveal

$$m_1 m_2 \left(\frac{d^2 \mathbf{r}_1}{dt^2} - \frac{d^2 \mathbf{r}_2}{dt^2} \right) = m_2 \mathbf{F}_{12} - (-m_1 \mathbf{F}_{12}) = (m_1 + m_2) \, \mathbf{F}_{12}.$$

The rules of calculus and equation (16.1) allow us to write this as

$$\mu \left(\frac{d^2 \mathbf{r}_{12}}{dt^2} \right) = \mathbf{F}_{12}(r_{12}),$$

(16.6)

where the "reduced mass" μ is defined by

$$\mu = \frac{m_1 m_2}{m_1 + m_2} = \frac{m_1 m_2}{M} .$$

(16.7)

Now we have it! Equation (16.6) is the equation of motion that we must solve to find $\mathbf{r}_{12}$. Then we know $\mathbf{R}_1$ and $\mathbf{R}_2$ from equation (16.3a), and we already know $\mathbf{R}_{cm}$ from equation (16.5); thus, **the whole problem is solved, as we know where the two bodies' center of mass is and their positions relative to that.**

16.4.3. Orbits and Periods

The various ways to look at the orbits are shown in figure 16.4. The orbit given by $\mathbf{r}_{12}(t)$ is shown in the left diagram of figure 16.4, and this is the one calculated using equation (16.6). It is the orbit of body 1 relative to body 2. If we assumed body 2 was extremely massive, $m_2 \gg m_1$, we would use exactly that equation to solve the whole problem, except that we would use m_1 instead of μ, and, of course, the center of mass would coincide with body 2.

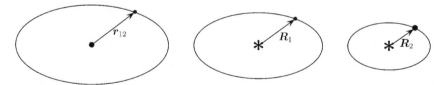

Figure 16.4. *Left*: This orbit is given by r_{12}. *Center and right*: These are orbits of the two bodies around the center of mass *, assuming body 2 is more massive than body 1.

(Obviously I am thinking about body 2 as the Sun and body 1 as one of the planets in our solar system. Newton is carefully not saying that because he is building up his general formalism; then, in *Book III*, he will use it for describing the solar system.)

What would be the effect of that assumption or approximation—using an infinitely massive body 2 instead of the proper center of mass formulation with a finite if very large m_2? Notice that the mass comes before the time derivatives in equation (16.6), and we can think of a scaled time to go from one case to another. We find that, if T is the period, the correct and approximate results are connected by

$$T_{correct} = \sqrt{\frac{\mu}{m_1}}\ T_{approx.} = \sqrt{\frac{m_2}{m_1 + m_2}}\ T_{approx.} \qquad (16.8)$$

For the solar system, these corrections are small; however, for some of the exoplanets or double star systems mentioned in section 6.1.1, the effect could be quite large. (Actually, we are getting a sneak preview of the fact that Kepler's third law of planetary motion linking orbit sizes and periods is not exactly correct. More on that when we get to *Book III*.)

The diagrams in the center and at the right of figure 16.4 show the orbits of the two bodies around the center of mass. Their scales follow from equation (16.3a). If we substituted those equations into equation (16.6), we would get equations of motion that looked like each body moving according to a force center at the center of mass, which is exactly what Newton says in *Proposition LXI*, quoted above in section 16.2.1.

Notice that in each case, because we are dealing with centripetal forces, the orbit is such that the position vector sweeps out equal areas in equal times, as Newton points out in *Corollary 3* to *Proposition LVIII*.

16.5. WHAT HAS BEEN DONE AND WHY IT WORKS

This is such an important step that it is worthwhile making a few general comments. **The general two-body problem has been broken up into two** separate problems: (1) the motion of their center of mass and (2) an equivalent single-body problem that tells us about their motion around that center.

The result for the center-of-mass motion, equation (16.5), is related to the fact that there are no external forces acting and a resulting conservation of linear momentum law, as explained in sections 6.6.1 and 6.6.2.

The single-body problem is relatively easily solved because the swept-out-area law, the conservation of angular momentum, and the conservation of energy allow us to integrate the differential equations of motion, as discussed in chapters 11 and 13.

In summary, to put it in modern terms, it is the fact that the theory for the types of centripetal forces being used leads to conservation laws for linear momentum, angular momentum, and energy that allows us to completely solve the equations of motion in the one- and two-body cases; the necessary integrations, as explained in section 13.1.1, are greatly simplified. Newton does not speak in those terms, but he too essentially uses that approach, if not explicitly.

16.5.1. What If There Are External Forces?

Newton actually considered this case in a *Corollary* to the laws of motion (see section 6.4.1). He asks us to consider the case where equal accelerative forces are acting, and the force on each body is proportional to its mass. Thus we modify the equations of motion (16.4b) to get

$$m_1 \frac{d^2\mathbf{r}_1}{dt^2} = \mathbf{F}_{12} + m_1\mathbf{E} \quad \text{and} \quad m_2 \frac{d^2\mathbf{r}_2}{dt^2} = -\mathbf{F}_{12} + m_2\mathbf{E}, \quad (16.9)$$

where $\mathbf{E}$ gives us the details of the external force. **Equation (16.6) for the relative motion of the two bodies is unchanged** because, in the process of deriving it, the external-force terms will cancel out. However, the center-of-mass equation of motion now becomes

$$\frac{d^2\mathbf{R}_{cm}}{dt^2} = \mathbf{E}. \qquad (16.10)$$

The center of mass will no longer move uniformly in a straight line or be at rest, and its motion now depends on the nature of $\mathbf{E}$. However, **it is important to note that the relative motion of the two particles is unchanged by this external force, and the two-body problem is still broken up into those two separate problems**, the center-of-mass motion and the bodies' relative motion.

16.6. THE MANY-BODY PROBLEM

Newton has solved the two-body problem, and perhaps he initially thought he could find a similar procedure for the case of more than two interacting bodies. That he could not do, but he does discuss cases involving linear forces (force directly proportional to distance) and inverse-square-law forces. Of course, the result for the center of mass remains valid no matter how many bodies are involved; it is determining the relative motions of those bodies that causes the problem.

16.6.1. Why Those Forces—Again?

By now, you will realize that those two types of forces (linear and inverse square) come up in the *Principia* over and over again. But look in a modern textbook on classical mechanics and you will find it is the same. There are two reasons for this. First, both types of force lead to linear differential equations (see sections 11.2.2 and 11.3.4) for the one- and two-body situations, and that means we can handle the mathematics without too much trouble.

Second, luckily for us, that mathematical tractability goes along with two forces that are of enormous practical importance. The linear force gives rise to oscillatory motion. The classical harmonic oscillator is ubiquitous in physics from pendulum motion onward. The force due to a stretched spring is a linear force, investigated by Hooke and sometimes named for him. The inverse-square-law force enters the theory of gravity and the Coulomb interaction between electric charges. Hence, a large fraction of physics can be studied using those two forces.

16.6.2. Examples

Newton begins with the linear force case in the following:

PROPOSITION LXIV. PROBLEM XL.

Supposing forces with which bodies mutually attract each other to increase in a simple ratio of their distances from the centres; it is required to find the motions of several bodies among themselves.

Remarkably, Newton has homed in on the only many-body problem that can be readily solved. You can find the details in almost any advanced textbook on classical mechanics, usually under a heading such as "Multiple or Coupled oscillators." In modern terms, we get a set of coupled linear equations for the equations of motion, but, because of that linearity property, they can be solved using some simple calculus and results from linear algebra (see the "Further Reading" section at the end of this chapter). Newton does not use that approach. He develops an iterative method that allows him to go from the two-body case to the three-body case, and then to continue adding more bodies as required (Chandrasekhar gives a full explanation—see this chapter's "Further Reading" section).

Newton's second example comes in here:

PROPOSITION LXV. THEOREM XXV.

Bodies, whose forces decrease as the square of their distances from their centres, may move among themselves in ellipses; and by radii drawn to the foci may describe areas proportional to the times very nearly.

Notice that Newton says *"very nearly."* He cannot solve this problem exactly, but he does understand that good approximations can be made. This may seem almost trivially obvious to us, with all our modern experience, but we should remember that for Newton and his readers this is classical mechanics set out for the very first time.

In *Case I*, Newton says, *"imagine several lesser bodies to revolve about some very great one at different distances from it."* He notes that we still know about the center-of-mass motion, and that *"the lesser will revolve about the great one in ellipses, and by radii drawn thereto will describe areas proportional to times."* Of course, we must *"except the errors"* of *"the mutual actions of the lesser bodies upon each other."* Obviously Newton is preparing the way for his treatment of the solar system in *Book III*.

Case II concerns two bodies revolving around each other with another larger body a great distance away. Newton explains that this can be dealt with using the external-force theory discussed above in section 16.3.1. If the distance is great enough, the force exerted by the massive body is essentially parallel for the bodies, so equations (16.9) may be used. He then points out that the argument may be extended from two to any number of bodies revolving around each other. It may not be as obvious, but this is also something Newton will need when he comes to *Book III*.

Naturally Newton thinks about turning the argument around so that something may be deduced about an observed physical situation:

> COR 3. Hence if the parts of this system move in ellipses or circles without any remarkable perturbation, it is manifest that, if they are all impelled by accelerative forces tending to any other bodies, the impulse is very weak, or else is impressed very nearly equally and in parallel directions upon all of them.

16.7. THE THREE-BODY PROBLEM

Another system of vital importance to Newton is that of the Earth, the Moon, and the Sun. In Newton's time, the need for navigational aids was becoming ever greater, and naval disasters gave added urgency to the problem of finding positions at sea, especially longitude. Newton hoped that his lunar theory would be of help. Notice that Newton is now looking for a causal theory to explain the Moon's motion, rather than data fitting and the construction of lunar tables. Today we know that the theory of three interacting bodies also has applications in astrophysics with such things as star triplets and several exoplanets.

We can be clear at the outset: Newton cannot solve the three-body problem. He gives three propositions for the problem, the first accompanied by twenty-two *Corollaries* that have become famous for their thoroughness and inventiveness. They are the beginning of perturbation theory. The Moon's orbit has intricate small variations in it; the question is how the Earth and the Moon together orbit around the Sun, and how the Sun perturbs the Earth-Moon orbiting system.

I will return to the problem of the Moon's motion when discussing *Book III*, but to show how much it dominated Newton's thinking about the three-body problem, here is the start of that great first *Proposition* and the diagram that accompanies it. The labels T (for *Terra*) and S (for *Sun*) are obvious.

PROPOSITION LXVI. THEOREM XXVI.

If three bodies, whose forces decrease as the square of the distances, attract each other mutually; and the accelerative attractions of any two towards the third be between themselves inversely as the squares of the distances; and the two least revolve around the greatest . . .

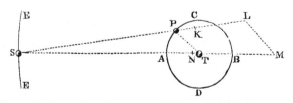

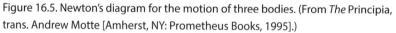

Figure 16.5. Newton's diagram for the motion of three bodies. (From *The* Principia, trans. Andrew Motte [Amherst, NY: Prometheus Books, 1995].)

16.7.1. The Most Famous Problem

In his classic text (see the "Further Reading" section at the end of this chapter), The eminent mathematician Sir Edmund Whittaker says, "the most celebrated of all dynamical problems is known as the Problem of Three Bodies."[1] From Newton onward, the problem has attracted the greatest mathematicians and scientists, intense efforts were made to solve it, prizes were offered, and a vast literature was produced—and is still being added to today.

We might expect to find solutions under very special restrictions. Newton hinted at such cases in his discussion of *Proposition LXV* when he wrote:

> *neither is it possible that bodies attracting each other according to the law supposed in this Proposition should move exactly in ellipses, unless by keeping a certain proportion of distances from each other.*

I do not know whether those superb theorists Leonard Euler (1707–1783) and Joseph Louis Lagrange (1736–1813) took that hint, but they did find exact solutions for the three-body problem. In Euler's case, the three bodies are always in a line, and the ratio of their distances apart remains constant. For Lagrange, the bodies are always at the corners of an equilateral triangle. The resulting orbits are shown in figure 16.6.

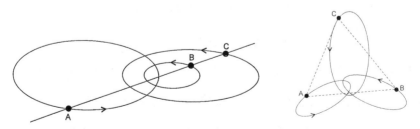

Figure 16.6. The motion of three bodies in the solutions given by Euler (*left*) and Lagrange (*right*). (Diacu, Florin; *Celestial Encounters.* © 1996 Florin Diacu and Philip Holmes, published by Princeton University Press. Reprinted by permission of Princeton University Press.)

16.7.2. An Intractable Problem

Recall that a full solution for a dynamical problem will give the position $\mathbf{r}(t)$ (or $x[t]$, $y[t]$, and $z[t]$) and the velocity $\mathbf{v}(t)$ (or its x, y, and z components) for all times t for every particle involved in the motion. To find that solution, we must solve or integrate the relevant equations of motion, which are second-order differential equations, and build in the initial conditions by appropriately setting the constants of integration (see section 13.1.1 for a simple example). As explained in chapter 13, the conservation laws for momentum, angular momentum, and energy help us to immediately do some of the required integrations. That is why we can readily deal with the one- and two-body problems.

For the three-body problem, there are three coupled equations of motion to be solved to give positions and velocities for all three particles. However, there are no extra conservation laws and expressions that we can use to carry out the required integrations. In fact, in 1887 mathematician and astronomer Ernst Heinrich Bruns proved that there are no further algebraic integrals or conservation laws to help us. Of course, there is a vast amount mathematically known about the three-body problem, and a solution in terms of an infinite series was given by Sundman in 1912. That solution is not of great practical value. Today, high-speed computers can be used to track three-body motions (see this chapter's "Further Reading" section).

16.8. A REMINDER FOR THE DOUBTERS

No doubt Newton would have felt pleased with what he achieved in *Section XI* of the *Principia*, but he would also be well aware of the fact that it involves forces acting over distances and without any specific mechanisms given for how those forces actually physically act. This has been a theme throughout the *Principia* (and, in fact, it continues right to the very last page). Newton obviously felt in was necessary yet again to set out his working approach and to answer possible criticisms. This he does in a final *Scholium*.

After a little lead into the next section, Newton states:

I here use the word attraction in general for any endeavor of what kind soever, made by bodies to approach each other, whether that endeavor arise from the action of the bodies themselves, as tending mutually to or agitating each other by spirits emitted; or whether it arises from the action of the aether or of the air, or of any medium whatsoever, whether corporeal or incorporeal, any how impelling bodies placed therein towards each other. In the same general sense I use the word impulse, not defining in this treatise the species or physical qualities of forces, but investigating the quantities and mathematical proportions of them.

I am not giving the origin of forces or the mechanisms behind them, Newton is saying, but rather a way of tackling the mathematical problems and then their applications. He very clearly sets out his plan:

In mathematics we are to investigate the quantities of forces with their proportions consequent upon any conditions supposed; then, when we enter upon physics, we compare those proportions with the phenomena of Nature, that we may know what conditions of those forces answer to the several kinds of attractive bodies. And this preparation being made, we argue more safely concerning the physical species, causes, and proportions of forces.

Newton is insisting the following: first I do the mathematics, then the physics.

16.9. CONCLUSION

It would be hard to overestimate the significance of *Section XI* of the *Principia*. An enormous part of the formalism of classical mechanics has now been drawn together to show how to solve the most important problems for the motion and interaction of bodies. The results are even more astounding when we remember back to the work of Galileo and even Huygens. Newton now has a general formalism for classical mechanics and a mathematical framework for using it to solve the great problems of physics and astronomy. Not only did Newton do all that, but he also set the traditions and the path for future researchers, right through to the present time (see chapter 28).

16.10. FURTHER READING

As always, the books listed in section 6.9.1 should be consulted. The book by Chandrasekhar carefully explains what Newton is doing and how he works in his own mathematical way. All the books on classical mechanics are good on this topic; I like the treatment of Barger and Olsson as well as that of Thornton and Marion. For readable accounts of the history and later modern developments, see:

Diacu, F., and P. Holmes. *Celestial Encounters*. Princeton, NJ: Princeton University Press, 1996.

Grossman, N. *The Sheer Joy of Celestial Mechanics*. Boston: Birkhäuser, 1996.

Linton, C. M. *From Eudoxus to Einstein: A History of Mathematical Astronomy*. Cambridge: Cambridge University Press, 2004.

Peterson, I. *Newton's Clock*. New York: W. H. Freeman, 1993.

A classic reference for the three-body problem is the book by Whittaker. Also see Goldstein and the books just cited. The literature is vast, but here are a few more suggestions if you wish to go further:

Baittin, R. H. *An Introduction to the Mathematics and Methods of Astrodynamics*. New York: American Institute of Aeronautics and Astronautics, 1987. [see chs. 2 and 8]

Barrow-Green, June. *Poincaré and the Three Body Problem*. Providence, RI: American Mathematical Society, 1997. [a good introduction to the problem and the groundbreaking work by Poincaré and others]

Diacu, F. "The Solution of the n-body Problem." *The Mathematical Intelligencer* 18 (1996): 66–70.

Saari, D. G. "Orbits of All Sorts." *Nature* 395 (1998): 19–20.

Styer, D. F. "Simple Derivation of Lagrange's Three-Body Equilibrium." *American Journal of Physics* 58 (1990): 917–19.

Valtonen, M., and S. Mikkola. "The Few-Body Problem in Astrophysics." *Annual Reviews in Astronomy and Astrophysics* 29 (1991): 9–29.

Whittaker, E. T. *A Treatise on the Analytical Dynamics of Particles and Rigid Bodies*. 4th ed. New York: Dover, 1944. First published 1904 by Cambridge University Press.

The literature on exoplanets is rapidly expanding as new discoveries flood in. Here are some references for those wishing to explore further:

Boss, A. P. "Extrasolar Planets." *Physics Today* (September 1996): 32–38.

Lunine, J. I., B. Macintosh, and S. Peale. "The Detection and Characterization of Exoplanets." *Physics Today* (May 2009): 46–51.

Jones, H. R. A. "Exoplanets and Their Properties." In *Astrophysics Update*. Edited by John W. Mason. Berlin: Springer, 2004.

Mayor, M., and P.-Y. Frei. *New Worlds in the Cosmos: The Discovery of Exoplanets*. Cambridge: Cambridge University Press, 2003. [the history and discovery story told by one of the discoverers of the first-known exoplanet]

Seager, S., ed. *Exoplanets*. Tucson: University of Arizona Press, 2010. [everything you want to know about the subject!]

Southworth, J. "Research News and Views: A New Class of Planet." *Nature* 481 (2012): 448.

Material on binary stars and other star systems can be found in most astronomy textbooks, for example:

Freedman, R. A., and W. J. Kaufmann III. *Universe*. 7th ed. New York: W. H. Freeman, 2005.

Karttunen, H., et al. *Fundamental Astronomy*. Berlin: Springer-Verlag, 1987.

Motz, L., and A. Duveen. *Essentials of Astronomy*. London: Blackie and Sons, 1977.

BIG BODIES AND SUPERB THEOREMS

We now come to *Sections XII* and *XIII* in the *Principia*, in which Newton adds the final steps to the formalism he will use when discussing gravity and the solar system. This is how Newton motivates and introduces these sections in the *Scholium* at the end of the previous *Section*:

> *It is reasonable to suppose that forces which are directed to bodies should depend upon the nature and quantity of those bodies . . . we are to compute the attractions of the bodies by assigning to each of their particles its proper force, and then collecting the sum of them all. . . . Let us see, then, with what forces spherical bodies consisting of particles endued with attractive powers in the manner above spoken of must act mutually upon one another.*

Newton is recognizing that, so far, he has been dealing with what today we call a point particle—a body so small that we can ignore its size effects and specify its position using a single set of coordinates. It is now time to go beyond that notion. Interestingly, over two hundred years later, Einstein came to emphasize the importance of this very point:

> One remark, however, appears indispensable. The notion "material point" is fundamental for mechanics. If now we seek the mechanics of a bodily object which itself cannot be treated as a material point—and strictly speaking every object "perceptible to our senses" is of this category—then the question arises: How shall we imagine the object to be built up out of material points, and what forces must we assume as acting between them? The formulation of this question is indispensable, if mechanics is to pretend to describe the objects completely.[1]

So Newton is about to take one of the great steps in mechanics, and it will require all his ingenuity and mathematical skills to tackle it. We know that the question of how to deal with large bodies worried Newton, and it probably held up the writing of parts of the *Principia* (although the delay was nothing like the twenty-five years sometimes claimed). In a letter written to Edmond Halley in 1686 (mostly concerning his arguments with Hooke), Newton wrote:

> I never extended the duplicate proportion lower than superficies of the earth & before a certain demonstration I found the last year have suspected it did not reach accurately enough down so low. . . . There is so strong an objection against the accurateness of this proportion, yet without my Demonstrations . . . it cannot be believed by a judicious Philosopher to be any where accurate.[2]

Newton returns to these difficulties in *Book III*. Incidentally, the day on which Newton wrote that letter, June 20, was obviously not a good day for him because he fired off a second letter to Halley again raging about Hooke's claims. Hooke purported to have first put forward various ideas, but, as we know, he lacked the mathematical skills to develop them as Newton did. Thus we find in the second letter Newton's famous broadside:

> For 'tis plain, by his words he knew not how to go about it. Now is not this very fine? Mathematicians, that find out, settle, and do all the business, must content themselves with being nothing but dry calculators and drudges; and another, that does nothing but pretend and grasp at all things, must carry away all the invention, as well as those that were to follow him, as of those that went before.[3]

Certainly none of Newton's peers were capable of producing the outstanding results discussed in this chapter.

17.1. PROBLEMS, PLANS, AND METHODS

Newton has said that we must sum up over all the small particles forming a larger body, and both he and we do that by using a continuous distribution of matter and integrating. However, as usual, for some cases he uses a variety of geometrical and limiting techniques. I will reflect on the mathematics in these sections in this chapter's discussion section (17.7). We should also note that in summing over all the particles, Newton is using a fundamental property of forces: if two or more forces are acting at the same time, they may be simply added together. This is the linearity or superposition property, which Newton recognized, as I discussed in section 6.2.

Newton has often been criticized for poor organization and lapses in logic, but in this case, he does follow a reasonably obvious path. He begins with a particle attracted by a spherical shell, moves on to the solid-sphere case, and concludes with interacting spheres, including the nonuniform density case. Nonspherical bodies are treated in *Section XIII*.

As far as forces are concerned, Newton concentrates on the inverse-square-law force, as that is the vital case needed when presenting the theory of gravity. He does, as usual, give some neat results for the linear force and gives a truly remarkable discussion of the general-force case, with power-law forces given as specific examples.

17.2. FORCE DUE TO A SPHERICAL SHELL

Newton begins by finding the effect of a spherical shell on a particle, or *corpuscle*, to use his word. You might be surprised to learn that his first case has the corpuscle inside the shell:

PROPOSITION LXX. THEOREM XXX.

If to every point of a spherical surface there tend equal centripetal forces decreasing as the square of the distance from those points, I say, that a corpuscle placed within that surface will not be attracted by those forces any way.

This is a case where no integration is required. Newton considers all the paired solid angles or cones, as shown in figure 17.1. Now if the area on the shell is distance d from the corpuscle P, its area on the shell is proportional to d^2. However, the force exerted on P by that little piece of the shell will be proportional to $1/d^2$. The dependence on d cancels out, and then you can see that the two areas at KL and IH, being directly opposite, will exert equal but opposite forces on P, that is, no net force at all. (Newton fills in the geometrical details.) The whole shell can be split up in that way and, thus, we get the (perhaps strange at first sight) result that no force is exerted at any point inside the shell, not just at the center, where symmetry gives the obvious result.

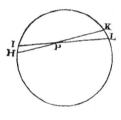

Figure 17.1. Newton's diagram for *Proposition LXX*. (From *The* Principia, trans. Andrew Motte [Amherst, NY: Prometheus Books, 1995].)

Now moving outside:

PROPOSITION LXXI. THEOREM XXXI.

The same things supposed as above, I say, that a corpuscle placed without the spherical surface is attracted towards the centre of the sphere with a force inversely proportional to the square of its distance from the centre.

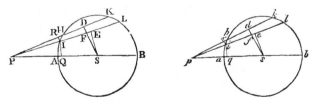

Figure 17.2. Newton's diagram for *Proposition LXXI*. (From *The* Principia, trans. Andrew Motte [Amherst, NY: Prometheus Books, 1995].)

The diagram shows us that Newton is using one of his ingenious comparison techniques for a "geometrical proof which must have left its readers in helpless wonder," as the eminent mathematician J. E. Littlewood put it in 1948.[4] You can see the calculus proof and Newton's working compared in Littlewood's paper "Newton and the Attraction of a Sphere" (see the "Further Reading" section at the end of this chapter).

While these results are theoretically interesting, their main value is their use when discussing the more practically important configurations.

17.3. FORCE DUE TO A SPHERE

With the results for a spherical shell in hand, it is comparatively easy to move on to solid spheres. The modern reader immediately thinks in calculus terms ("integrating over the shells"), and Newton gives us a similar little calculus hint in a *Scholium*:

> *By the surfaces of which I here imagine the solids composed, I do not mean surfaces purely mathematical, but orbs so extremely thin, that their thickness is as nothing; that is, the evanescent orbs of which the sphere will at last consist, when the number of the orbs is increased, and their thickness diminished without end.*

This takes us to:

PROPOSITION LXXIII. THEOREM XXXIII.

If to the several points of a given sphere there tend equal centripetal forces decreasing as the square of the distance from the points, I say, that a corpuscle placed within the sphere is attracted by a force proportional to its distance from the centre.

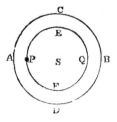

Figure 17.3. Newton's diagram for *Proposition LXXIII*. (From *The* Principia, trans. Andrew Motte [Amherst, NY: Prometheus Books, 1995].)

If you have not previously seen this result, you may find it very strange, but first remember as Newton explains:

> It is plain (by Prop. LXX) that the concentric spherical surfaces of which the difference AEBF of the spheres is composed [see figure 17.3], have no effect at all upon the body P.

So as we move inside the sphere, only the inner part EQF of radius r, say, has any effect, and that reduces as r decreases, as it depends on the volume $(4/3)\pi r^3$ of the material acting on P; this balances out the rise in strength due to the inverse-square law at small distances, giving a final force acting on P, which is directly proportional to r. Thus, inside the sphere, it is a linear force directed toward the center, which results when the summations or integrations are completed. Now we can understand why Newton chose that seemingly overcomplicated cycloidal pendulum inside a sphere (see section 15.3).

Newton covers the case of a particle outside a solid sphere in the following:

PROPOSITION LXXIV. THEOREM XXXIV.

The same things supposed, I say, that a corpuscle situate without the sphere is attracted with a force inversely proportional to the square of its distance from the centre.

The proof, of course, follows directly from *Proposition LXXII*, and today we naturally begin as Newton does by saying "*For suppose the sphere to be divided into innumerable concentric spherical surfaces . . .*"

17.3.1. Time for a Pause

It is easy to be blasé about the results given above, but they are monumentally important. Newton has shown how to go from small bodies, the idealized "point particles," to large bodies built up from many particles, as is the Earth, the planets, and the Sun. The theory can now be applied not just to those small experimental systems, like bodies moving down planes, projectiles, pendulums, and colliding bodies (as in Newton's double pendulums), but also to bodies on the astronomical scale. Future workers, as we will see in chapter 28, developed Newton's theory so that it could be applied to large everyday bodies and their rotations, as for spinning tops.

However, it is the amazing simplicity of the results that impresses. For that most important case, the inverse-square-law force, the balances and summations tell us that the effect is a force acting as if from the center of the sphere; it follows an inverse-square law if we are outside the sphere, and a linear law if we are inside the sphere. Beautiful!

For some situations, we know that Newton's mechanics must be replaced by Einstein's relativistic mechanics. However, there is still a theorem relating influences of matter inside and outside spherically symmetric regions. It is called Birkhoff's theorem (see the "Further Reading" section at the end of this chapter).

We should also be careful to distinguish between what Newton actually says and what he does not say. Note that in *Proposition LXXIV* he talks about "*a force inversely proportional to the square of its distance from the centre.*" As usual, Newton deals with proportions, and nowhere does he give the magnitude or strength of the total force. He does use all sorts of phrases, like "*the particles are as the spheres, that is, as the cubes of the diameters,*" but the final step is never made. Thus it may be a little unfair (and certainly Robert Weinstock, the old controversy stirrer, thought so—see this chapter's "Further Reading" section) when we read statements like physicists Vernon Barger and Martin Olsson's:

> Newton's theorem follows: the gravitational force of any spherically symmetric distribution of matter at a distance R from the center is the same as if all the mass within the sphere of radius R were concentrated at the center.[5]

Newton did not explicitly make that final step, which sums to get the total magnitude to go along with the spatial behavior of the force. In some ways, it seems so obvious that it is hard to think that he was not aware of the comprehensive total result.

17.4. TWO OR MORE SPHERES

Newton can now move to the practical case of two spheres, and, at the same time, he allows for radial density variations (as occur in the Earth and other planets):

> *PROPOSITION LXXVI. THEOREM XXXVI.*
>
> *If spheres be however dissimilar (as to density of matter and attractive force) in the same ratio onward from the centre to the circumference; but everywhere similar, at every given distance from the centre, on all sides and round about; and the attractive force of every point decreases as the square of the distance of the body attracted: I say, that the whole force with which one of these spheres attracts the other will be inversely proportional to the square of the distance of the centres.*

The variations in density must depend on only the distance from the sphere center, and it must be "*the same on all sides and round about,*" as Newton somewhat strangely puts it (see figure 17.4).

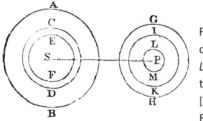

Figure 17.4. Newton's diagram for *Proposition LXXVI*. (From *The* Principia, trans. Andrew Motte [Amherst, NY: Prometheus Books, 1995].)

Newton has now produced a result of great generality. Along with the propositions in this section, there are numerous *Corollaries* about the forces in certain circumstances and other deductions that may be made. For example, for this *Proposition*:

> COR. 1. *Hence if many spheres of this kind, similar in all respects, attract each other mutually, the accelerative attractions of each to each, at any equal distances of the centres, will be as the attracting spheres.*

. . .

> COR. 8. *All those truths above demonstrated, relating to the motion of bodies about the foci of conic sections will take place when an attracting sphere, of any form and condition like that above described is placed in the focus.*

As Newton tells us here and elsewhere, all the results about bodies moving under inverse-square-law forces can still be used when the small bodies are replaced by finite spheres.

17.5. OTHER FORCES

Of course, the inverse-square-law force is the most important (and we know that Newton will introduce it into his theory of gravity in *Book III*), but by now we know that Newton is always thorough and keen to add as much generality as possible into his results. In the next two *Propositions*, *LXXVII* and *LXXVIII*, he extends the result in the previous *Proposition* to the linear-law force (force directly proportional to the distance). Basically he shows that if the spheres are made up of particles attracting a corpuscle with a force directly proportional to the distance between them, then whole spheres attract each other with a force depending directly on the distance between their centers.

In a *Scholium*, Newton gives us his plan:

> *It would be tedious to run over the other cases, whose conclusions are less elegant and important, so particularly as I have done these* [previous cases]. *I choose rather to comprehend and determine them all by one general method as follows.*

So the reader is led to the truly remarkable eightieth *Proposition*. I will quote it in full and give you the diagram, so you can see how carefully Newton sets up the problem and how he presents his solution. (And

I expect many of you at first will be as bewildered as were Newton's contemporaries!)

PROPOSITION LXXX. THEOREM XL.

If to the several equal parts of a sphere ABE described about the centre S there tend equal centripetal forces; and from the several points D in the axis of the sphere AB in which a corpuscle, as P, is placed, there be erected the perpendiculars DE meeting the sphere in E, and if in those perpendiculars the lengths DN be taken as the quantity DE² × PS/PE, and as the force which a particle of the sphere situate in the axis exerts at the distance PE upon the corpuscle P conjunctly: I say, that the whole force with which the corpuscle P is attracted towards the sphere is as the area ANB, comprehended under the axis of the sphere AB, and the curved line ANB, the locus of the point N.

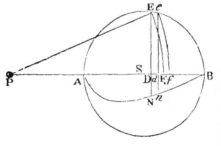

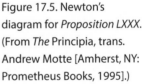

Figure 17.5. Newton's diagram for *Proposition LXXX*. (From *The* Principia, trans. Andrew Motte [Amherst, NY: Prometheus Books, 1995].)

Newton is telling us how to construct the area bounded by the line *AB* and the curve *ANB* in terms of the geometric quantities *DE*, *PS*, and *PE*, together with "*the force which a particle of the sphere situated in the axis exerts at the distance* PE *upon the corpuscle* P." Find that area, and you have the total force exerted by the sphere on the corpuscle at *P*. Clearly Newton is telling us to find the value of a certain integral, and that will give us the total force. I will return to the mathematical contortions in section 17.7. If you wish to follow Newton's mathematical manipulations in detail, I suggest Chandrasekhar as your guide (see his work listed in the "Further Reading" section at the end of this chapter).

Let the corpuscle in figure 17.5 be a distance *R* from the sphere center, so $PS = R$, and let the sphere have radius a. Then in modern terms, if we say that the attractive force exerted on a corpuscle at distance y from particles in the sphere is $f(y)$ per unit volume of material forming the sphere, Newton's prescription agrees with this formula for the total force $F(R)$ exerted by the sphere on the corpuscle:

$$F(R) = \pi \int_{R-a}^{R+a} f(y) \left[y^2 - \frac{(y^2 - a^2 + R^2)^2}{4R^2} \right] dy. \qquad (17.1)$$

To discuss Newton's power-law examples, I will write

if $f(y) = \dfrac{\kappa}{y^n}$, then the sphere force is $F(R) = \left(\dfrac{4}{3}\pi a^3 \right) \dfrac{\kappa}{R^n} C$. (17.2)

So the total force exerted by the sphere is written in terms of the original force multiplied by the volume of the sphere and a factor C, which we might expect to depend on R. For the inverse-square-law force, $n = 2$, Newton discovered the amazing result that C is a constant, $C = 1$. As his examples of using *Proposition LXXX*, Newton took inverse-power-law forces with $n = 1, 3,$ and 4. Here are his results, with C expressed for neatness in terms of the corpuscle distance measured in terms of the sphere radius, so $C = C(\rho)$ where $\rho = R/a$:

$n = 1$ $\qquad C = (3/8)(\rho^2 + 1) - \dfrac{3(\rho^2 - 1)^2}{16\rho} \ln\left(\dfrac{\rho+1}{\rho-1} \right),$

$n = 2$ $\qquad C = 1,$

$n = 3$ $\qquad C = (3/8)\rho(\rho^2 + 1)\ln\left(\dfrac{\rho+1}{\rho-1} \right) - (3/4)\rho^2,$

$n = 4$ $\qquad C = \rho^2/(\rho^2 - 1).$

Of course, if you look in the *Principia*, you will not find those formulas but rather a description of them in geometric terms. For example, instead of the logarithm function, you will find the term *hyperbolic area*. The formulas do emphasize just how special the result is for the inverse-square-law force, $n = 2$; for other force laws, the total force given by integrating over the sphere becomes a complicated function of R.

17.5.1. A Mathematical Gem for inside the Sphere

Newton writes:

> By the same method one may determine the attraction of a corpuscle situated
> within the sphere, but more expeditiously by the following Theorem.

He then proceeds to show (in *Proposition LXXXII. Theorem XLI*) that, if
the corpuscle outside the sphere is at a distance R_1 from the sphere center, and
the one inside is a distance R_2 from the center, and $R_1 R_2 = a^2$ where a is the
sphere radius, then for forces proportional to the mth power of the distance,

$$\frac{F_1}{F_2} = \frac{\sqrt{R_1^{m+1}}}{\sqrt{R_2^{m+1}}}.$$

For the inverse-square-law force, $m = -2$, and the formula can be used
to check the result given in section 17.3.

This amazing result is related to the method of images, used later in
electrostatics problems, as shown by Chandrasekhar (see this chapter's
"Further Reading" section).

17.6. NONSPHERICAL BODIES

Newton deals with segments of spheres and then concludes: "*The attractions
of spherical bodies being now explained, it comes next in order to treat of the laws
of attraction in other bodies consisting in like manner of attractive particles.*"
Thus, we come to *Section XIII: "Of the Attractive Forces of Bodies That Are
Not of a Spherical Figure."*

Newton gives some general results, and neat results for the linear-
force case—basically, the composite bodies in question attract with a total
linear force directed toward the center of gravity or mass (*Propositions
LXXXVIII and LXXXIX*).

He then turns to the case of a thin disk composed of particles attracting
a corpuscle outside the disk. I have already discussed this problem in
section 7.4, where I used it to illustrate Newton's way of working. Armed
with this result, he can now deal with bodies that may be broken up into a

stack of such disks, and an integration process can be used as in the earlier work. So we find:

PROPOSITION XCI. PROBLEM XLV.

To find the attraction of a corpuscle situated in the axis of a round solid, to whose several points there tend equal centripetal forces decreasing in any ratio of the distances whatsoever.

Figure 17.6 shows two of Newton's diagrams, and the spheroid case is obviously of importance when we get to considerations of the shape of the earth in *Book III*.

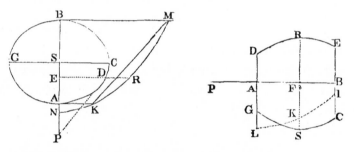

Figure 17.6. Newton's diagrams for *Proposition XCI*. (From *The* Principia, trans. Andrew Motte [Amherst, NY: Prometheus Books, 1995].)

Finally, Newton considers some limiting cases that take us to planar geometries. He generalizes the uniform, constant force that Galileo effectively assumes and shows how Galileo's parabola will be replaced with other curves. (He can do that because he generalized Galileo's falling-body result to one-dimensional motion under any force in *Section VII*.)

17.7. DISCUSSION

Sections XII and *XIII* of the *Principia* are, without doubt, a mathematical tour de force. Newton demonstrates his skill in formulating and solving mathematical problems and in producing results of great physical significance. It would appear that he was (justifiably) proud of this work. In the *Scholium* where he summarizes the results, he writes, "*which is very remarkable*"—this is the sort of statement found rarely in the book. There is also the evidence of his letter to Halley, quoted earlier. We can also turn to his portrait painted when he was an old man of eighty-three (see figure 1.3). It appears that he is holding the *Principia* open at pages dealing with *Proposition LXXXI*, where he discusses the evaluation of the area or integral needed for the general-force case. Surely that must have been the deliberate choice of a man reflecting on his greatest achievements. We would all agree; these are truly superb results.

The adjective *superb* to describe Newton's work comes from the mathematician J. W. L. Glaisher (1848–1928). Glaisher gave an address to mark the bicentenary of the publication of the *Principia*, and it is still worth quoting as a summary of this work and its relevance:

> No sooner had Newton proved this superb theorem [they are all superb, but perhaps he is picking out *Theorem XXXI*]—and we know from his own words that he had no expectation of so beautiful a result till it emerged from his mathematical investigation—than all the mechanism of the universe at once lay spread before him. When he discovered the theorems that form the first three sections of Book I, when he gave them in his lectures of 1684, he was unaware that the sun and earth exerted their attractions as if they were but points. How different must these propositions have seemed to Newton's eyes when he realised that these results, which he had believed to be only approximately true when applied to the solar system, were really exact![6]

17.7.1. Mathematical Considerations

Newton is at his most penetrating and mathematically creative best in these sections. Few, if any, of his contemporaries would have been able to cope with that level of mathematics, and it remains challenging to this very day. This area of mechanics is now tackled using potentials rather than forces (see chapter 28), and potential theory is a major subject in applied mathematics.

We also have the old problem of geometric working and presentation.

There is doubt about Newton's claims for using analytical methods, as I have discussed earlier in sections 7.5 and 13.6. However, in this case, it is beyond belief that the results discussed above could have been generated without using certain calculus techniques. I refer you to the recent book by Niccolò Guicciardini (see the "Further Reading" section at the end of this chapter) for extensive analysis of Newton's mathematical approaches. In terms of the material covered in this chapter, it is worth quoting Guicciardini's authoritative summary:

> I wrote at length about these two corollaries to present some examples of how quadratures occurring in the Principia were discussed by Newton and his acolytes. These are not a reconstruction of what Newton might have been able to do but exactly how he analytically resolved certain problems in the Principia. There are, indeed, a number of problems in the Principia that are resolved "granting the quadrature of curvilinear figures". Newton tackled these problems by squaring the curves but gave no details in the printed text about how such quadratures could be performed. Newton's contemporaries were aware of the fact that Newton was hiding the analysis of these problems. As Fontenelle stated
>
>> Furthermore, it is a justice due to the learned M. Newton, and that M. Leibnitz himself accorded to him: That he has also found out something similar to the differential calculus, as it appears in his excellent book entitled [sic] *Philosophiae Naturalis Principia Mathematica*, published in 1687, which is almost entirely about calculus.
>
> In some cases, as with Corollary 3, Proposition 41, Book 1, and Corollary 2, Proposition 91, Book 1, it is possible to recover Newton's analysis on its own terms.

The analysis that Newton was "*hiding*" in *Sections XII* and *XIII* is brilliant, and whenever I look at the results, I am always amazed at what he achieved, especially after slogging through some of the integrations to check for myself!

17.8. FURTHER READING

Almost all the classical-mechanics books listed in section 6.9.1 introduce the material in this chapter and the basics of modern potential theory. A detailed guide to Newton's methods and results is given by:

Chandrasekhar, S. *Newton's* Principia *for the Common Reader*. Oxford: Clarendon Press, 1995.

For expert analysis of Newton's mathematical workings, see:

Guicciardini, N. *Isaac Newton: On Mathematical Certainty and Method*. Cambridge, MA: MIT Press, 2011.

For the general-relativity equivalent result, Birkhoff's theorem, see the following:

Weinberg, Steven. *Gravitation and Cosmology*. New York: John Wiley, 1972.

———. *The First Three Minutes*. Glasgow, UK: Fontana/Collins, 1981. [see pp. 44–46]

Other references:

Littlewood, J. E. "Newton and the Attraction of a Sphere." *Mathematical Gazette* 32 (1948). Reprinted in *Littlewood's Miscellany*. Edited by B. Bollobas. Cambridge: Cambridge University Press, 1982, pp. 179–81.

Schmid, C. "Newton's Superb Theorem: An Elementary Proof." *American Journal of Physics* 79 (2011): 536–39.

Stein, S. K. "Inverse Problems for Central Forces." *Mathematics Magazine* 69 (1996): 83–93.

Weinstock, R. "Newton's *Principia* and the External Gravitational Field of a Spherically Symmetric Mass Distribution." *American Journal of Physics* 52 (1984): 883–90.

HERE ENDETH *BOOK I*

We are now at the end of *Book I*, and it is time to summarize and evaluate. However, before I do that, there is one final piece of work to comment on.

18.1. MECHANICS FOR OPTICS

The final section in *Book I* is something of an oddity as its title reveals: *SECTION XIV*: *"The Motion of Very Small Bodies When Agitated by Centripetal Forces Tending to the Several Parts of Any Very Great Body."*

Newton studies the motion of these small bodies in planar regions with specified forces. The selection of *Principia* diagrams shown in figure 18.1 quickly tells what sort of problems are involved as the paths bend and change as they move through various regions.

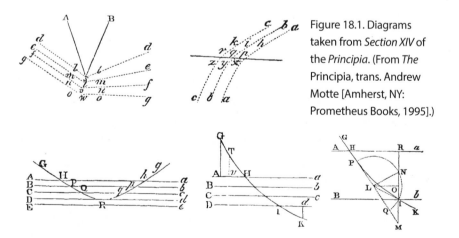

Figure 18.1. Diagrams taken from *Section XIV* of the *Principia*. (From *The Principia*, trans. Andrew Motte [Amherst, NY: Prometheus Books, 1995].)

Newton explains why he has carried out this investigation:

These attractions bear a great resemblance to the reflections and refractions of light made in a given ratio of secants, as was discovered by Snell; and consequently in a given ratio of the sines, as was exhibited by Descartes. For it is now certain from the phenomena of Jupiter's satellites, confirmed by the observations of different astronomers, that light is propagated in succession, and requires about seven or eight minutes to travel from the sun to the earth.

Therefore because of the analogy there is between the propagation of the rays of light and the motion of bodies, I thought it not amiss to add the following propositions for optical use; not at all considering the nature of the rays of light, or inquiring whether they are bodies or not; but only determining the trajectories of bodies which are extremely like the trajectories of rays.

Newton favored a description of light as a stream of corpuscles, and in this section he shows how a mechanical picture can be created that reveals light paths as seen in reflection, refraction, total internal reflection, diffraction, and the bending of light at curved surfaces, as used in lens design. He is again being careful to state that he is thinking of an *analogy*, rather than buying into any discussion of the true nature of light.

18.2. NEWTON'S DEVELOPMENT OF CLASSICAL MECHANICS

By the end of *Book I*, Newton has given us a whole scheme for classical mechanics. Figure 18.2 lets us see how it develops and fits together, and it is worthwhile standing back and reviewing some of the major steps.

Newton begins with conceptual matters and the basic quantities that are used in mechanics and that he builds into his three laws of motion. It is a central feature of Newton's mechanics that it is set out in general; no longer are we looking at specific problems, as Galileo and Huygens mostly did, but we have the basics given for dealing with any mechanical system. To emphasize this generality, we find Newton deducing things that will apply in any system: momentum is conserved, and the motion of the center of mass is specified. This move to general laws and the deduction of general properties, like the conservation of momentum, are features that Newton introduced into the theory, and they have provided the model for the scientific method ever since.

Newton then defines the force that dominates mechanics: the centripetal or central force. In modern terms, if the force center is chosen as the origin, then the force on a particle at position **r** is given by $\mathbf{F}(\mathbf{r}) = F(r)\hat{\mathbf{r}}$, so $F(r)$ gives the strength of the force (which will be negative for an attractive force), and the unit vector tells us that the force acts directly along the line from the force center to the particle. This produces the most basic one-body problem: find the motion of a particle moving under the influence of a given centripetal force.

Once again, he discovers generality: in this case, the line from force center to the particle sweeps out equal areas in equal times, or, in modern terms, angular momentum is conserved. That property depends on only the direction of the force and not $F(r)$. The form of the force leads in modern terms to the conservation of energy, which is a general property that Newton does not discover. However, his results about the path independence of speed changes (*Proposition XL*) do take him to general results that today we most simply obtain using the energy-conservation law.

Using the angular-momentum property, Newton now shows how to divide the orbital-motion problem into two parts: first find the orbit shape, then find the position on the given orbit for any specified time. He also shows how to use the initial conditions to fix the orbit details—for example, will the orbit be an ellipse, a parabola, or a hyperbola when an attractive inverse-square law is specified? Newton gives examples for a variety of forces and also discusses some general properties of orbits, such as the degree to which they close back on themselves.

Finally, Newton prepares for applications to real-world situations. He shows that the formalism he has developed can be used for the problem of two interacting bodies; that problem splits into a one-body problem plus a center-of-mass motion problem, both of which are already solved. He also shows that real-world large bodies composed of many particles may be brought into his scheme by summing over the forces. For spherical bodies, he can replace the large body by one force with its center at the large body's center.

The layout in figure 18.2 serves as a guide to the development of mechanics, and it is essentially what is found in mechanics textbooks from the *Principia* right up to those modern texts listed in section 6.9.1.

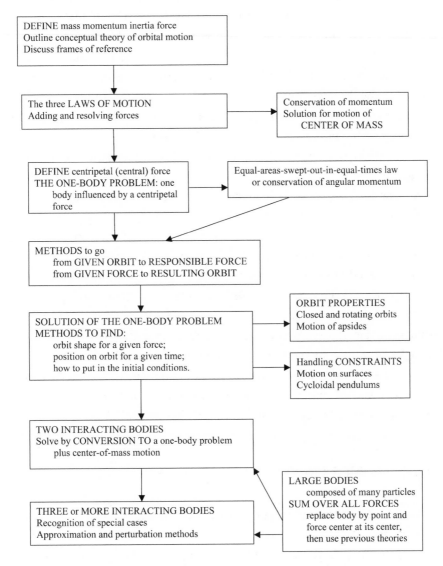

Figure 18.2. Newton's development of the basics of classical mechanics.

18.3. EXAMPLES AND THE INVERSE-SQUARE-LAW FORCE

Newton gives examples to show how his general theory is used, most often with $F(r) = $ constant $\times r^n$ or with the power-law forces. The simplest case has $n = 1$ to give the linear force, which corresponds to the ubiquitous harmonic oscillator. This force is used for localized systems, like pendulums, but is obviously not useful for very large systems, as the force increases without bound as the distances involved increase. It is a force for confinement, and interestingly such forces have recently been used to describe the confinement of quarks within protons and other elementary particles.

The inverse-square-law force with $n = -2$ is physically meaningful for large distances, and, as Newton will show in *Book III*, it is the appropriate force law for gravity. So it is that Newton uses this force law in a great many of his examples, ready for future applications.

Newton discovers many important properties of motion under an attractive inverse-square law, and a summary is given in figure 18.3. I find it quite incredible that he already came to two of the most amazing of those properties. First, of all the force laws that give diminishing force as distance increases, only the inverse-square law leads to bounded orbits that are all closed (a result now known as Bertrand's theorem—see section 14.4.1). Second, a large spherical body formed from particles each exerting an inverse-square-law force on an external body may be replaced by one force acting at its center, and that force too follows an inverse-square law. Even Newton had to comment "*which is very remarkable.*"

18.4. CONCLUSION

An outline of Newton's development of classical mechanics can be gained by looking at the tables and flowcharts in figures 5.5, 6.5, 10.7, 18.2, and 18.3.

I will conclude this chapter by making four points. First, Newton has now given us the basic theory for classical mechanics that has stood the test of time and retains its validity today, needing only extensions or replacements when extremes of scale or time are involved, as when we enter the domains of relativity and quantum theory. It is true that some people detect wooly thinking and difficulties as Newton struggles with the under-lying conceptual matters and some mathematical niceties, but we should remember that he created this formalism; there is nothing approaching it in the writings of Galileo, Descartes, or even the brilliant Huygens. It is also true that much needed to be added to Newton's mechanics to increase the range of physical situations that can be covered (Newton began that in *Book II*, and others followed, as I will outline in chapter 28), but the basics are now in place.

Second, Newton has set out a methodology for us to follow. He has developed a theory showing how to state general laws and deduce general properties of dynamical systems. By introducing a range of forces, like the power-law forces, he has shown us how to mathematically explore that formalism and exhibit its possible outcomes.

Third, Newton has taken us to a new level of thoroughness and sophis-tication. Right from his insistence on separately considering not just a result but also its converse (in *Section II, Propositions I* and *II*), through to his proof that the Kepler equation has no algebraic solution (in *Section VI*), and on to his stunning results for rotating orbits (*Section IX*) and composite bodies (*Sections XII* and *XIII*), we see work of the greatest inventiveness that only a genius like Newton can produce.

Fourth, we must remember that Newton was also inventing a whole new mathematical approach for mechanics. It is true that he couched much of it in geometrical terms that are almost incomprehensible to most of us today and that we need to translate it into a more modern form. Nevertheless, he showed that the subject was amenable to a mathematical treatment, the power of which has evoked a sense of wonder ever since.

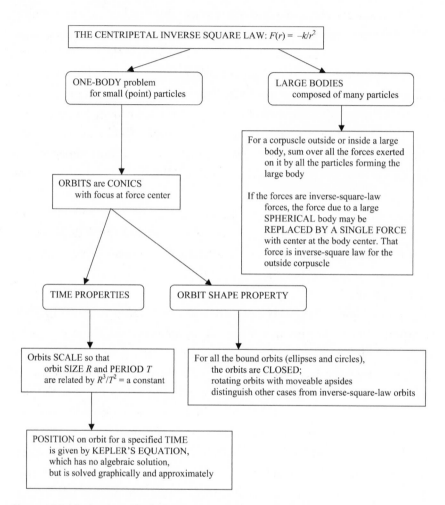

Figure 18.3. Mechanics with the inverse-square-law centripetal force.

PART 5

ABOUT *PRINCIPIA BOOK II*

Book II of the *Principia* is often ignored, and that is something of a tragedy because it shows Newton at his diverse best. Here we have the great mathematical physicist, the experimentalist, and the theoretical physicist displaying his powers of invention and analysis. The material concerns resisted motion, such as air resistance on projectiles, and the specification, measurement, and inquiry into the origins of the forces involved. This leads him to ideas about fluids and the propagation of disturbances through fluids, such as sound waves. Newton has reached the limits of his mathematical and physical background, but he still manages to initiate whole areas of mathematics, physics, and engineering with a remarkable display of clever techniques and analogies.

Reading plans: It would be a shame to miss out on Book II, but if you are anxious to see how Newton develops the theory of gravitation and how he applies it to the solar system, then do move ahead to part 6. As an alternative, just read the introductory remarks and conclusion for each chapter, plus section 19.1. That will give you some idea about Book II and perhaps inspire you to make a mental note to return here later.

STARTING ON *BOOK II*

I n some ways, *Book II* of the *Principia* is an enigma. In *Book I*, Newton has presented the theory of dynamics and is now all ready for *Book III* and claiming the great prize: the theory of the solar system. Very little of *Book II* is even mentioned in *Book III*, so why should we bother with it? The simple answer is that, if you are only interested in astronomical phenomena and large-scale terrestrial properties, then you may well ignore *Book II*. However, you might change your mind after reading this famous evaluation by the highly respected mathematician and historian Clifford Truesdell:

> For ingenuity and insight, and, above all, for sense of problem, Book II is the most brilliant and fertile work ever written in mechanics; it was so received at once by the dozen men who could understand it; and so it remains. For over one hundred years the finest geometers devoted themselves to criticism, development, and correction of the ideas and arguments sketched in it, to solution of the problems it attacked unsuccessfully, to exploration and conquest of the ranges of mechanics it opened.[1]

Here is a short introduction to this "most brilliant and fertile work ever written in mechanics."

19.1. QUESTIONS ABOUT *BOOK II*

19.1.1. What Is *Book II* About?

Book II is concerned with the motion of bodies in cases where the medium in which they are situated retards that motion, as in the well-known example of air resistance. Newton is at his thorough best, dealing with all the classic problems of dynamics: force-free motion and falling bodies; projectiles; pendulums; and central-force motion. He sets out the dynamical theory, reports on an experimental program, and begins a theory of fluids and gases in order to understand their basic properties and how they create resistance forces. There are nine sections; the full contents details are given in figure 19.1.

BOOK II. OF THE MOTION OF BODIES (continued)

SECTION

I. *Of the motion of bodies that are resisted in the ratio of velocity*

II. *Of the motion of bodies that are resisted in the duplicate ratio of their velocities*

III. *Of the motion of bodies that are resisted partly in the ratio of velocities, and*

    partly in the duplicate of the same ratio

IV. *Of the circular motion of bodies in resisting mediums*

V. *Of the density and compression of fluids; and of hydrostatics*

VI. *Of the motion and resistance of funependulous bodies*

VII. *Of the motion of fluids and the resistance made to projected bodies*

VIII. *Of motion propagated through fluids*

IX. *Of the circular motion of fluids*

Figure 19.1. The contents of the *Principia Book II*.

Scattered throughout *Book II* are sections giving ideas and results about such things as the definition of mass (one of the high points in the *Principia*—see section 19.7) , the calculus, new analytical techniques that have become standard in applied mathematics, new physical concepts, and so on in a breathtaking display of inventiveness.

Book II gives us a fine example of the extent and versatility of Newton's genius. We see the mathematical physicist at work as he demonstrates how to use his formalism for dynamics. We see the innovative experimentalist in action as he matches theory and data to check force parameters. We see the theoretical physicist probing the underlying physical world to gain an understanding of the origin of the forces controlling macroscopic phenomena.

19.1.2. Why Did Newton Write *Book II?*

The first answer that comes to mind is Newton's desire to destroy Descartes's rival philosophy of nature. In part III of Descartes's *Principia Philosophiae*, we find:

> *24. That the heavens are fluid.*

Descartes goes on to present his vortex theory of planetary motion (discussed in section 2.6.2 of this book), and in *Book II* Newton aims to discredit that theory. Even if the moving-fluid-and-vortex theory is dismissed, it still leaves the possibility of planets having to move through a fluid, and this is another topic Newton considers (and also effectively deals with in *Book III*).

Second, how to understand the effects of air resistance on falling bodies and projectiles, like cannonballs, was already an old problem going back to Aristotle and questions about vacuums. It was natural for Newton to show that his theory of dynamics could tackle this complex area, perhaps particularly so, since Galileo had written of it that "it is not possible to give any exact description" and suggested the way out was "to discover and demonstrate the theorems in the case of no impediments,"[2] and so look at only those cases where it could be argued that air resistance could be ignored. Descartes in his *Principia Philosophiae* stated that "it is obvious, moreover, that they [projectiles] are gradually slowed down, either by the air itself or by some other fluid bodies through which they are moving,"[3] but he gave no idea about how such effects were to be more exactly described or explained. The effect of air resistance on falling bodies still intrigues us today—who can forget those images of the *Apollo 15* astronauts dropping a hammer and a feather together on the Moon? Robert Boyle carried out similar experiments (observed by Newton) in a vacuum created by his air pump.

Third, it also seems likely that as Newton became involved in this work, he saw it as an opportunity to demonstrate the power of those methodological ideas he put forward in his *Preface* and his hope for the future: "*I wish we could derive the rest of the phenomena of nature by the same kind of reasoning from mechanical principles.*"

19.1.3. What Is Successful about *Book II*?

Newton had the thing Galileo lacked: the concept of force and a law telling how forces change a body's motion. Thus he could theoretically explore the effects of various types of resistance forces. What Newton began can be found in virtually every modern textbook, such as those listed in section 6.9.1. He also devised a series of experiments to link his theory to physical situations (see chapter 20). Finally, he introduced many of the basic ideas of hydrodynamics and showed how wave motion and other phenomena might be explained (see chapter 21). In short, he gave what Truesdell called "the most brilliant and fertile work ever written in mechanics."

19.1.4. Where Does Newton Fail and Why?

Recall what Newton wrote in the *Scholium* at the end of *Book I, Section XI*:

> In mathematics we are to investigate the quantities of forces with their proportions consequent on any conditions supposed; then, when we enter upon

physics, we compare those proportions with the phenomena of Nature, that we may know what conditions of those forces answer to the several kinds of attractive bodies. And this preparation being made, we argue more safely concerning the physical species, causes, and proportions of the forces.

That is Newton's standard approach in the *Principia*, and now he applies it with forces describing "*attractive bodies*" replaced by those describing a medium's resistance to motion.

The problem is that Newton has now moved into a more difficult area of science: the mathematics is more difficult (see section 19.2); the experiments and the fitting of theory to applications is more difficult; and, finally, the search for the underlying "*species, causes, and proportions of the forces*" takes him into a completely new and complex part of physics. Newton has reached the limits of what he can achieve with his level of mathematics and physics. It is, then, remarkable just how far he was able to go with a stunning array of arguments, analogies, and innovations.

19.1.5. Following Newton's Approach

I will cover the three strands of Newton's approach in the rest of this chapter (specifying model forces and developing the theory), in chapter 20 (observing Newton's experimental work), and in chapter 21 (discussing Newton's attempts to link the previous work to properties of fluids and to develop some of the theory of those fluids as it is required elsewhere in physics). As always, I wish to communicate my admiration for the concepts and results that Newton develops in *Book II* rather than struggle through the mathematical details as he provided them. Naturally there must be a little mathematics in these chapters, but readers should understand that it is mostly in the modern form of equations and results that are important and that can be appreciated without becoming mired in the details. (**In summary, read on and do not be put off by the sight of a few equations!**)

19.2. DYNAMICS IN GENERAL

It is probably simplest if I use the modern formulation (see section 6.6) in which Newton's second law tells us that a body of mass m has velocity $\mathbf{v}$ governed by the equation

$$m\frac{d\mathbf{v}}{dt} = \mathbf{F} + \mathbf{F_R}, \tag{19.1}$$

$$\text{where} \quad \mathbf{F_R} = f(v)\hat{\mathbf{v}}. \tag{19.2}$$

Here $\mathbf{F}$ is the force (if any) driving the motion and $\mathbf{F_R}$ is the resistance force, the assumed form of which is given in equation (19.2). The strength of the force depends on the speed v, and the direction is given by the unit vector in the direction of the velocity, meaning that $f(v)$ must be negative to give a force opposing the motion. Newton chooses to use

$$\mathbf{F_R} = -av\hat{\mathbf{v}} - bv^2\hat{\mathbf{v}}. \tag{19.3}$$

Newton's approach requires him to solve for possible motions in terms of the force parameters a and b, then to use observational or experimental data to find their practically relevant values. In chapter 21, we shall see how he tries to understand the force parameters using the theory of fluids.

One general point must be made at the outset. Resistance forces model the interaction of the moving body with the medium in which the motion takes place, and energy is transferred from the moving body to that medium. Therefore, we no longer have the conservation of energy law and, hence, a key tool in solving dynamical problems (see sections 11.3.6 and 13.2) is lost.

19.2.1. The Simplest Case

Newton recognizes (and all textbooks follow suit) that the simplest case is when the resistance force is proportional to the velocity, so $b = 0$ in equation (19.3). For motion in one dimension, we must solve

$$m\frac{dv}{dt} = F - av, \tag{19.4}$$

which is a **linear** in the speed v. In two dimensions, this linear force property translates into the two equations

$$m\frac{dv_x}{dt} = F_x - av_x \quad \text{and}$$

$$m\frac{dv_y}{dt} = F_y - av_y. \tag{19.5}$$

We have a separate equation like equation (19.4) for each component. Since $v_x = dx/dt$, a second integration will then give the position $x(t)$. If F depends on position $F = F(x)$ in equation (19.4), and we will need to go to the differential equation

$$m\frac{d^2x}{dt^2} = F(x) - a\frac{dx}{dt}. \tag{19.6}$$

For the important case of a harmonic oscillator or pendulum, F depends only on x, and so we still have a linear differential equation that is relatively easy to solve.

19.2.2. How the Difficulties Enter

Physically important cases do not have $b = 0$, in which case, if we set $a = 0$, equations (19.4) and (19.5) become

$$m\frac{dv}{dt} = F - bv^2, \tag{19.7}$$

$$m\frac{dv_x}{dt} = F_x - b\sqrt{v_x^2 + v_y^2}\; v_x, \quad \text{and}$$

$$m\frac{dv_y}{dt} = F_y - b\sqrt{v_x^2 + v_y^2}\; v_y. \tag{19.8}$$

Two things have changed. Equation (19.7) is no longer linear in v, so we must use nonlinear differential equations, although the constant F case is still simple. Second, for motion in two (or three) dimensions, we now have coupled differential equations (the equation for v_y involves v_x)—compare this to equations (19.8) and (19.5).

Recall that Galileo pointed out that, for projectiles, we can combine motions in the horizontal and vertical directions after considering them separately. The modern version is given in section 11.2.1. That separability property is lost if we have resistance forces other than those depending linearly on the velocity, and so even projectile motion becomes a challenging problem.

With that background in place, we can move to Newton's results for the first major problems in dynamics with resistance forces.

19.3. FREE MOTION AND FALLING BODIES

These are the simplest cases, with either no force or a constant one driving the motion. The equations of motion can be integrated—for more on this, see the textbooks listed in section 6.9.1 (the one by Fowles is commonly used). I will give a few details for the linear case ($b = 0$) so we can see how Newton presents his results in *Section I: "Of the Motion of Bodies That Are Resisted in the Ratio of the Velocity."*

19.3.1. Linear Resistance Force and Free Motion

For free motion, we set $F = 0$ in equation (19.4). Assuming motion starts at $x = 0$ with initial speed v_0, integration of the equation of motion gives

$$v = v_0 e^{-(a/m)t} \text{ and } mv_0 - mv = ax. \tag{19.9}$$

The first of the equations in (19.9) tells us how the resistance force slows down the body (exponentially, in this case), and the second one shows how the force reduces the momentum as the body travels a distance x, with the force constant a controlling the magnitude of the reduction. This physical emphasis is not so prevalent in modern textbooks, but it was beautifully stated by Newton in:

PROPOSITION I. THEOREM I.

If a body is resisted in the ratio of its velocity, the motion [momentum] *lost by resistance is as the space gone over in its motion.*

In *Proposition II*, Newton gives us the result for the first equation in (19.9), but he gives it in terms of what happens to the speed in succeeding equal time intervals. The result is

$$\frac{v_1}{v_0}, \qquad \frac{v_2}{v_0} = \left(\frac{v_1}{v_0}\right)^2, \qquad \frac{v_3}{v_0} = \left(\frac{v_1}{v_0}\right)^3, \qquad \frac{v_4}{v_0} = \left(\frac{v_1}{v_0}\right)^4, \qquad \cdots$$

We tend to look for a complete formula (as in equation [19.9]) rather than interpretative results like those just given (which are perhaps reminiscent of Galileo's approach). Newton could calculate the complete details, as he explains in a *Corollary* to the second *Proposition*. If you read Newton's result, you will find not exponentials but mention of areas under a hyperbola. The link is in the mathematical relationships as follows:

hyperola $xy = 1$ area $\int (1/x)dx = \ln(x)$ if $w = \ln(x)$, then $x = \exp(w)$.

Newton was a great calculator, and it is amusing to see that, strangely, he once calculated the area under a hyperbola, a logarithm, to fifty-five decimal places (see Fauvel's *Let Newton Be!* referenced in the "Further Reading" section at the end of this chapter).

19.3.2. Linear Resistance Force and Falling Bodies

Assume a vertical x axis and a body released from rest at a height $x = h$. The force is now $F = mg$, and equation (19.4) gives

$$v = -v_T \left\{ 1 - e^{-(a/m)t} \right\}, \quad \text{where} \quad v_T = \frac{mg}{a}. \tag{19.10}$$

The speed v increases until it reaches v_T, the "*greatest speed that the body can acquire by falling*" (as Newton puts it), and known today as the terminal speed. The fall time t and the fall height h are related by

$$h = v_T t - \left(\frac{v_T m}{a} \right) \left\{ 1 - e^{-(a/m)t} \right\}. \tag{19.11}$$

Thus we can calculate all details for a falling body, and so can Newton, although in the *Principia* he has no formulas like ours, but he gives results expressed as in the example in section 19.3.1.

The above results are neat and easy to interpret, so you might wonder why we do not stop here. Newton supplies the answer.

19.3.3. Choice of Force Laws

Newton explains in a *Scholium* why we must go further:

> But, yet, that the resistance of bodies is in the ratio of the velocity, is more a mathematical hypothesis than a physical one. In mediums void of all tenacity, the resistances made to bodies are in the duplicate [squared] ratio of the veloci- ties. For by the action of a swifter body, a greater motion in proportion to a greater velocity is communicated to the same quantity of the medium in a less time; and in an equal time, by reason of greater quantity of the disturbed medium, a motion is communicated in the duplicate ratio greater; and the resistance (by Law II and III) is as the motion communicated. Let us, there- fore, see what motions arise from this law of resistance.

Newton is giving physical reasons for the form of the resistance force. This is a complicated topic, and I return to it in the next chapter.

19.3.4. Further Results

So, following that *Scholium*, Newton repeats the above analysis but now with $a = 0$ in equation (19.3). He presents *Section II*: *"Of the Motion of Bodies That Are Resisted in the Duplicate* [Squared] *Ratio of Their Velocities."*

He then gives the general case so that he will have the two free parameters a and b at his disposal when he turns to fitting experimental data, that is, *Section III*: *"Of the Motion of Bodies That Are Resisted Partly in the Ratio of the Velocities, and Partly in the Duplicate of the Same Ratio."*

The modern details are given by Fowles and in other textbooks cited in section 6.9.1.

19.3.5. Summary

Newton has shown that he can solve for free and falling bodies for a variety of resistance forces. His results are often expressed in forms giving more emphasis to the physical outcomes than a "cold" formula. However, it is clear that he can reduce the results to integrals (as we do), although he talks about areas under curves rather than integrals. He has identified the important concept of terminal speed for a falling body—modern discussions often use raindrops as examples.

The groundwork has now been done: describe the motion for given force parameters a and b. Newton is ready for the reverse process: use observed motions to find the practical values of those force parameters, and that I will cover in chapter 20.

19.4. A STRANGE INTERLUDE

In the middle of *Section II*, we find:

> LEMMA II. *The moment of any genitum is equal to the moments of each of the generating sides drawn into the indices of the powers of those sides, and into their co-efficients continually.*

The language is totally opaque to a modern reader, but in this *Lemma*, Newton presents rules for calculus that he needs for the topics being considered. Today we would expect to see results like

$$\frac{dx^n}{dx} = nx^{n-1} \quad \text{and} \quad \frac{d(x^n x^p)}{dx} = nx^{n-1}x^p + px^n x^{p-1}.$$

They are there in Newton's *Lemma* but expressed rather differently.

A reader might well ask why the *Lemma* comes here and not in *Section I* of *Book I*, where Newton sets out his mathematical methods. Or why even introduce it at all rather than give some relevant reference? Part of the answer comes back to his dispute with Leibnitz (see section 7.5). Here is Newton's *Scholium* that comes after the *Lemma*:

> *In a letter of mine to Mr J. Collins, dated December 10, 1672, having described a method of tangents, which I suspected to be the same with Slusius' method, which at that time was not made public, I subjoined these words: This is one particular, or rather a Corollary, of a general method, which extends itself, without any troublesome calculation, not only to the drawing of tangents to any curve lines, whether geometrical or mechanical, or any how respecting right lines or other curves, but also to the resolving other abstruser kinds of problems about the crookedness, areas, lengths, centres of gravity of curves, &c. . . . . This method I have interwoven with that other of working in equations, by reducing them to infinite series. So far that letter. And these last words relate to a treatise I composed on that subject in the year 1671. The foundation of that general method is contained in the proceeding lemma.*

Quite clearly, Newton is using this opportunity to reinforce his claims as the discoverer of calculus. Recall that we are reading the 1726 third edition of *Principia*. If we go back to the earlier editions, the *Scholium* is quite different and talks about "*in correspondence which I carried on ten years ago with the very able geometer G. W. Leibnitz,*" going on to how they both had the same methods in calculus. The battle has progressed! (See this chapter's "Further Reading" section to follow the whole story.)

19.5. PROJECTILES

This is another topic with a long history and with much debate about the exact nature of a projectile's trajectory (see the "Further Reading" section at the end of this chapter). Galileo finally showed that if air resistance is ignored, then the trajectory is a parabola; the derivation of that result using Newton's laws was given in section 11.2.1. Naturally, Newton reworked the problem including resistance forces.

For the linear case ($b = 0$), the equations to be solved are as in equation (19.5) with $F_x = 0$ and $F_y = -mg$. The equations again separate, and each can be solved to replace the resistance-free solution in equation (11.6) with

$$x = \frac{V\cos(\alpha)}{\gamma}\left\{1 - e^{-\gamma t}\right\}, \qquad \text{where } \gamma = \frac{a}{m},$$

and

$$y = \left\{\frac{V\sin(\alpha)}{\gamma} + \frac{g}{\gamma^2}\right\}\left\{1 - e^{-\gamma t}\right\} - \frac{gt}{\gamma}.$$

$$(19.12)$$

The projectile is launched from the origin at an angle α to the horizontal x axis and with speed V. These results are considerably more complicated than those for the resistance-free case, equation (11.6), and it is immediately clear that experimental work based on projectiles is not a good way to investigate resistance forces. Newton does give a diagram (see figure 19.2) so the reader can see the effect on trajectory shape and projectile range when the launch angle is changed. The non-parabolic nature of the trajectories is clear.

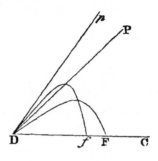

Figure 19.2. Projectile paths for the same launch speed but different launch angles. (From *The* Principia, trans. Andrew Motte [Amherst, NY: Prometheus Books, 1995].)

As an illustration of the complication introduced even in the linear-resistance case, we can no longer find a simple exact formula for the range, and of course Galileo's optimum range condition, $\alpha = 45°$, will now change to something depending on γ. Newton went on to consider the speed-squared case ($a = 0$ in equation [19.3]), and, in *Proposition X*, he also introduced the density as a variable in the resistance-force strength. That proposition has become famous because Newton made a mistake in his original version. This was reported to Newton by Niklaus Bernoulli, whose uncle, Johann Bernoulli, had detected it. Newton scrambled to find the error and managed to make a very-last-minute correction as the *Principia's* second edition was going to press. In typical style, Newton failed to give any credit to the Bernoullis; this is yet another example of Newton's ego at work and his propensity to alienate people by his lack of generosity.

19.6. DECAYING ORBITS

Given Newton's hostility toward Descartes's ideas (see section 19.1.2), it is no surprise to find him laying the groundwork for combating some of them in *Section IV: "Of the Circular Motion of Bodies in Resisting Media."*

This is not an easy problem, and Newton also realizes that for astronomical applications he must allow for the density of the medium decreasing as the distance from the attracting body increases. His major result is given here:

PROPOSITION XV. THEOREM XII.

If the density of a medium in each place thereof be reciprocally as the distance of the places from an immovable centre, and the centripetal force be in the duplicate ratio of the density [giving the usual inverse-square-law force]; I say that a body may revolve in a spiral which cuts all the radii drawn from that centre in a given angle.

Newton is saying that the resistance force will cause the orbit to decay, and in certain cases it may take the form of a particular spiral (see figure 19.3).

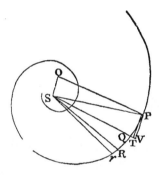

Figure 19.3. Newton's diagram for *Proposition XV*. (From *The* Principia, trans. Andrew Motte [Amherst, NY: Prometheus Books, 1995].)

The orbit *PQR* is such that the radii *SP* (and the like) from the force center *S* all cut the spiral at the same angle.

The proof is involved, and it has been suggested—both in Newton's time and recently—that there are errors to be fixed. Notice that the *Proposition* uses the word *may*; it does not say all orbits are of that particular form. References to discussion papers are given in the "Further Reading" section at the end of this chapter; Chandrasekhar gives a modern version of Newton's proof. In this case, the resistance force is proportional to the density and the speed squared.

However, the important point is that Newton has recognized that resistance forces will cause orbits to change from circles or ellipses to some sort of spiral motion, and that the density of the medium is an important variable. He gives three further *Propositions* exploring various aspects of the problem.

This topic became important again when artificial satellites were launched to orbit the Earth. One approach to the problem is to use perturbation theory, so the orbit is assumed fixed for one cycle and its form is used to calculate how the resistance force changes the orbit parameters over that cycle. That gives a new orbit, and the process repeats. The interesting physics involved is discussed by Blitzer, for example. Over time, the orbit reduces in size and changes shape. The density variation in the Earth's atmosphere is more like an exponential than Newton's form, and suitable parameters are now available (see references to King-Hele). Modern satellites tend to have very high orbits, so the resistance forces have only minor effects.

19.7. THE PENDULUM AND THE DEFINITION OF MASS

In *Proposition XXIV*, Newton returns to a question that he raised in the very first page of the *Principia*: How to define mass? He recognizes that mass ("*the quantity of matter*") is used in two different ways: first, when considering inertia; second, when considering weight or the effect of gravitation on a body. To explore further, he must use the second law of motion, and he chooses to do that now. (This seems like a strange place to insert this work, as it has nothing to do with resistance forces and surely belongs in *Book I*. There will be more on this topic in *Book III*, so Newton is not making life too easy for his readers!)

In modern terms, we consider a simple pendulum consisting of a "light string" of length l supporting a bob of mass m. If oscillations are through small angles θ, Newton's second law gives

$$m\frac{d^2\theta}{dt^2} = -\left(\frac{mg}{l}\right)\theta \quad \text{or, cancelling the } m, \quad \frac{d^2\theta}{dt^2} = -\left(\frac{g}{l}\right)\theta. \quad (19.13)$$

You will find that in virtually every textbook you care to consult. However, it was the genius of Newton that he recognized how the masses in equation (19.13) are not necessarily the same, and their cancellation should not be made without justification. On the left-hand side, we should write m_I to indicate that this mass describes the inertia of the pendulum bob. On the right, we have the force or weight of the bob and we should use the gravitational mass m_G. Thus, for the equation of motion and the resulting period T, we should write

$$m_I \frac{d^2\theta}{dt^2} = -\left(\frac{m_G g}{l}\right)\theta, \quad \text{giving period} \quad T = 2\pi\sqrt{\frac{m_I l}{m_G g}}. \qquad (19.14)$$

If we square the period formula, we get the results for one pendulum and for two of the same length:

$$\frac{m_I}{m_G} = \frac{T^2 l}{4\pi^2 g}, \quad \text{and for comparing pendulums 1 and 2:} \quad \frac{m_{I1}}{m_{I2}} = \frac{m_{G1}}{m_{G2}}\frac{T_1^2}{T_2^2}. (19.15)$$

Newton sets out these results, obtained using an incremental form of the second law, in *Corollaries 1* to *4*, where he uses weights instead of gravitational mass ratios. Here is the stunning summary expressed as:

PROPOSITION XXIV. THEOREM XIX.

The quantities of matter in funependulous bodies [those forming the bob in a pendulum] *whose centres of oscillation are equally distant from the centre of suspension, are in a ratio compounded of the ratio of the weights and the duplicate ratio of the times of oscillation in vacuo.*

(Incidentally, Newton is not assuming a point particle here and is referring to Huygens's work on composite pendulums and their "*centre of oscillation*." Nothing if not thorough!)

Newton sums up the vital importance of these results in:

Cor. 7. And hence appears a method of comparing bodies one among another, as to the quantity of matter in each; and of comparing the weights of the same body in different places, to know the variation of its gravity [g varies at different places on the Earth, as we will see in *Book III*]. *And by experiments made with greatest accuracy* [to be reported when discussing *Proposition VI* in *Book III*] *I have always found the quantity of matter in bodies to be proportional to their weight.*

This must be taken as one of the high points in the *Principia*. As mentioned in chapter 5, this was the profound result that made such an impact on Albert Einstein and was a key to his general theory of relativity.

19.8. DAMPED PENDULUMS

In *Section VI*, *"Of the Motion and Resistance of Funependulous Bodies* [Pendulums]," *Propositions XXV* to *XXXI* deal with pendulums when various resistance forces are present to damp down their oscillations. For the cycloidal pendulum (see chapter 15), Newton shows that the oscillations remain isochronous (independent of the amplitude of the swings) even when a resistance force proportional to the velocity is present. An important motivation for this work is to provide the theoretical basis for analyzing experiments that Newton made to measure the parameters specifying the resistance forces in various mediums (see the next chapter). In particular, *Propositions XXX* and *XXXI* show how the fractional decrease in the amplitude of the swings may be related to the resistance force at the bottom of the swing. These are technical and intricate calculations, but Newton does produce some straightforward relationships that may be applied, as explained in the next chapter.

19.9. CONCLUSION

It is a measure of Newton's vision and thoroughness that the work described in this chapter still forms the basis for an introduction to the effects of resistance forces as presented in modern textbooks. In fact, in some cases, Newton goes much further. He has tackled the four great problems of dynamics (free fall, projectiles, pendulums, and inverse-square-law-force orbits), producing significant results in all cases, often eloquently expressed in physically meaningful ways. Furthermore, he has now established the background for experimental work and investigations of the fundamental origins of resistance forces.

As ever, we have also seen Newton's quirky side, with sections on the methods of calculus and on the fundamental properties of mass slipped in to bewilder the unwary reader!

19.10. FURTHER READING

The reader is referred to the books listed in section 6.9.1. For dynamics, the book by Fowles, as well as the one by Thornton and Marion, is recommended. Sommerfeld nicely treats the cycloidal pendulum. Chandrasekhar explains Newton's working for a few of the most important propositions, and Cohen's *Guide* is a good place to start for historical details.

Fauvel, J., ed. *Let Newton Be!* Oxford: Oxford University Press, 1988. [articles on the *Principia*; Jon Pepper shows examples of Newton's calculations]

Hall, A. R. *Ballistics in the Seventeenth Century.* Cambridge: Cambridge University Press, 1952.

On orbits and spiral motion:

Blitzer, L. "Satellite Orbit Paradox: A General View." *American Journal of Physics* 39 (1971): 882–86.

Chandrasekhar, S. "The Effects of Air-Drag on the Descent of Bodies." *Newton's* Principia *for the Common Reader.* Oxford: Clarendon Press, 1995.

Erlichson, H. "Newton's Solution to the Equiangular Spiral Problem." *Historia Mathematica* 19 (1992): 402–13.

King-Hele, D. *Theory of Satellite Orbits in an Atmosphere*. London: Butterworths, 1964.

King-Hele, D., and D. M. C. Walker. "The Effects of Drag on Satellite Orbits: Advances in 1687 and 1987." *Vistas in Astronomy* 30 (1987): 269–89.

Weinstock, R. "Newton's *Principia* and Inverse Square Orbits in a Resisting Medium: A Spiral of Twisted Logic." *Historia Mathematica* 25 (1998): 281–89.

Wilson, C. "Newton on the Equiangular Spiral: An Addendum to Erlichson's Account." *Historia Mathematica* 21 (1994): 196–203.

NEWTON THE EXPERIMENTALIST

For those people thinking of Newton as a mathematician and the *Principia* as an impenetrable mathematical treatise, it will come as a shock to know that over twenty pages in *Book II* are devoted to reports of experiments. However, we must remember that Newton was in the vanguard of the Scientific Revolution, and, as explained in chapter 2, one of its central tenets was the necessity of finding out about nature by doing experiments (see section 2.5). As a young man, Newton met the pioneering experimentalist Robert Boyle, and during their friendship, Newton was greatly influenced by him (see the "Further Reading" section at the end of this chapter).

It is also obvious that even in his boyhood, Newton took a great delight in practical matters. This delightful excerpt from Stukeley's memoir tells us about

> his strange inventions, uncommon skill & industry in mechanical works. They tell us that instead of playing among the other boys, when from school, he always busied himself at home, in making knickknacks of diverse sorts, & models in wood, of whatever his fancy lead him to. For which purpose he furnished himself with little saws, hatchets, hammers, chisels, & a whole shop of tools which he used with much dexterity.[1]

We must also again refer to his *Preface*, where Newton states his aim: "*from the phenomena of motions to investigate the forces of nature.*" He has done the theory and is ready to do the experiments. We are reminded of Einstein's comment that "it is the theory which decides what we can observe."[2]

Newton's objective is to find parameters for the resistance forces, the values of a and b in equation (19.3). He carried out a large number of experiments using pendulums and observations of falling bodies. The forces involved are now usually called drag, and before going to Newton's experimental details, it is worth looking at modern information on this topic.

20.1. DRAG: DEFINITION AND MEASUREMENTS

The term *drag* is now used for the resistance force D, and it is usual to define it in terms of a nondimensional drag coefficient C_D according to

$$D = C_D \tfrac{1}{2} A \rho v^2. \tag{20.1}$$

A is the relevant area for the body moving through the fluid (πa^2 for a sphere of radius a), ρ is the fluid density, and v is the body's speed through the fluid. The drag coefficient depends on the dimensionless Reynolds number R_e,

$$C_D = C_D\left(R_e\right), \quad \text{where} \quad R_e = \rho v l / \mu. \tag{20.2}$$

Here l is a characteristic length—a sphere's diameter, say—and μ is the fluid viscosity. The Reynolds number is sometimes interpreted as giving a measure of different contributions to the drag expressed as a ratio of inertial to viscous effects. Notice that if C_D is a constant, then we have the resistance force proportional to the speed squared, as Newton suggested.

Graphs of C_D versus R_e are given in many books, such as that by Binder, as well as the resource by Daugherty and Franzini, and in the papers by Achenbach, Frohlich, and George E. Smith (see the "Further Reading" section at the end of this chapter). Figure 20.1 shows the form of the drag versus the Reynolds-number curve for a sphere.

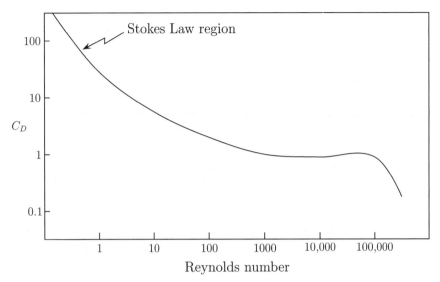

Figure 20.1. A sketch of the drag coefficient versus the Reynolds number for a sphere. Note the logarithmic scales.

It is immediately obvious that drag is a complex phenomenon with an intricate dependence on fluid-flow properties in the wake of the sphere. For a small Reynolds number, the logarithmic curve is linear with a negative slope, and equations (20.1) and (20.2) combine to give the result that drag depends linearly on the speed, which was the first case investigated by Newton in *Section I* of *Book II*. In his discussion *Scholium*, he said this was "*more a mathematical hypothesis that a physical one*," but we see that it is actually appropriate in a limited regime. This linear-drag law is now called Stokes's law.

The C_D-versus-R_e curve soon changes as the Reynolds number increases and, for a region of Reynolds numbers between 1,000 and 100,000, is very roughly flat, indicating a drag depending roughly on speed squared. Beyond that, there is a rapid major decline in the drag coefficient as more complex fluid-motion effects kick in. (See, for example, the discussion in Binder's section 5-29 and the wonderful photographs in Van Dyke's *Album*). The paper by Frohlich and the book by de Mestre give interesting examples taken from sport, where Reynolds numbers around 100,000 are commonly involved. Newton's experiments involved Reynolds numbers all the way up to around 80,000.

It is clear that Newton has moved into a difficult area of science. In fact, this topic has become very significant in engineering, where a vast amount of empirical and experimental work has been done in such fields as aeronautics.

20.2. EXPERIMENTS WITH PENDULUMS

The pendulum is an interesting dynamical system and also one that can be used as an experimental tool. We saw Newton using it to test predictions of his laws (see section 6.5) and noted that he empirically corrected his measurements for the effect of air resistance. Now Newton uses the pendulum to find the resistance-force parameters, as in equation (19.3). He has established the required theory in *Propositions XXX and XXXI*, and now comes the *General Scholium*, which begins:

> From these propositions may be found the resistance of mediums by pendulums operating therein. I found the resistance of the air by the following experiments. I suspended a wooden globe or ball weighing 57.7/22 ounces troy, its diameter 6.7/8 London inches by a fine thread on a firm hook, and the centre of the oscillation of the globe was 10.1/2 feet. I marked on the thread a point 10 feet and 1 inch distant from the centre of suspension; and even with that point I placed a ruler divided into inches, by the help whereof I observed the lengths of the arcs described by the pendulum. (See figure 20.2.)

Notice that Newton is using a very long pendulum so that for small enough swings, it is isochronal, like a cycloidal pendulum, a point he discusses in detail. (See the source by I. Bernard Cohen listed in the "Further Reading" section at the end of this chapter for a detailed discussion of what Newton implied by the word *pendulum*.) As ever with Newton, there is a cleverly conceived underlying concept and some brilliant analysis to make use of it.

20.2.1. The Basic Idea

The *Scholium* continues: "*Then I numbered the oscillations in which the globe would lose 1/8 part of its motion.*" Newton goes on to find the reduction in amplitude per swing and shows how that may be used to find resistance-force parameters.

To explain, consider the simplest case with a resistance force proportional to the velocity, so in equation (19.3), $b = 0$ and a is to be found. In this case, the pendulum equation of motion can be solved exactly (see any modern textbook, such as Fowles's section 3.4) and

$$\theta = \theta_0 \cos(\omega t)e^{-(a/2m)t} \qquad (20.3)$$

The pendulum bob has mass m and has been released from rest with $\theta = \theta_0$ as in figure 20.2. The angular frequency ω is close to the unperturbed value $\sqrt{\frac{g}{l}}$. After one swing or half a period, the pendulum fails to reach amplitude θ_0 by an amount

$$\Delta\theta = \theta_0 \left(1 - e^{-(a/2m)(T/2)}\right) \approx \theta_0 (aT/4m). \qquad (20.4)$$

If we use the approximation 22/7 for π, we can write this in Newton's form as

$$\left(\frac{7}{11}\right)\frac{l\Delta\theta}{l} = \frac{\text{(maximum resitance force)}}{\text{(pendulum weight)}}. \qquad (20.5)$$

This is the result Newton gives in *Proposition* XXX. If we measure the reduction in swing $\Delta\theta$, we can find the force parameter a.

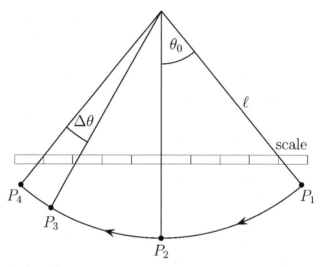

Figure 20.2. Notation for Newton's pendulum experiment (not to scale). The pendulum is released from P_1 and would swing to P_4 if there were no resistance force. When resistance is present, it swings to P_3 falling short by the angular amount $\Delta\theta$.

20.2.2. The Full Method

Newton uses a resistance force with magnitude

$$F_R = av + dv^{3/2} + bv^2, \tag{20.6}$$

where he has introduced (without any justification) an extra term proportional to the speed to the power 3/2, presumably to model the complex drag phenomena that we saw in section 20.1. He thus needs to determine the three force parameters a, d, and b, and it is here that we see the Newton genius in action.

Newton sets

$$\frac{l\Delta\theta}{l} = AV + BV^{3/2} + CV^2, \tag{20.7}$$

where V is the maximum speed for the particular release condition (occurring at or very close to the bottom of the swing), so V is known and depends on the release angle: $V = V(\theta_0)$. Of course V gives the maximum resistance force, and from *Proposition XXX*, Newton has results like equation (20.5), which he uses to produce a total result that we can express as

$$\frac{l\Delta\theta}{l} = \left\{ \left(\frac{11a}{7}\right)V + \left(\frac{10d}{7}\right)V^{3/2} + \left(\frac{4b}{3}\right)V^2 \right\} \Big/ mg . \quad (20.8)$$

Comparing equations (20.7) and (20.8), we see that our force parameters a, d, and b may be obtained from A, B, and C, and the resistance force is given in terms of the weight mg.

In this case, Newton must measure $\Delta\theta$ for three values of θ_0 and thus find three equations like (20.7), which he can solve as simultaneous equations for A, B, and C.

20.2.3. Implementation

Newton displaced the pendulum by 2, 4, 8, 16, 32, and 64 inches and counted the oscillations for the amplitude to have diminished by an eighth in each case. (Incidentally, those larger displacements were surely pushing the simple, linear-pendulum approximation.) This gave him six sets of data (given in the *Principia*), and, hence, more than the requisite three simultaneous equations for finding A, B, and C. He chose to use the second, fourth, and sixth cases, and after doing the mathematics, he concluded that:

> The resistance of the globe will be to its weight as $0.0000583V +$ $0.00007593V^{3/2} + 0.0022169V^2$ to the length of the pendulum between the centre of suspension and the ruler, that is, 121 inches.

Given the nature of the method, it is amusing to see the number of figures Newton uses in his result—surely 0.0022169 should be 0.0022, at best. In fact, choosing other data sets will not give consistent results, and Newton realized that there were difficulties with this method.

The first edition of the *Principia* contained different results; the full story of this work is described by George E. Smith (see the "Further Reading" section at the end of this chapter).

20.2.4. Further Work

Newton did a great deal of research using his pendulum method and, after giving more data, says, "*I leave the calculation to such as are disposed to make it*"! He became aware of the difficulties, for example the resistance from the motion of the suspending thread. Using some choices of data sets will produce negative values for the force parameters. He tried different bobs and tried to conclude that the resistance force was proportional to the square of the diameter of the bob. (He noted a worry that one of the bobs was not exactly spherical.)

In a fascinating series of experiments, Newton measured pendulum oscillations in water and quicksilver (mercury). Comparing oscillations in air and water, he concluded that:

> *the resistance of the pendulum oscillating in water, that is, that part that is proportional to the square of the velocity, and which only needs to be considered in swift bodies, is to the resistance of the whole same pendulum, oscillating in air with the same velocity, as about 850 to 1, that is as, the density of water to the density of air, nearly.*

That result agrees with the modern form in equation (20.1).

Newton's work in this area covers nine pages in the *Principia*, and I leave it to interested readers to read them and marvel again at his industry and inventiveness, even if the final results are none too satisfactory. One final topic is worthy of note.

20.2.5. Detecting the Aether

It is typical of Newton's respect for the importance of experiments that he investigates the possibility of an aether existing by devising a pendulum experiment. It is interesting to read his own words, *since it is the opinion of some that there is a certain aethereal medium extremely rare and subtile, which freely pervades the pores of all bodies; and from such a medium, so pervading the pores of bodies, some resistance must needs arise; in order to try whether the resistance, which we experience in bodies in motion, be made upon their outward superficies [surfaces] only, or whether their internal parts meet with any considerable resistance upon their superficies, I thought of the following experiment.*

Basically, he uses a box as the pendulum bob, experimenting with it empty and filled with lead as comparison cases. (It is a measure of Newton's care that he notes that the lead case will stretch the supporting thread and he must correct for that.) He detects no great difference (citing a ratio of 78:77), and notes that *"this experiment is related from memory, the paper being lost,"* so he does not have all details. Even so, the experiment was carefully and intricately designed, and about one and a half pages are devoted to the subject, so he must have thought his no-aether-detected result to be of significance.

20.2.6. Modern Experiments

Some beautiful modern experiments by Bolster, Hershberger, and Donnelley allow us to see the change in fluid behavior around a pendulum as it changes from the low Reynolds number, Stokes flow regime, to a critical-change region with a Reynolds number around 700. The oscillations are made in water with dissolved Thymol blue, allowing fluid motions to be observed. In the first case, there is laminar flow, and then we can see the shedding of vortex rings. No doubt Newton would have delighted in these experiments, which clearly show how resistance-force changes are correlated with fluid-flow properties. The interested reader is encouraged to look at the figures in the references given in the "Further Reading" section at the end of this chapter.

20.3. FALLING BODIES

The observation of falling bodies must surely be the oldest of all experiments and also the one causing the most speculation. I. Bernard Cohen (see this chapter's "Further Reading" section) gives a good history, including a discussion of Galileo's exploits. The fact that different masses fall at the same rate is always clouded by the influence of air resistance, and now Newton uses that variability to investigate the resistance force itself. Based on his previous theoretical work, he concludes that the principal relevant resistance force is proportional to speed squared, so we have $a = 0$ in equation (19.3), and b is the parameter controlling the force magnitude. Newton suggests that, for a sphere of diameter d moving in a fluid with density ρ,

$$F_R \propto \rho d^2 v^2, \quad \text{so} \quad F_R = -bv^2. \tag{20.9}$$

His results for this study are given in *PROPOSITION XL. PROBLEM IX*: "*To Find by Phenomena the Resistance of a Globe* [Sphere] *Moving through a Perfectly Fluid Compressed Medium.*"

The necessary theory is given, then fourteen experiments are described and discussed in a following *Scholium*.

20.3.1. The Modern Theory

Taking the x axis as vertically downward, the equation of motion

$$m\frac{d^2x}{dt^2} = mg - bv^2 = mg - b\left(\frac{dx}{dt}\right)^2 \tag{20.10}$$

must be solved with the starting condition $x = 0$ and speed $v = 0$ at $t = 0$. The solution may be found in textbooks such as *Dynamics* by Horace Lamb, and

$$v(t) = V \tanh\left(\frac{gt}{V}\right), \quad \text{where } V \text{ is the terminal velocity } V = \sqrt{\frac{mg}{b}} \tag{20.11}$$

$$x(t) = \frac{V^2}{g}\log\left\{\cosh\left(\frac{gt}{V}\right)\right\} \quad \text{or} \quad x = \frac{V^2}{2g}\log\left\{\frac{V^2}{V^2 - v^2}\right\}. \quad (20.12)$$

Thus x, the distance fallen, is known in terms of the elapsed time t or the speed v at that point in the fall. These results were known to Newton (although they are not given in that form), as shown by checking the figures he gives with eight-figure accuracy for theoretical fall results in his table.

20.3.2. Experiments in Water

Newton's *Scholium* begins:

> In order to investigate the resistances of fluids from experiments, I procured a square wooden vessel, whose length and breadth on the inside was 9 inches English measure, and its depth 9 feet ½; this I filled with rainwater; and having provided globes made up of wax and lead included therein, I noted the time of the descents of these globes, the height through which they descended being 112 inches.

He had "seconds" and "half-seconds" pendulums available for timing. He reports on three experiments like that, and then for experiments four to twelve, he used a tank about fifteen feet deep. For experiments one, two, and four, he measured the time of fall, calculated the theoretical fall distance for that time, and compared it with the actual distance fallen. In each case, the theory agreed with experiment to better than 1 percent.

For the other experiments, he recorded the distance fallen, calculated the theoretical time for falling that distance, and compared it with the measured times. He did several measurements for each case, and the calculated times generally fitted with the spread of measured times. For the heavier bodies, his times tended to be a little low, suggesting his theory overestimated the resistance in these cases.

Newton recognizes that there are problems with these experiments, such as interaction of the fluid motion with the tank sides. He also notes that, for some releases, the globes oscillated a little, and that would complicate the motion. The discussion is very detailed, and Newton comes over as a careful and perceptive experimenter.

20.3.3. Free Fall

Experiments thirteen and fourteen investigate the fall of bodies in air using the newly restored St. Paul's Cathedral. Experiment thirteen was carried out in June 1710 and used a number of glass globes variously filled with air, water, or quicksilver (mercury). Newton describes a release mechanism by which the globe was released from a hinged table and, at the same moment, a "seconds" pendulum was set oscillating so the time of fall could be measured. (Later he discusses this mechanism and errors it may introduce.) The theory and experiment appear to agree with about 2 to 5 percent accuracy according to the table of results.

Experiment fourteen was carried out in July 1719 by John Desaguliers, who was one of Newton's confidants and Curator and Operator of Experiments in 1714 for the Royal Society. This time, bladders shaped into spheres and balls of lead *"were let fall from the lantern on top of the cupola of the same church, namely from a height of 272 feet."* Thus a good distance of fall was obtained, and Newton describes the complicated release and timing schemes, the latter using *"a pendulum vibrating four times in a second."*

The bladders could be filled with liquids, and the whole experiment was clearly very carefully carried out. Newton even comments about the manner in which *"the fifth bladder was wrinkled, and by its wrinkles was a little retarded."* It is interesting to look through his table of results, reproduced here in figure 20.3.

The weights of the bladders	The diameters	The times of falling from a height of 272ft	The spaces which by the theory should have been fallen	The difference between theory and experiment
128 grains	5.28 inches	19"	271 feet 11 inch	−0 ft 1 inch
156	5.19	17	272 ½	+0 0½
137½	5.3	18	272 7	+0 7
97½	5.26	22	277 4	+5 4
99⅛	5	21⅛	282 0	+10 0

Figure 20.3. Newton's table of results from experiment fourteen.

These results are very good, partly because the larger fall would leave the experiment just a little less sensitive to timing errors.

20.3.4. How Good Are These Results?

I can do no better than to quote the summary of George E. Smith, the expert on this topic:

> How good were these vertical fall data? The best way to determine this is to ignore Newton's theory and instead determine the values of the drag coefficient C_D implied by the results. [Smith then refers to a figure much like figure 19.1.] The two results in air, with a Reynolds number of 40,000 and C_D ranging from 0.499 to 0.518 for Desaguliers's and a Reynolds number of 78,000 and C_D ranging from 0.504 to 0.538 for Newton's, show remarkably good agreement with modern values. The results in water are a little on the high side, perhaps due to sidewall effects from the fairly narrow troughs or from free-surface effects at the beginning of the descent. Regardless, the C_D values range between 0.462 and 0.519. Even more significant, the lowest value, 0.462, occurs at a Reynolds number where the modern measured value reaches a minimum. The experiments themselves were therefore of high quality. One can make a case that no better data for resistance of spheres over this range of Reynolds numbers were published before the twentieth century.

That is quite a compliment for Newton the experimentalist!

20.4. CONCLUSION

Newton's experimental work typifies what we find in *Book II*: ingenious methods and a care and thoroughness that were surely a model for his times. His pendulum experiments were a clever attempt to evaluate the different components in the resistance force, but the method was beset by difficulties, which Newton, to his credit, clearly recognized. The vertical-fall experiments show us Newton at his best, even if his underlying theoretical basis is problematical, as we will see in the next chapter.

20.5. FURTHER READING

For more on experiments and Boyle's part in the Scientific Revolution, see, for example:

Dear, Peter. *Revolutionizing the Sciences: European Knowledge and Its Ambitions, 1500–1700*. Houndmills, UK: Palgrave, 2001.

Hunter, M. *Boyle: Between God and Science*. New Haven, CT: Yale University Press, 2009.

Shapin, Steven. *The Scientific Revolution*. Chicago: University of Chicago Press, 1996. [includes an extended, comprehensive bibliographic essay]

For detailed discussion of Newton's experiments and connected matter, see:

Bertoloni Meli, Domenico. *Thinking with Objects: The Transformation of Mechanics in the Seventeenth Century*. Baltimore: Johns Hopkins University Press, 2006.

Cohen, I. Bernard. *The Birth of a New Physics*. New York: Norton, 1985.

———. "A Guide to Newton's *Principia*." In *Isaac Newton: The* Principia, by I. Bernard Cohen and Anne Whitman. Berkeley: University of California Press, 1999.

Gauld, C. F. "Newton's Investigation of the Resistance to Moving Bodies in Continuous Fluids and the Nature of 'Frontier Science.'" *Science and Education* 19 (2010): 939–61.

———. "Newton's Use of the Pendulum to Investigate Fluid Resistance: A Case Study and Some Implications for Teaching about the Nature of Science." *Science and Education* 18 (2009): 383–400.

Fowles, G. R. *Analytical Mechanics*. 4th ed. Philadelphia: Saunders College Publishing, 1986.

Lamb, Horace. *Dynamics*. Cambridge: Cambridge University Press, 1945.

Smith, George E. "The Newtonian Style in Book II of the *Principia*." In *Isaac Newton's Natural Philosophy* . Edited by J. Z. Buchwald and I. Bernard Cohen. Cambridge, MA: MIT Press, 2001.

————. "Newton's Study of Fluid Mechanics." *International Journal of Engineering Science* 36 (1998): 1377–90.

For information on modern pendulum experiments, see:

Bolster, D., R. E. Hershberger, and R. J. Donnelly. "Dynamic Similarity, the Dimensionless Science." *Physics Today* (September 2011): 42–47.
————. "Oscillating Pendulum Decay by Emission of Vortex Rings." *Physical Review E* 81 (2010): 046317-1-046317-6.

For fluid mechanics and drag:

Achenbach, E. "Experiments on the Flow Past Spheres at Very High Reynolds Numbers." *Journal of Fluid Mechanics* 54 (1972): 565–75.
Binder, R. C. *Fluid Mechanics*. Englewood Cliffs, NJ: Prentice-Hall, 1943.
Daughterty, R. L., and J. B. Franzini. *Fluid Mechanics with Engineering Applications*. New York: McGraw-Hill, 1965.
De Mestre, N. *The Mathematics of Projectiles in Sport*. Cambridge: Cambridge University Press, 1990.
Frohlich, C. "Aerodynamic Drag Crisis and Its Possible Effect on the Flight of Baseballs." *American Journal of Physics* 52 (1984): 325–34.
Rouse, Hunter. *Elementary Mechanics of Fluids*. New York: John Wiley, 1946.
Van Dyke, Milton. *An Album of Fluid Motion*. Stanford, CA: Parabolic Press, 1982.

21

WHAT LIES BENEATH

Recall that *Book II* of the *Principia* concerns motion when there is resistance from the medium in which that motion takes place. In the previous two chapters, we have seen how Newton describes this situation using his theory of dynamics and then explores the form of the assumed resistance forces using experiments with pendulums and falling bodies. In this chapter, I will discuss Newton's third approach to this motion in which he asks: How do resistance forces arise? What form do they take? And how do they depend on the properties of the medium? This takes him into the theory of fluids (liquids and gases) and naturally extends beyond resistance forces to motions in the fluids themselves.

For me, this is one of the most remarkable parts of the *Principia*. While it is clear that he has reached the limits of his theoretical formalism, nevertheless, Newton still manages to introduce concepts and methods that remain of importance for any mathematical scientist. He manages to produce results using a stunning variety of innovative approaches and analogies. This is Newton at his best!

To deal with a fluid, one needs a theory or equations extending dynamics from the realm of particles to small elements of the fluid and how they move and interact with each other. Alternatively, one may consider the fluid as a collection of particles and try to use their motions to describe how the whole collection, the fluid, behaves. These are vast areas of scholarship and research, and Newton produces pioneering work in both approaches. I will discuss some of his most significant steps to give you a flavor of this part of *Book II*. To go further, consult the references in the "Further Reading" section at the end of this chapter.

21.1. HYDROSTATICS

We have reached *Section V*: "*Of the Density and Compression of Fluids; and of Hydrostatics.*"

Newton begins by defining his terms and immediately identifying a consequence:

THE DEFINITION OF A FLUID.

A fluid is any body whose parts yield to any force impressed on it, and by yielding, are easily moved among themselves.

PROPOSITION XIX. THEOREM XIV.

All the parts of a homogeneous and unmoved fluid included in any unmoved vessel, and compressed on every side (setting aside the consideration of condensation, gravity and all centripetal forces), will be equally pressed on every side, and remain in their places without any motion arising from that pressure.

That sums up our basic concept of a fluid. Newton goes on later to recognize what today we call Boyle's law when he routinely considers fluids with the "*density of a fluid proportional to the compression.*" Today we express the relationship between pressure P, volume V, and density ρ as

$$PV = \kappa, \text{a constant} \quad P = \kappa/V = K\rho \quad \text{or} \quad P \propto \rho. \quad (21.1)$$

Newton then shows how gravity will change the pressure in a column of fluid and comments on the pressure in the Earth's atmosphere. As always, he covers a range of force and geometry examples.

Thus Newton has introduced some continuum ideas that he can use in the static case.

21.2. SCALING AND BOYLE'S LAW

The application of Boyle's law is crucial in much of Newton's work, so it is not surprising that he delved into its origins. Furthermore, he has here the chance to demonstrate what he had in mind when he wrote in his *Preface*:

> *I wish we could derive the rest of the phenomena of nature by the same kind or reasoning from mechanical principles* [as he has used for gravity and the solar system]; *for I am induced by many reasons to suspect that they may all depend upon certain forces by which the particles of bodies by some causes hitherto unknown, are either mutually impelled towards each other, and cohere in regular figures, or are repelled and recede from each other.*

So it is here that we find the impressive

PROPOSITION XXIII. THEOREM XVIII.

If a fluid be composed of particles mutually flying [about] *each other, and the density be as the compression* [that is, Boyle's law applies], *the centrifugal forces of the particles will be reciprocally proportional to the distances of their centres. And, vice versa, particles flying* [about] *each other, with forces that are reciprocally proportional to the distances of their centre, compose an elastic fluid, whose density is as the compression.*

Newton is claiming (within the limits of his proof and its assumptions, it must be noted) that Boyle's law holds if and only if the fluid comprises particles with repulsive forces depending inversely on the distance between them. That is a remarkable achievement, and equally remarkable is the technique Newton introduces into applied mathematics to make the proof. He gives a scaling argument based on the cubes of fluid, as in figure 21.1. The ratio of the sides AB and ab, which today we might use in terms of the parameter x, with $ab = xAB$, can be used to scale the areas, volumes, and densities (for example, the area of side $abcd = x^2ABCD$). Newton is then able to argue that if Boyle's law holds, the pressures lead to interparticle repulsive forces varying inversely on the distance. He also shows how to start with those forces and come to Boyle's law. This is an extremely original analysis, and the interested reader will find Newton's description of it quite easy to follow, if a little lengthy. Scaling arguments are now routinely used in science and have led to important relationships in biology (see the book by Schmidt-Nielsen listed in this chapter's "Further Reading" section). The later growth of the molecular theory of gases, the kinetic theory, is covered in chapter 28.

Newton gives a *Scholium* in which he enumerates other possible laws and their forces and concludes with his standard methodological warning:

> *But whether elastic fluids do really consist of particles so repelling each other, is a physical question. We have here demonstrated mathematically the property of fluids consisting of particles of this kind, that hence philosophers [scientists] may take occasion to discuss that question.*

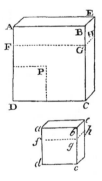

Figure 21.1. Newton's diagram in support of *Proposition XXIII*. The small cube is a scaled version of the larger one. (From *The* Principia, trans. Andrew Motte [Amherst, NY: Prometheus Books, 1995].)

Newton's particle approach to gases is not correct, but the basic idea of understanding a macroscopic law by considering microscopic effects is valid and gave rise to the kinetic theory of gases (see chapter 28).

21.3. FLUID MOTION AND RESISTANCE FORCES

From the hydrostatics case, we now move to *Section VII*: "*Of the Motion of Fluids, and the Resistance Made to Projected Bodies.*"

As mentioned earlier, Newton does not have a mathematical theory for fluid motion, so he must use various fluid models and comparisons of different flow situations. He begins with the idea of comparing similar systems, and this idea, now known as the principle of similitude, has proved to be of great value in science and engineering. Here is Newton's own formulation:

PROPOSITION XXXII. THEOREM XXVI.

Suppose two systems of bodies consisting of an equal number of particles, and let the correspondent particles be similar and proportional, each in one system to each in the other, and have a like situation among themselves, and the same given ratio of density to each other; . . . I say, that the particles of those systems will continue to move among themselves with like motions in proportional times.

E. T. Whittaker, in his classic *Treatise*, summed it up: "If any system of connected particles and rigid bodies is given, it is possible to construct another system exactly similar to it but on a different scale."[1] These ideas are behind the use of the Reynolds number, as discussed in section 20.1; two systems may seem quite different but have parameters scaling to give the same Reynolds number. (Readers might be surprised to learn that pendulum motions of 100 micron spheres in liquid helium and one-inch steel spheres in water have similar behavior, as shown in the paper by Bolster, Hershberger, and Donnelly.)

Newton uses models of fluids to conclude that "*the resistance of the globe* [sphere] *is in a ratio compounded of the duplicate* [square] *ratio of the velocity, and the duplicate ratio of the diameter, and the ratio of the density of the medium.*" We would write

$$\text{Resistance force} = kv^2d^2\rho. \qquad (21.2)$$

Newton considers the constant k for an incompressible rarefied medium (with a dependence on whether or not the particle collisions are elastic) and for an incompressible continuous medium. He also compares the sphere and cylinder, finding resistance doubled for the cylinder of same diameter.

During these calculations, Newton introduces yet another idea that has proved to be of lasting value. He writes (when discussing *Proposition XXXIV*, comparing sphere and cylinder motions):

> For since the action of the medium upon the body is the same (by Cor. 5 of the Laws) whether the body move in a quiescent medium, or whether the particles of the medium impinge with the same velocity upon the quiescent body . . .

As a consequence of the laws of motion, we may think of the body moving in a stationary fluid, or a fluid flowing onto a fixed body. This is a conceptually useful device for theorizing, but it is also of enormous practical importance for things like wind tunnels and the testing of airplane and car shapes.

There is one more problem that Newton considers in this section that reminds us that we are reading the work of a master theorist.

21.3.1. The Shape of Least Resistance

In a *Scholium* following *Proposition XXXIV*, Newton writes:

> By the same method other figures may be compared together as to their resistance; and those may be found that are most apt to continue their motions in resisting mediums.

Newton is finding the optimum shapes for moving through a fluid (on the particle-collision model) so that the effects of resistance are minimized.

He first considers a cone (see figure 21.2) with a circular base $CEBH$, a center O, and a height OD. Surprisingly, the optimum shape is not a cone with a vertex at D; it is a frustum obtained by slicing through at D a cone with vertex S given by the construction $QS = QC$, where Q is the point midway between O and D.

Newton now produces a breathtaking result: he gives the general optimum shape (see figure 21.2). I will not give details (it is far too complicated), but the solution is in the form of a curve to be rotated around an axis to create the required optimum shape. Interested readers should consult the resources by Goldstine and Chandrasekhar (see the "Further Reading" section at the end of this chapter) for a complete exposition.

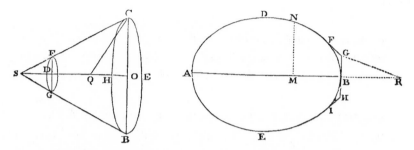

Figure 21.2. Newton's diagrams for least-resistance shapes. *Left*: The optimum cone with axis *OD* and moving to the left. *Right*: The general shape for a body moving to the right and with an axis along the line *AMBR*. (From *The* Principia, trans. Andrew Motte [Amherst, NY: Prometheus Books, 1995].)

This is a problem in the calculus of variations (see the book by Lemons listed in the "Further Reading" section at the end of this chapter for a simple introduction to the subject). Classic problems in this branch of mathematics are: Given a closed curve of fixed length, what shape should it take to enclose the maximum area? (Answer: A circle.) What shape should a smooth curve joining two fixed points in a vertical plane take if a particle slides down it in the least time? (Answer: A cycloid.) According to Goldstine, "the first genuine problem"—that is, the least-resistance-shape problem considered here—"of the calculus of variations was formulated and solved by Newton in late 1685."[2] A look at Goldstine's discussion will convince you that this is a stunning piece of work by Newton. Chandrasekhar calls it "the most sophisticated mathematical problem treated in the *Principia*."[3]

There is no proof in the *Principia*, and I. Bernard Cohen may be consulted for historical commentary on that matter as well as references to Whiteside's definitive study. Newton states "*which Proposition may be of use in the building of ships*," and, according to Cohen, it may be that one John Craig suggested the problem to Newton in that form. As it turns out, the conditions assumed by Newton have no relevance to shipbuilding but may be of interest for very high speed flight, as Newton himself indicates in *Corollaries* 2 and 3 to *Proposition XXXIII*. For more on this topic, see the book edited by Miele listed in the "Further Reading" section at the end of this chapter.

21.5. PROPAGATION OF DISTURBANCES IN FLUIDS

Although not immediately part of Newton's investigation into the motion of bodies through fluids, we can surmise that he discovered that motion of the fluid itself was physically relevant and something he could investigate. Thus we come to *Section VIII: "Of Motion Propagated through Fluids."*

Newton again uses his discrete particle model to discuss how a pressure disturbance is propagated in a fluid. He gives the following:

PROPOSITION XLI. THEOREM XXXII.

A pressure is not propagated through a fluid in rectilinear directions unless where the particles of a fluid lie in a right line.

PROPOSITION XLII. THEOREM XXXIII.

All motion propagated through a fluid diverges from a rectilinear progress into the unmoved spaces.

These *Propositions* are supported by the diagrams in figure 21.3, which show the spreading and diffraction of disturbances in fluids. Newton then explains how a "*tremulous body*" will transmit a series of compressions and impulses through an elastic medium.

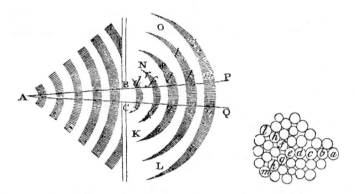

Figure 21.3. Newton's diagrams in support of *Propositions XLI* and *XLII*. (From *The Principia*, trans. Andrew Motte [Amherst, NY: Prometheus Books, 1995].)

In a later *Scholium*, Newton makes an important comment revealing his ideas about sound and light:

> *The last Propositions respect the motions of light and sounds; for since light is propagated in right lines it is certain that it cannot consist in action alone (by Prop. XLI and XLII).*

Newton is ruling out a wave theory for light. Of course, later experiments revealing the diffraction and interference effects explained by wave theory would prove him wrong.

21.5.1. Oscillations and Waves: Water Waves

Newton does not have the equations for continuum mechanics and derivations of wave equations, so he relies on ingenious (as ever) physical constructions and analogies. His investigative tool is set out in a *Proposition* and its supporting diagram:

PROPOSITION XLIV. THEOREM XXXV.

If water ascend and descend alternately in the erected legs KL, MN, of a canal or pipe; and a pendulum be constructed whose length between the point of suspension and the centre of oscillation is equal to half the length of the water in the canal; I say, that the water will ascend and descend in the same times in which the pendulum oscillates.

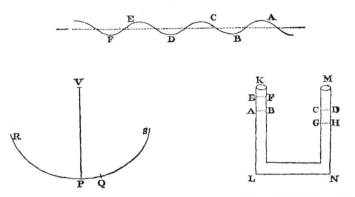

Figure 21.4. Newton's diagrams supporting *Propositions XLIV* and *XLVI*. (From *The Principia*, trans. Andrew Motte [Amherst, NY: Prometheus Books, 1995].)

Newton discusses how the change in water level *AB* to *EF* and the pendulum arc *PQ* are related. Although he talks about a cycloidal pendulum, he draws a simple pendulum. We must assume small oscillations, as in a *Corollary* he talks about the oscillations of the water "*all performed in equal times, whether the motion be more or less intense.*" We have the notion of isochronal oscillations and the transfer of that concept to waves, thus giving the modern idea of simple linear harmonic waves.

He goes on to define the "*breadth of the waves,*" what today we call their wavelength λ in terms of the distance between hollows or between "*tops of the ridges*" (*FD* or *EC* in figure 21.4). He is then able to use his pendulum model to show that "*the velocity of waves is in the subduplicate ratio* [square root] *of the breadths.*" The equivalent modern formula for waterwave speed c is

$$c = \sqrt{\frac{g\lambda}{2\pi}}, \quad \text{when } \lambda << \text{water depth.}$$

In an interesting aside, the ever-thorough Newton shows that he has thought about the motion of water under waves and the trajectories followed:

These things are true upon the supposition that parts of water ascend or descend in a right line [see his model in figure 21.4]; but, in truth, that ascent and descent is rather performed in a circle; and therefore I propose the time defined by this proposition as only near the truth.

21.5.2. Oscillations and Waves: Sound Waves

Propositions XLVII to *L* deal with waves in compressible fluids, and sound waves in gases are the obvious model. Newton deals with the propagation of oscillations or "*pulses*" and links the detailed dynamics to the equivalent motion around a circle. (I leave it to the interested reader to follow up on the details—the arguments given in prose are, as always, rather long and tedious. See this chapter's "Further Reading" section.)

Newton discovers the basic relationship for wave motion: the speed is equal to the wavelength multiplied by the frequency. He also relates the speed of sound waves to properties of the medium in which they propagate:

PROPOSITION XLVIII. THEOREM XXXVIII.

The velocities of pulses propagated in an elastic fluid are in a ratio compounded of the subduplicate ratio of the elastic force directly, and the subduplicate ratio of the density inversely [the square root of elastic force divided by the density]; *supposing the elastic force of the fluid proportional to its condensation.*

Newton had the properties of this elastic fluid summed up in Boyle's law, equation (21.1). Using that formula, he could evaluate the square root of P divided by density ρ and come up with his value for c, the speed of sound. Newton calculated that $c = 979$ feet per second. This was a problem for Newton because a variety of experiments (some of which he made himself) gave larger values, around 1,140 feet per second. However, rather than admit his theory was lacking in some way, he tried to come up with arguments to correct his result, things we might today call fudge factors (see the paper by Westfall listed in the "Further Reading" section at the end of this chapter). Again the ego of the man is evident, and such unworthy episodes detract from his reputation.

The problem remained for about a hundred years, until Pierre-Simon Laplace corrected Newton's theory in 1816. The problem lies in the use of Boyle's law and Newton's assumption that there was no heat exchange in the rapid changes of pressure in a sound wave. Instead of P/ρ in Boyle's law, we should use P/ρ^γ, where γ is the specific-heat constant. Correcting Newton's work in this way gives a satisfactory result for the speed of sound.

21.6. CIRCULAR MOTION AND VORTEX THEORY

Finally we come to what many people see as the rationale for the whole of *Book II*, that is, *SECTION IX: "Of the Circular Motion of Fluids."*

It is here that Newton attacks Descartes's vortex theory of planetary motion (see section 2.6.2). Descartes gave no explanation for how his vortex theory could lead to Kepler's laws for planetary motion. Newton took up the challenge—with the aim, of course, of discrediting Descartes's theory.

This is a difficult area of fluid dynamics, and Newton was well beyond his capabilities to discuss the problem in mathematical detail. It is interesting to note that here, for the first time, he makes some assumptions about the way viscosity enters in order to transmit fluid motion from one region to another. He introduces a *Hypothesis*:

> *The resistance arising from the want of lubricity in parts of a fluid, is, caeteris paribus* [other things being equal], *proportional to the velocity with which parts of the fluid are separated from each other.*

This has now become the standard way to relate shearing stress, viscosity, and velocity gradient; fluids behaving this way are, in fact, known as Newtonian fluids. Newton needs the idea here to see how a revolving cylinder or a revolving sphere can spread out rotations across the fluid. (He never used this idea when discussing resistance forces, and it was Stokes who showed how viscous effects operate at very low Reynolds numbers.)

Proposition LI and *LII* consider the circular motion generated around rotating cylinders and spheres. Newton claims that, as the distance from the center increases, the periodic times for the fluid circles increase directly as the distance or as the square of the distance, respectively; neither case will lead to Kepler's third law. He also points out the dynamical difficulties associated with a body moving around in a circle—its tendency to fly outward. While Newton (and his readers) were satisfied that he had shown how Descartes's scheme is not viable, there are difficulties with Newton's work, and the whole area is one of great technical difficulty (see this chapter's "Further Reading" section).

21.7. CONCLUSION

There is a tendency to skip over *Book II* of the *Principia*, but just the parts covered in this chapter show how much is lost by such a decision. While it is true that Newton had reached the limit of his mathematical and physical backgrounds and skills, he still manages a dazzling array of achievements. He has begun the process of finding microscopic, or lower scale, explanations for macroscopic phenomena, the development of which constitutes a large part of modern physics. (In fact, Newton had already used this concept when considering forces between large bodies, as discussed in chapter 18.) There are the techniques: scaling theory, the principle of similitude, and the calculus of variations. There is the pioneering physics: particle theory for the properties of gases, the equivalence of a moving body in a quiescent fluid and a fluid flowing past a fixed body, the idea of optimum minimum-resistance shapes, the beginnings of wave theory, the calculation of the speed of sound from first principles, and ideas about resistance forces and viscosity. We can now understand why in chapter 19 we heard Clifford Truesdell call *Book II* "the most brilliant and fertile work ever written in mechanics."[4] We shall find even more examples of Newton's inventiveness and fluids theory in *Book III*, when he deals with the theory of tides and the shape of the Earth.

We are now ready to move on to the final book of the *Principia*, as Newton tells us at the very end of *Book II*:

The hypothesis of vortices is utterly irreconcilable with astronomical phenomena, and rather serves to perplex than explain the heavenly motions. How these are performed in free spaces without vortices, may be understood by the first book; and I shall now more fully treat of it in the following book.

21.10. FURTHER READING

Boas, M., and R. Hall. "Newton's 'Mechanical Principles.'" *Journal of the History of Ideas* 20 (1959): 167–78. [discusses how Newton uses the mechanical hypothesis and microscopic-particle theory]

Cohen, I. Bernard. "A Guide to Newton's *Principia*." In *Isaac Newton: The Principia*, by I. Bernard Cohen and Anne Whitman. Berkeley: University of California Press, 1999.

Darrigol, O. *Worlds of Flow: A History of Hydrodynamics from the Bernoullis to Prandtl*. Oxford: Oxford University Press, 2005.

Rouse, Hunter, and Simon Ince. *History of Hydraulics*. New York: Dover, 1957.

Smith, George E. "The Newtonian Style in Book II of the *Principia*." In *Isaac Newton's Natural Philosophy*. Edited by J. Z. Buchwald and I. Bernard Cohen. Cambridge, MA: MIT Press, 2001.

For more on the principle of similitude:

Binder, R. C. *Fluid Mechanics*. Englewood Cliffs, NJ: Prentice-Hall, 1943. [see ch. 4]

Bolster, D., R. E. Hershberger, and R. J. Donnelly. "Dynamic Similarity, the Dimensionless Science." *Physics Today* (September 2011): 42–47.

Huntley, H. E. *Dimensional Analysis*. London: Macdonald, 1952.

Rayleigh, Lord. "The Principle of Similitude." *Nature* 95 (1915): 66–68. [a classic paper with examples given by one of the great scientists]

Schmidt-Nielsen, K. *Scaling: Why Is Animal Size So Important?* Cambridge: Cambridge University Press, 1984.

Whittaker, E. T. *A Treatise on the Analytical Dynamics of Particles and Rigid*

Bodies. New York: Dover, 1944. [this edition is the first American printing; see sec. 33, "Similarity in Dynamical Systems"]

For shapes of least resistance:

Chandrasekhar, S. *Newton's* Principia *for the Common Reader*. Oxford: Clarendon Press, 1995. [see ch. 26]
Goldstine, Herman H. *A History of the Calculus of Variations*. New York: Springer-Verlag, 1980. [see sec. 1.2]
Lemons, D. S. *Perfect Form: Variational Principles, Methods, and Applications in Elementary Physics*. Princeton, NJ: Princeton University Press, 1997. [lovely introduction]
Miele, A., ed. *Theory of Optimum Aerodynamic Shapes*. New York: Academic Press, 1965. [see especially part 4, "Newtonian Hypersonic Flow"]

For wave theory and the speed of sound:

Cannon, J. T., and S. Dostrovsy. *The Evolution of Dynamics: Vibration Theory from 1687 to 1742*. New York: Springer-Verlag, 1981.
Chandrasekhar, S. *Newton's* Principia *for the Common Reader*. Oxford: Clarendon Press, 1995. [see ch. 28]
Finn, B. S. "Laplace and the Speed of Sound." *Isis* 55 (1964): 7–19.
Westfall, R. S. "Newton and the Fudge Factor." *Science* 179 (1973): 751–58.

Viscosity and vortices:

Dobson, G. J. "Newton's Errors with the Rotational Motion of Fluids." *Archive for the History of the Exact Sciences* 54 (1999): 243–54.
Langlois, W. E. *Slow Viscous Flow*. New York: Macmillan, 1964.
Tritton, D. J. *Physical Fluid Dynamics*. New York: Van Nostrand Reinhold, 1977.

PART 6

THE MAJESTIC
PRINCIPIA BOOK III

At last Newton is ready to introduce his theory of universal gravity. He can then explain a whole breathtaking variety of earthly and astronomical phenomena. In the final *General Scholium*, he summarizes his main achievements, notes some remaining difficulties, and tells us about the role of God in his worldview.

Chapter 22. *BOOK III* AND GRAVITY
Chapter 23. THEORY OF THE SOLAR SYSTEM
Chapter 24. EARTHLY PHENOMENA
Chapter 25. CHALLENGES: THE MOON AND COMETS
Chapter 26. THE CONCLUDING *GENERAL SCHOLIUM*

Reading plans: These chapters cover the final great triumphs of the Principia. *They are largely nonmathematical and should not be missed! You must read the final chapter; it takes you through Newton's thought-provoking conclusions and some of his most celebrated pronouncements.*

BOOK III AND GRAVITY

Book III is about the force of gravity, how Newton argued for the form of the force, and then how he lays out the consequences of that force in nature. He is clearly following the approach set out in the *Preface*:

> For all the difficulty of philosophy [what we might call natural philosophy or science] *seems to consist in this—from the phenomena of motions to investigate the forces of nature, and then from these forces to demonstrate the other phenomena.*

In his introduction to *Book III*, Newton reminds us that he has been developing the mathematical formalism and principles, although "*lest they should have appeared of themselves dry and barren, I have illustrated them here and there with some philosophical Scholiums.*" Now for the big finale: "*it remains that, from the general principles, I now demonstrate the frame of the System of the World.*" He tells us that, for the presentation, he "*chose to reduce the substance of this book into the form of Propositions (in the mathematical way).*" (Refer back to section 7.5 for the story behind that decision.) Newton also specifies the bare minimum requirement for reading *Book III*:

> It is enough if one carefully reads the Definitions, the Laws of Motion, and the first three Sections of the first book. He may then pass on to this book, and consult each of the remaining Propositions of the first two books, as references in this, and his occasions, shall require.

That is the material discussed in chapters 5, 6, 8, 9, and 10 of this book.

In this chapter, I will examine how Newton makes the step "*from the phenomena of motions to investigate the forces of nature"* for gravity. In the following chapters, I discuss how he goes *"from these forces to demonstrate the other phenomena."*

22.1. WHAT IS REQUIRED?

The force of gravity is usually defined in terms like this: any two bodies a distance r apart and of mass m_1 and m_2 attract each other with a force acting directly along the line joining them and with a magnitude of

$$F = \frac{Gm_1m_2}{r^2}, \quad \text{where G is the gravitational constant.} \quad (22.1)$$

Thus, Newton must establish four things: (1) gravity is an attractive centripetal (central) force, (2) it depends inversely on the square of the distance between the two bodies, (3) it depends on the product of the masses of the bodies, and (4) it operates for any two bodies, implying that it is a universal force.

We will not find an equation like (22.1) in the *Principia*, as we know Newton uses proportions and comparisons in his working. He does not introduce the constant G, about which I will make a few comments later.

As he has told us, the necessary mathematical formalism is in place, and now he must make contact with the physical world and fit theory and observational data together. That is no simple task, and to explain his approach and support his arguments, he first sets out a basis for proceeding.

22.2. NEWTON'S "RULES OF REASONING IN PHILOSOPHY"

This is one of the most famous and dissected parts of the *Principia*. Newton sets out four *Rules* that he can appeal to in support of his working.

> RULE I. *We are to admit no more causes of natural things than such as are both true and sufficient to explain their appearances.*

Some readers will be reminded of Ockham's razor. The idea here is that "*Nature does nothing in vain, and more is in vain when less will serve; for Nature is pleased with simplicity.*"

Notice that Newton mentions "*simplicity,*" a point I return to in section 22.7.2.

RULE II. *Therefore to the same natural effects we must, as far as possible, assign the same causes.*

It is worth checking Newton's examples: "*As to respiration in man as in beast; the descent of stones in Europe and America; the light of our culinary fire and of the sun; the reflection of light in the earth, and in the planets.*"

Although he is wrong about the mechanism for the Sun's output, he is correct about reflection of light being the same everywhere, and a much deeper and far-reaching assumption is involved here. Newton is saying, to put it in modern terms, that the physical effects and laws of physics that we observe on Earth should be taken as valid everywhere in the universe. This is a major change from Aristotle and his theory of perfect heavens; it forms the basis for everything we do in astronomy and astrophysics.

RULE III. *The qualities of bodies, which admit neither intension nor remission of degrees, and which are found to belong to all bodies within the reach of our experiments, are to be esteemed the universal qualities of all bodies whatsoever.*

This follows on from *Rule II* by saying that those constant properties of bodies that we observe in our experiments, like their inertia, should be taken to be present universally. So it is, for example, that our modern knowledge of atoms and spectral lines developed here on Earth may be used to tell us about the properties of stars everywhere in the universe.

RULE IV. *In experimental philosophy we are to look upon propositions collected by general induction from phenomena as accurately or very nearly true, notwithstanding any contrary hypotheses that may be imagined, till such time as other phenomena occur, by which they may either be made more accurate, or liable to exceptions.*

We use induction to come up with our theories, and we should stick with them until there is a clash with new experimental data. As Newton says, *"this rule we must follow, that the argument of induction may not be evaded by hypotheses."*

22.2.1. Discussion of the *Rules*

Newton's *Rules* have attracted much attention and questioning: How sound are they, and do they stand up to a logical and philosophical analysis? How closely does Newton actually follow his rules? The "problem of induction" is a venerable philosophical topic. The interested reader should consult the references listed in the "Further Reading" section at the end of this chapter.

In some ways, this is a personal topic. Some modern scientists (possibly Weinberg or Feynman) are alleged to have said that "the philosophy of science is about as much use to scientists as ornithology is to birds."[1] I expect they, like me, find the collection of *Rules* perceptive and a reasonable guide to some of the assumptions we make and the methods we use in science. I believe it is also useful to heed the words of Albert Einstein:

> If you want to find out anything from the theoretical physicists about the methods they use, I advise you to stick closely to one principle: don't listen to their words, fix your attention on their deeds.[2]

So, let us see how Newton uses (or attempts to use, depending on how philosophical you wish to be) his *Rules of Reasoning*. I will go through his working and reserve discussion until section 22.7.

22.3. DATA

Newton chooses to work with the following six pieces of information, or *Phenomena*, as he calls them.

PHENOMENON I.

That the circumjovial planets [the moons of Jupiter] *by radii drawn to Jupiter's centre describe areas proportional to the times of description; and that their periodic times, the fixed stars being at rest, are in the sesquiplicate* [3/2 power] *proportion of their distances from its centre.*

Note the fixed stars as reference. You will probably say, "Yes, Jupiter's moons obey Kepler's laws."

PHENOMENON II.

That the circumsaturnal planets [the moons of Saturn] *by radii drawn to Saturn's centre describe areas proportional to the times of description; and that their periodic times, the fixed stars being at rest, are in the sesquiplicate* [3/2 power] *proportion of their distances from its centre.*

Saturn's moons behave like Jupiter's. Tables of numerical data are given for these two phenomena.

PHENOMENON III.

That the five primary planets, Mercury, Venus, Mars, Jupiter and Saturn, with their several orbits, encompass the sun.

Newton mentions that Mercury and Venus have phases (like Earth's Moon) so it is clear that they do orbit the Sun. Also that for Jupiter and Saturn, "*the shadows of their satellites* [moons] *that appear sometimes upon their discs make it plain that the light they shine with is not their own, but borrowed from the sun.*"

PHENOMENON IV.

That the fixed stars being at rest, the periodic times of the five primary planets, and (whether of the sun about the earth, or) of the earth about the sun are in the sesquiplicate [3/2 power] *proportion of their mean distances from the sun.*

Numerical data from Kepler and others is given.

PHENOMENON V.

Then the primary planets, by radii drawn to the earth, describe areas nowise proportional to the times; but that the areas which they describe by radii drawn to the sun are proportional to the times of description.

We must refer planetary orbits to the Sun, not the Earth, for Kepler's second law to hold.

PHENOMENON VI.

That the moon, by a radius drawn to the earth's centre, describes an area proportional to the time of description.

We are now ready for the *Propositions* that use those data to argue for the nature of the forces involved.

22.4. THE FORM OF THE FORCES

Newton can now call upon *Propositions* from *Book I*, specifically *Propositions I, II,* and *IV,* and their *Corollaries* to arrive at the following:

PROPOSITION I. THEOREM I.

That the forces by which the circumjovial planets are continually drawn off their rectilinear motions, and retained in their proper orbits, tend to Jupiter's centre; and are reciprocally as the squares of the distances of the places of those planets from the centre.

The same thing applies for Saturn's moons.

Notice that Newton treats these systems of planets around Jupiter and Saturn as independent systems with no account taken of the action of the Sun on them. He can do that because of the remarkable fourth *Corollary* to the laws of motion, discussed in section 6.4.1. Surprisingly, even the thorough Newton fails to remind the reader of that.

PROPOSITION II. THEOREM II.

That the forces by which the primary planets are continually drawn off their rectilinear motions, and retained in their proper orbits, tend to the sun; and are reciprocally as the squares of the distances of the places of those planets from the sun's centre.

As an extra piece of evidence in this case, Newton notes that there is no detected motion of the line joining the apsides, and so his wonderful result in *Proposition XLV* (see section 14.3 to review) indicates an inverse-square law in operation. Any deviation from inverse square "*would produce a motion of the apsides sensible enough in every single revolution, and in many of them enormously great.*"

PROPOSITION III. THEOREM III.

That the force by which the moon is retained in its orbit tends to the earth; and is reciprocally as the square of the distance of its place from the earth's centre.

The forces· involved in these orbits are centripetal (central) and satisfy an inverse-square law. Incidentally, we see now why Newton was so careful about checking that the area law is an if-and-only-if law, as it is the test for a central force (see section 8.3 for review).

22.5. BUT IS IT GRAVITY?

Newton has characterized the force maintaining moons and planets in their orbits, but he has not yet identified it as the force of gravity that we know from our experiences on Earth. This is one of the great discoveries in science, and Newton takes the first step in the following:

PROPOSITION IV. THEOREM IV.

That the moon gravitates towards the earth, and by the force of gravity is continually drawn off from its rectilinear motion, and retained in its orbit.

By his famous "Moon Test" Newton links the force keeping the Moon in its orbit with the gravitational force responsible for falling bodies and pendulum motions here on Earth. This must surely be one of Newton's most inspired pieces of work.

Look at figure 22.1. The Moon is assumed to travel through the angle θ in one minute, and in order to remain in its orbit, Newton says, it must "fall" toward the Earth a distance D away from the straight-line motion it would follow if no force is acting on it. Newton gives the following data:

Moon's orbit: period is 27 days, 7 hours, 43 minutes (= 39,343 minutes);
Moon's mean radius: approximately sixty times the Earth's radius;
Earth's circumference: 123,249,600 Paris feet.
Using those data, he finds D to be 15 1/12 Paris feet.

(I find 15.009 rather than 15.08, but I do not know what Newton used for π, for example.) If you want to see how it works, you should note that θ is $2\pi/39{,}343$ and $D = R\theta^2/2$, where R is the radius of the Moon's orbit, which is sixty times the Earth's radius, which in turn, when multiplied by 2π, gives the circumference of 123,249,600 Paris feet.)

If the same force is responsible for the Moon's "fall" and a body's fall on Earth, because of the inverse-square law, it must be 60^2 stronger on the Earth's surface than at the Moon's orbit. To compensate, we can reduce the fall time on Earth by 60 (so it becomes one second instead of the one-minute fall he took for the Moon); then the fall distances should be the same. (Remember the $\frac{1}{2}gt^2$ formula to understand why the time squared is involved.) **Thus we get the result: if the Moon in its orbit falls 15 1/12 Paris feet in one minute, a body on Earth (maybe an apple!) should fall the same distance in one second. Newton quotes Huygens's measurements to confirm that is actually the case.** Newton then concludes "*and therefore (by Rule I and II) the force by which the moon is retained in its orbit is the very same force which we commonly call gravity.*"

Much has been written about these calculations, and there is no doubt that Newton shamefully fudged parts of them to get such good agreement (see this chapter's "Further Reading" section, especially the paper by

Westfall.) The idea was correct, and the figures are enough to make the correspondence between gravity as observed here on Earth and the force acting on the Moon without any fiddling in the calculations. Newton did not wish to leave open any grounds for criticism (and more of the hated controversies), but a little more of his darker character is revealed here.

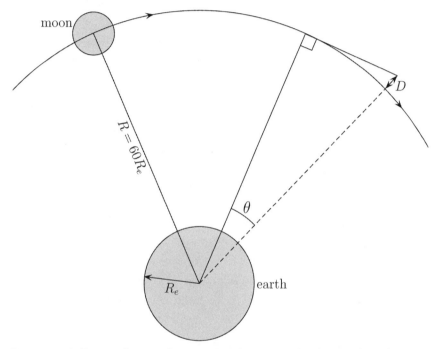

Figure 22.1. A diagram showing the Moon traveling an angular distance θ and falling through distance D, as calculated in Newton's Moon Test. The Earth has radius R_e, and the radius R of the Moon's orbit is approximately $60R_e$.

To back up this momentous conclusion, Newton attaches a *Scholium* in which he gives a supporting thought experiment. He considers moons orbiting the Earth at lower and lower orbits. Eventually, one is at the level of mountaintops and so is subject to the force of gravity as we observe it on Earth. If the force maintaining the orbit is not one and the same as that gravity, "*it would descend with twice the velocity, as being impelled by both these forces conspiring together.*" there is only one force acting, and it is Earth's gravity controlling both moon orbits and the fall of bodies on Earth.

Incidentally, there is some debate about exactly when Newton invented his Moon Test and whether observation of a falling apple played a part in his musings on the subject of gravity. Consult the book by Cohen or look to Westfall's biography to investigate further. We know that Newton apparently embellished his past in his old age, but, just for fun, here is the famous Stukeley recollection of this profound piece of thinking:

> After dinner, the weather being warm, we went into the garden, & drank tea under the shade of some appletrees, only he and myself. Amidst other discourse, he told me, he was in just the same situation, as when formerly, the notion of gravitation came into his mind. "Why should that apple always descend perpendicularly to the ground?" thought he to himself: occasioned by the fall of an apple as he sat in a contemplative mood: "why should it not go sideways or upwards? But constantly to the earth's centre? Assuredly the reason is that the earth draws it. There must be a drawing power in matter and the sum of the drawing power in the matter of the earth must be in the earth's centre," not in any side of the earth.[3]

22.5.1. Gravity Everywhere

Newton is now ready to compare the Earth and its Moon with Jupiter and its moons as similar systems, suggesting we invoke *Rule II* and say that gravity is the common factor. He also mentions the use of the third law of motion to show how gravity acts mutually—Jupiter attracts its moons, and they attract Jupiter, both with the force of gravity. All of this is bound up in the following:

PROPOSITION V. THEOREM V.

That the circumjovial planets gravitate towards Jupiter; the circumsaturnal towards Saturn; the circumsolar towards the sun; and by the forces of their gravity are drawn off from rectilinear motions, and retained in curvilinear orbits.

In what must be one of the most important and powerful statements in the *Principia*, Newton sums up and states what is to be done from now on in a *Scholium*:

The force which retains the celestial bodies in their orbits has been hitherto called centripetal force; but it being now made plain that it can be no other than a gravitating force, we shall hereafter call it gravity. For the cause of the centripetal force which retains the moon in its orbit will extend itself to all the planets, by Rule I, II, and IV.

22.6. GRAVITY AND MASS

Finally, Newton discusses how the strength of the gravitational inverse-square-law force is given in terms of the masses of the bodies involved, expressed as the product $m_1 m_2$ in the modern form of equation (22.1). His conclusions are summed up in two *Propositions*:

Proposition VI. Theorem VI.

That all bodies gravitate towards every planet; and that the weights of bodies towards any the same planet, at equal distances from the centre of the planet, are proportional to the quantities of matter which they severally contain.

Proposition VII. Theorem VII.

That there is a power of gravity tending to all bodies, proportional to the several quantities of matter which they contain.

Newton gives two *Propositions*, VIII and IX, covering interacting composite bodies and the linear behavior of the gravitational force inside a composite body (using the results discussed in chapter 17).

22.6.1. Gravitational and Inertial Mass

Body 2 exerts a gravitational force ("*a power of gravity*"—*Proposition VII*) proportional to its mass m_2, and body 1 responds to that force ("*weight of a body towards*"—*Proposition VI*) proportional to its mass m_1. We can reverse the situation to ask what force body 1 exerts on body 2, and we get the same magnitude by the third law of motion. Thus we deduce that the same masses are used in both situations, that is, for finding both the origin of the force and a body's response to it.

Notice that in his *Propositions*, Newton talks about "*quantities of matter*," which is how he introduced mass at the very start of the *Principia* (see chapter 5). It was there that he first used mass when discussing inertia and added "*the same is known by the weight of each body; for it is proportional to the weight, as I have found by experiments on pendulums, very accurately made, which shall be shown hereafter.*"

There are two masses: the inertial mass m_I and the gravitational mass m_G. The motion of body 1 under the gravitational influence of body 2 is calculated using

$$m_{I1} \frac{d^2 \mathbf{r}_1}{dt^2} = \frac{Gm_{G1}m_{G2}}{\left| \mathbf{r}_1 - \mathbf{r}_2 \right|^2} \hat{\mathbf{r}}_{12}. \tag{22.2}$$

If we assume that $m_I = m_G$, the mass of body 1 cancels out in equation (22.2), and so it is that we get the result that free fall is the same for all bodies, irrespective of their individual masses. The same sort of result holds for pendulum motion, as discussed in section 19.7, where I gave the theory for Newton's "*experiments on pendulums.*"

Newton shows that he clearly understood the situation when he introduces the topic while discussing *Proposition VI*:

> It has been, now for a long time, observed by others, that all sorts of heavy bodies (allowance being made for the inequality of retardation which they suffer from a small power of resistance in the air) descend to the earth from equal heights in equal times; and that equality of times we may distinguish to a great accuracy, by the help of pendulums.

So now Newton is discussing this topic for the third time, and here he tells us that with his two-identical-pendulums experiment (see section 19.5), he "*tried the thing in gold, silver, lead, glass, wood, water, and wheat.*" He concludes that the two masses, inertial and gravitational, are the same, and "*I could manifestly have discovered a difference of matter less than one thousandth part of the whole, had any such been.*"

This is one of the great results in physics, and by 1906 Baron Roland von Eötvös had reduced the uncertainty to 5 parts in 10^8. The equality of the two masses is now called the principle of equivalence, and the accuracy is now around the 1 part in 10^{12} level (see the "Further Reading" section at the end of this chapter).

Newton goes on to discuss properties of mass and how they relate to the possibility of an aether, but there is no speculation about the origin of that vital equality of the masses. That came some two hundred years later with Einstein:

> The general theory of relativity owes its existence in the first place to the empirical fact of the numerical equality of the inertial and gravitational mass of bodies, for which fundamental fact classical mechanics provided no interpretation.[4]

Einstein was able to show that "our extension of the principle of relativity [from the special to the general] implies the necessity of the law of the equality of inertial and gravitational mass."[5]

22.7. WHAT HAS BEEN DONE?

Newton's working in this part of the *Principia* can seem somewhat contorted, and it has been much analyzed (see this chapter's "Further Reading" section, especially the articles by Harper and Smith and the book by Crowe). There appears to be an underlying methodology that Newton does not spell out in detail. We need to follow Einstein's dictum and "fix our attention on his deeds."

Newton presents data about planetary motions and moons. He then effectively says that if you use his laws of motion and the results flowing from them, as presented in *Book I*, you will see that the theory is matched to those data when you use a centripetal inverse-square-law force with strength dependent on the masses of the interacting bodies. Furthermore, on the basis of the Moon Test and the *Rules of Reasoning*, we should call that force gravity and take the theory as holding universally until conflicting experimental data are produced.

We can only admire the way in which those wonderful if-and-only-if results given in *Book I* are now used in such a simple but powerful way. Newton has followed his program and gone "*from the phenomena of motions to investigate the forces of nature.*" However, there is a confusing problem: the data that Newton uses are not exact. He created a simplified version of the known observations about the motion of planets and moons. He was obviously aware of that, and he even gave what we might take as warnings, such as the comment in the third *Corollary* to *Proposition V*:

> All the planets do mutually gravitate towards one another, by Cor. 1 and 2. And hence it is that Jupiter and Saturn, when near their conjunction, by their mutual attractions sensibly disturb each other's motion. So the sun disturbs the motions of the moon.

Thus the data he uses could not possibly be the exact, whole story.

It appears that Newton has used this work basically as a suggestive argument to come up with a theory of universal gravity that may be used and continually tested according to *Rule of Reasoning III*. So, as I discuss in the next chapter, he now turns to a more genuine theory of the solar system and then on to other gravitational effects. In fact, this process has continued and his theory has stood the test of time.

It is true that there are contortions and confusions, but we are now learning about one of the greatest pieces of scientific work ever produced. Already by 1837, Whewell was writing in his great *History of the Inductive Sciences*:

> It is indisputably and incomparably the greatest scientific discovery ever made, whether we look at the advance which it involved, the extent of the truth disclosed, or the fundamental and satisfactory nature of this truth.[6]

Our modern acceptance of the theory might be amusingly summed up by *Apollo 8* astronaut William Anders's reply to a question about who was controlling the capsule: "I think Isaac Newton is doing most of the driving now."[7]

Whatever methods he used, Newton did get to the correct theory of universal gravity. On the way, he showed his usual amazing innovative powers with his Moon Test and with his recognition and deep appreciation for the importance of basic fundamental physical phenomena in his exploration of the equality of inertial and gravitational masses. By the Moon Test and *Rule of Reasoning II*, he linked terrestrial and extraterrestrial phenomena and set the course for astrophysics as we know it today.

22.7.1. What Has Not Been Done?

Newton has established the features of the gravitational force that we express as in equation (22.1), except for the overall strength constant G. His way of working with proportions and ratios means that he avoids the need to give a numerical value for G. The Moon Test provides a good example of that. In fact, it is extremely difficult to measure G, and of all the fundamental constants, it is one of the most poorly known.

The gravitational force is extremely weak, compared with the electric force, for example. However, it is always positive and linearly additive, so it builds up for composite, large bodies, as Newton showed with his "superb theorems" (see chapter 17). Consequently, there is no way to screen out gravitational effects (as can be done with a Faraday cage in the electromagnetic case), and that makes the problem of isolating effects in the laboratory very difficult.

The gravitational interaction of two small bodies compared with their attraction by the whole Earth is, according to Newton, *"far less than to fall under the observation of our senses."* Certainly, very refined experimental techniques are required. The first recognized experiment seems to be that by British scientist Henry Cavendish in 1798—but note that he framed his results in terms of a determination of the Earth's density rather than a measurement of G. There are now various fascinating modern experiments, some at the atomic level, and I give a few examples in the "Further Reading" section at the end of this chapter.

22.7.2. Einstein, Newton, and Simplicity

This is how Richard Feynman opens his lecture on gravitation:

> In this chapter we shall discuss one of the most far-reaching generalizations of the human mind. While we are admiring the human mind we should take some time off to stand in awe of a nature that could follow with such completeness and generality such an elegantly simple principle as the law of gravitation.[8]

While paying homage to Newton, Feynman also introduces something new here: an appreciation of the simplicity of the law of gravitation. One wonders how much Newton appreciated that simplicity. Surely he did; certainly he mentioned the word *simplicity* when discussing the *Rules of Reasoning*. He clearly knew the very special properties of the inverse-square-law force as shown by his brilliant results in *Book I* (see figure 18.3 for a summary). In fact, he might well have used the simplicity of his results to argue for their validity, as Einstein does:

> Our experience hitherto justifies us in believing that nature is the realization of the simplest conceivable mathematical ideas.[9]

Einstein used the simplicity argument when formulating his theory of general relativity, and certainly he never used arguments from data as we have seen Newton doing in this chapter. In fact, one of Einstein's constraints on his theory pays a great compliment to Newton: he required his theory to reduce to Newton's in cases of "weak gravitational" effects. The checking against experimental observations seems to be almost an afterthought and actually demanded new observations for comparisons. I have a feeling that Isaac Newton would have greatly enjoyed Einstein's response to the question of what he would say if God had not made the universe so that light was bent near the Sun as he predicted: "Then I would have to pity the dear Lord. The theory is correct anyway."[10]

22.8. FURTHER READING

Recommended discussion for this topic:

Aoki, S. "The Moon Test in Newton's *Principia*: Accuracy of the Inverse Square Law of Universal Gravitation." *Archive for the History of the Exact Sciences* 44 (1992): 147–90.

Cohen, I. Bernard. "A Guide to Newton's *Principia*." In *Isaac Newton: The Principia*, by I. Bernard Cohen and Anne Whitman. Berkeley: University of California Press, 1999.

———. "Newton's Discovery of Gravity." *Scientific American* (1981): 167–79. [an expert's "popular" introduction]

———. "Newton's Third Law and Universal Gravity." *Journal of the History of Ideas* 48 (1987): 571–93.

Harper, W. "Newton's Argument for Universal Gravitation." In *The Cambridge Companion to Newton*, edited by I. Bernard Cohen and George E. Smith. Cambridge: Cambridge University Press, 2002.

Smith, George E. *Closing the Loop: Testing Newtonian Gravity, Then and Now*. Isaac Newton Lectures at the Suppes Center. Available for download at http://www.stanford.edu/dept/cisst/visitors.html (accessed May 20, 2013).

Westfall, R. S. *Never at Rest: A Biography of Isaac Newton*. Cambridge: Cambridge University Press, 1980.

———. "Newton and the Fudge Factor." *Science* 179 (1973): 751–58.

On Newton's *Rules of Reasoning* and related matters:

Achinstein, P. "Newton's Inductivism and the Law of Gravity." In *Science Rules: A Historical Introduction to Scientific Methods*. Baltimore: Johns Hopkins University Press, 2004.

Crowe, M. J. *Mechanics from Aristotle to Einstein*. Santa Fe, NM: Green Lion Press, 2007. [see pp. 209–20]

Davies, E. B. "The Newtonian Myth." *Studies in History and Philosophy of Science* 34 (2003): 763–80. [disputes Popper's viewpoint and suggests he actually read little of the *Principia*]

Harper, W., and G. E. Smith. "Newton's Way of Inquiry." In *The Creation of Ideas in Physics*, edited by J. Leplin. Dordrecht, Neth.: Kluwer, 1995.

Okasha, S. *Philosophy of Science: A Very Short Introduction.* Oxford: Oxford University Press, 2002. [simple introduction with sections on induction]

Rodriguez-Fernandez, J. L. "Ockham's Razor." *Endeavour* 23 (1999): 121–25.

Spencer, Q. "Do Newton's Rules of Reasoning Guarantee Truth . . . Must They?" *Studies in History and Philosophy of Science* 35 (2004): 759–82.

On the principle of equivalence, see the relativity books by Einstein, Lambourne, Weinberg, and Will referenced in section 5.8. Also see chapters by Cook and Will in *Three Hundred Years of Gravitation*, ed. S. W. Hawking and W. Israel (Cambridge: Cambridge University Press, 1987). Furthermore:

Will, C. M. "The Confrontation between General Relativity and Experiment on the Living Relativity." *Living Reviews in Relativity* 9 (2006). Available online at http://relativity.livingreviews.org/Articles/lrr-2006 -3/download/lrr-2006-3Color.pdf (accessed May 20, 2013).

On the measurement of the gravitational constant G, see the cited chapter by Cook and:

Cavendish, Henry. "Experiments to Determine the Density of the Earth." *Philosophical Transactions of the Royal Society* 88 (1798): 469–526.

Clotfelter, B. E. "The Cavendish Experiment as Cavendish Knew It." *American Journal of Physics* 55 (1987): 210–13.

Ducheyne, S. "Testing Universal Gravitation in the Laboratory; or, the Significance of Research on the Mean Density of the Earth and Big G, 1798–1898." *Archive for the History of the Exact Sciences* 65 (2011): 181–227.

Fixler, J. B., et al. "Atom Interferometer Measurement of the Newtonian Constant of Gravity." *Science* 315 (2007): 74–77.

Gundlach, J. H., and S. M. Merkowitz. "Measurement of Newton's Constant Using a Torsion Balance with Angular Acceleration Feedback." *Physical Review Letters* 85 (2000): 2869–72.

Saulnier, M. S., and D. Frisch. "Measurement of the Gravitational Constant without Torsion." *American Journal of Physics* 57 (1989): 417–20.

Schlamminger, St. "A Measurement of Newton's Gravitational Constant." *Physical Review D.* 74, 082001 (September 7, 2006).

Schwarz, J. P., et al. "A Free-Fall Determination of the Newtonian Constant of Gravity." *Science* 282 (1998): 2230–34.

23

THEORY OF THE SOLAR SYSTEM

In the previous chapter, we saw how, at the start of *Book III*, Newton followed the first part of his program: *"from the phenomena of motions to investigate the forces of nature."* The rest of *Book III* is devoted to the second part: *"and then from these forces to demonstrate the other phenomena."* These *"other phenomena"* include the motion of the solar system, terrestrial phenomena, and the motion of the Moon and comets. Some of the problems involved are mathematically intricate and tough. Some of the effects are subtle and raise questions about the exactness of gravitational theory and its power to deal with them. There may be small effects, like tidal friction, changing the motion of the Earth and the Moon. However, even though it has taken centuries, we now know that Newton's gravitational theory has proved to be triumphant. In fact, each battle won in the quest to describe all the second- and third-order effects in the relevant phenomena has really been another step in confirming the validity of the theory of universal gravity.

What we find in the rest of *Book III* is Newton giving a good first attempt at applying his theory to *"the other phenomena."* However, perhaps more importantly, he is also defining problems and setting out challenges. For example: How does the massive Jupiter disturb the orbits of other planets? (This will be answered in this chapter.) What causes the tides and the precession of the equinoxes? (Next chapter.) How can we account for the motion of the Moon and comets? (Chapter 25.)

We know that Kepler's laws give a good first characterization of the solar system. We also know that Newton's theory with suitable assumptions will produce those laws. However, can that theory take us beyond that simplified model? That is the subject of this chapter.

23.1. BACKGROUND

Newton begins by considering the medium through which the planets move. He can call on his work in *Book II* and he concludes:

> *It is shown in the Scholium of Prop. XXII, Book II, that at the height of 200 miles above the Earth the air is more rare than it is at the superficies of the Earth in the ratio of 30 to 0.0000000000003998, or as 75000000000000 to 1 nearly. And hence the planet Jupiter, revolving in a medium with the same density with that superior air, would not lose by the resistance of the medium the 1000000th part of its motion in 1000000 years.*

Whatever the absolute accuracy of those figures, Newton reached the reasonable conclusion of *Proposition X*: "*That the Motion of the Planets in the Heavens May Subsist an Exceedingly Long Time.*"

Thus the theory in *Book I* dealing with motion when no resistance forces are present may be used. Newton also refers again to the experiments (presumably Boyle's) showing that a piece of gold and a feather fall together in a vacuum.

Newton is now ready to reveal the motion of the planets using his derivation of universal gravity, as discussed in the previous chapter:

> *We have discoursed above of these motions from the Phenomena. Now we know the principles on which they depend, from those principles we deduce the motions of the heavens a priori.*

Of course, this is where some people feel confused, or they worry about a circular argument and question the use of "*a priori.*" However, as we shall see, Newton goes beyond the simplified discussion given earlier. He has data from Kepler and other sources, but there is one thing he needs to establish before he can begin. Characteristically, he manages it with another stroke of genius.

23.2. MASSES OF THE PLANETS

The masses of the planets are important when considering their gravitational interactions and for determining the center of mass of the whole system. Going back to the *Corollaries* to *Proposition VIII*, we find that Newton has devised a method to calculate the masses of planets that have a satellite or moon. Suppose that the planet has mass m_p and the mass of the Sun is m_{sun}. Let the planet orbit around the Sun have period T_p and a semimajor axis a_p and its moon have orbit parameters T_m and a_m. Then the derivation of Kepler's third law tells us that

$$\frac{T_p^{\ 2}}{a_p^{\ 3}} = \frac{\left(4\pi^2/G\right)}{m_{sun}} \quad \text{and} \quad \frac{T_m^{\ 2}}{a_m^{\ 3}} = \frac{\left(4\pi^2/G\right)}{m_p} . \qquad (23.1)$$

Combining these two equations gives

$$\frac{m_p}{m_{sun}} = \frac{T_p^{\ 2}a_m^{\ 3}}{T_m^{\ 2}a_p^{\ 3}} . \qquad (23.2)$$

That elusive gravitational constant G has cancelled out, and we have a remarkable formula giving the mass of the planet in terms of its orbital details and its satellite's details. Newton can use these results for the Earth and its Moon and for Jupiter and Saturn with their moons. The results are:

Body:	Sun	Jupiter	Saturn	Earth
mass (*Principia*):	1	1/1,067	1/3,021	1/169,282
mass (modern):	1	1/1,047	1/3,500	1/332,480

(The modern results are from the book by Blanco and McKuskey, listed in the "Further Reading" section at the end of this chapter.)

We see that Jupiter and Saturn are massive planets. Newton also argues that the density of the planets decreases as their distance from the Sun increases. That conclusion is supported by the modern density data for the first six planets: 6.03; 5.11; 5.517; 4.16; 1.34; and 0.68 in units of *gm* per cubic centimeter.

Who would have thought that we could determine the masses of the planets? Here is another example of the genius Newton at work. Masses of other planets are now found by using data on spacecraft trajectories as they fly past a planet.

23.3. CENTER OF GRAVITY

Newton is now ready for the first general results that follow from the laws of motion (see section 6.3):

PROPOSITION XI.

That the common centre of gravity of the Earth, the sun, and all the planets, is immovable.

PROPOSITION XII.

That the sun is agitated by a perpetual motion, but never recedes far from the common centre of gravity of all the planets.

The center of gravity is always close to the center of the Sun because it is so massive when compared with all the planets. Newton points out that, for the Sun-Jupiter system, the center of gravity is just a little outside the Sun; for the Sun-Saturn system, it is a little inside. In the extreme case, if even all the planets (that is, those known to Newton—see section 23.7) were placed together on one side, "*the distance of the common centre of gravity of all from the centre of the sun would scarcely amount to one diameter of the sun.*" The conclusion is that, "*since that centre of gravity is in perpetual rest, the sun, according to the various positions of the planets, must perpetually be moved every way, but will never recede far from that centre.*"

At last, the ancient question about the center of the solar system—or of "the world," in old terminology—has been answered. Newton gives a famous *Corollary:* "*Hence the common centre of gravity of the Earth, the Sun, and all the planets, is to be esteemed the centre of the world.*" Neither the Earth nor the Sun should be taken as that center.

23.3.1. Being More Careful, and a Curious Hypothesis

The result that Newton is using (as discussed in chapter 6) tells us that, for a system of interacting bodies with no external forces acting on them, the center of mass (or gravity, here) is either at rest or moving uniformly relative to the inertial frame of reference used to describe the motions. Newton has chosen to take that center as at rest. To back up that, he has given the following:

HYPOTHESIS I. That the center of the system of the world is immovable.

There seems to be no reason for this hypothesis except that it had always been debated what, exactly, was at rest (the Earth? the Sun?) in the various approaches to the solar system. In fact, Newton finds that even in a heliocentric or Copernican system, the Sun does not form some immovable center. Modern astronomy has discovered motion of the solar system and rotations of our galaxy in the universe in general. In 1783, William Herschel suggested that the solar system was moving toward the constellation Hercules. What Newton might have said is that, to a high degree of accuracy for some time periods, it is possible to choose a local frame of reference in which the center of gravity of the solar system is at rest. There is no reference back to the discussion of absolute space, which came very early in the *Principia*.

23.4. THE MOTION OF THE PLANETS

Newton recognizes that the simple theory he gave in *Book I* and the resulting Kepler's laws give a very good description of the motion of the planets, and "*the mutual actions of the planets one upon another are so very small, that they may be neglected.*" (Kepler's third law holds at the tenth-of-a-percent level, for example. The period corrections given by equation [16.8] are at the one-twentieth-of-a-percent level for Jupiter.) However, he does go on to say that the case involving Jupiter and Saturn is significant; I will come to that in section 23.5 and to another important case in section 23.7.

Newton reminds us that the configuration of the orbits in space is basically fixed (see chapter 14 on rotating orbits), although again he suggests planetary interactions may cause the aphelions to move and he gives some estimates concluding that "*these motions are so inconsiderable, that we have neglected them in this Proposition [Proposition XIV].*" Of course, one of these "*inconsiderable*" motions for the planet Mercury played a key role in establishing Einstein's theory of general relativity (see section 14.4.2).

Newton then discusses how the orbital diameters and eccentricities may be found, along with the nature of the diurnal motions. Included in the latter is a discussion of the Moon and why we see only one face of it. Finally he discusses the shape of the planets themselves, a topic to be considered in the next chapter.

23.5. JUPITER AND SATURN

Since he has found their masses and knows their geometric configuration, Newton can evaluate the forces involved and feels compelled to state that "*it is true, that the action of Jupiter on Saturn is not to be neglected.*" He was also aware that long-term observations showed various deviations in their orbits away from constant Kepler ellipses. How to reconcile all of this with his theory of universal gravity must have worried Newton, as even in a manuscript predating the *Principia* he had written:

> But to consider simultaneously all these causes of motion and to define these motions by exact laws allowing of convenient calculation exceeds, unless I am mistaken, the force of the entire human intellect.[1]

In the discussion following *Proposition XIII*, Newton calculates that, at conjunction, the gravitational effect of Jupiter on Saturn is about 1/211 times that of the Sun, which is clearly not to be neglected. This maximum effect occurs every 19.86 years. The force of Saturn on Jupiter is comparatively less, but it is still important. The observations show that the variations for Saturn are indeed larger than those for Jupiter (see figure 23.1). It is clear that there are intricate mechanisms with variations revealing several periodic effects. This was known as the great inequality of Jupiter and Saturn. (Deviations from expected uniform motions were referred to as "inequalities.")

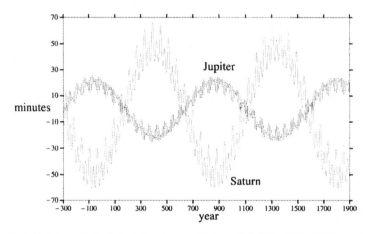

Figure 23.1. Major perturbations of Jupiter and Saturn, 251 BCE–1800 CE. The scale is minutes, and the graph records deviations from the mean positions. (From C. M. Linton, *From Eudoxus to Einstein: A History of Mathematical Astronomy* [Cambridge: Cambridge University Press, 2004], reprinted with the permission of Cambridge University Press.)

Newton has no way to tackle this problem, although he does suggest that an elliptical orbit for Saturn might have its focus at the center of gravity of Jupiter and the Sun. Showing how gravitational theory could account for these results became one of the great problems of celestial mechanics and thus attracted leading mathematicians. The problem was finally resolved in 1786 by Pierre-Simon Laplace, who saw it as a triumph for Newton's theory:

> *We shall see that this great law of nature represents all the celestial phenomena, down to their smallest details; there is not a single one of their inequalities that does not follow from it with admirable precision.*[2]

After correctly recognizing the problems for Jupiter and Saturn, Newton goes on: "*that perturbations of the other orbits are yet far less, except that the orbit of the Earth is sensibly disturbed by the Moon.*" However, that is a topic for chapter 25.

As an aside, I note that here, as throughout the *Principia*, when more than one force is involved, Newton simply takes a linear superposition of the component forces. Thus the theory of gravity is linear in the sources, making it simple to apply.

23.6. THE FIXED STARS

There is not too much about the stars in the *Principia*. They are mentioned in passing when discussing frames of reference (see section 5.5), and we shall discuss other mentions in chapters 25 and 26. So it is interesting to pick up on the two *Corollaries* to *Proposition XIV* (about the fixed aphelions and nodes of planetary orbits).

> COR. 1. *The fixed stars are immovable, seeing they keep the same position to the aphelions and nodes of the planets.*
>
> COR. 2. *And since these stars are liable to no sensible parallax from the annual motion of the Earth, they can have no force, because of their immense distance, to produce any sensible effect in our system. Not to mention that the fixed stars, every where promiscuously dispersed in the heavens, by their contrary attractions destroy their mutual actions, by Prop. LXX, Book I.*

Newton knew that the brightness of stars could change, and there were records of nova, but no motions had yet been detected. In fact, it was some time before star catalogues could be compared to check for changing positions, and it was not until 1837 that Bessel corrected Newton and reported parallax observations. He is able to conclude that the stars are at such "*immense distance*" that their gravitational effects on the solar system may be ignored.

Newton's final sentence is related to the stability of a system of fixed stars. He references *Proposition LXX* in *Book I*, which allows him to conclude that at a point inside a sphere surrounded by a spherically symmetric distribution of matter, there will be no gravitational effect due to that matter. A particle on the sphere will be influenced by the material inside the sphere, but not by that outside. It is fascinating to see that a similar argument can be made in relativistic cosmology where Newton's *Proposition* turns into Birkhoff's theorem (see the Weinberg reference listed in the "Further Reading" section at the end of this chapter).

23.7. SOLAR SYSTEM TRIUMPH: NEPTUNE

The six planets considered by Newton had been known since ancient times, and it came as a surprise when, in 1781, William Herschel discovered the seventh planet, Uranus. This triumph for observational astronomy provided a new test for Newton's theory of the solar system, and it proved to be a difficult task. The orbit of Uranus could not be obtained using Newton's theory, and there was speculation about the need to modify the force law at very large distances. Universal gravity and the simple inverse-square law were under attack.

Two people, Urbain Le Verrier in France and John Couch Adams in England, considered a different possibility: gravitational theory is correct, but there is an unknown eighth planet that is disturbing Uranus's orbit, just as Jupiter perturbs Saturn's orbit. In one of science's great stories, they separately calculated the possible orbit for such a planet, and on September 23, 1846, the German astronomer Johann Gottfried Galle used Le Verrier's predictions to locate and observe the new planet. It was eventually named Neptune. The story is fascinating because English astronomers failed to act on Adams's predictions, so the glory of finding Neptune passed to the French and the Germans. There was much argument about who did what, and the priority debate became quite bitter. (See this chapter's "Further Reading" section for the entertaining full story.)

This completed the main features of the solar system (as Pluto is no longer classed as a planet) and provided a most dramatic vindication of Newton's theory.

23.8. CONCLUSIONS

All the hard work of *Book I* has paid off; with a few *Propositions*, Newton has established the theory of universal gravity and demonstrated how it accounts for the solar system. As ever with Newton, there are some breathtaking steps reported in this chapter: the ancient argument about "what is the center of the world" has been settled, and the center of gravity is the appropriate answer; the seemingly impossible has been achieved with the calculation of the masses of Earth, Jupiter, and Saturn, relative to the mass of the Sun; challenges were set for future astronomers and mathematicians by the claim that all details of solar system motions, including the great inequalities of Jupiter and Saturn, could be explained by universal gravitation.

23.9. FURTHER READING

See the books recommended in section 6.9.1 and:

Blanco, V. M., and S. W. McCuskey. *Basic Physics of the Solar System.* Reading, MA: Addison-Wesley, 1961.

Cohen, I. Bernard. "Newton's Determination of the Masses and Densities of the Sun, Jupiter, Saturn, and the Earth." *Archive for the History of the Exact Sciences* 53 (1998): 83–95.

Linton, C. M. *From Eudoxus to Einstein: A History of Mathematical Astronomy.* Cambridge: Cambridge University Press, 2004.

Weinberg, S. *The First Three Minutes: A Modern View of the Origin of the Universe.* Glasgow, UK: Fontana/Collins, 1981. [see p. 45]

The writings of Curtis Wilson on the topics of this chapter are highly recommended:

Wilson, C. "The Great Inequality of Jupiter and Saturn: From Kepler to Laplace." *Archive for the History of the Exact Sciences* 33 (1985): 15–290.

———. "Newton and Celestial Mechanics." In *The Cambridge Companion to Newton*. Edited by I. Bernard Cohen and George E. Smith. Cambridge: Cambridge University Press, 2002.

———. "The Newtonian Achievement in Astronomy." In *The General History of Astronomy: Volume 2, Planetary Astronomy from the Renaissance to the Rise of Astrophysics—Part A: Tycho Brahe to Newton*. Edited by R. Taton and C. Wilson. Cambridge: Cambridge University Press, 1989.

On Neptune, see Linton (cited above), specifically chapter 10, and the following:

Grosser, Morton. *The Discovery of Neptune*. Cambridge, MA: Harvard University Press, 1962.

Hanson, N. R. "Le Verrier: The Zenith and Nadir of Newtonian Mechanics." *Isis* 53 (1962): 359–78.

Smith, Robert W. "The Cambridge Network in Action: The Discovery of Neptune." *Isis* 80 (1989): 395–422.

Standage, T. *The Neptune File*. London: Allen Lane, Penguin Press, 2000.

EARTHLY PHENOMENA

Newton next turns to problems directly concerning the Earth and motion on it. The Earth is spinning around a north–south axis. Now he can add gravitational theory to that in order to tackle four interconnected problems: What is the shape of the Earth? How does gravity vary over the surface of the Earth? What mechanism explains the tides? Why does the direction of the spin axis vary with time? These are old questions, but Newton has new methods in mechanics, especially the concept of force, to employ in devising new answers.

As ever, Newton shows us his unrivaled inventiveness in finding physical and mathematical models for attacking these problems. However, it becomes clear that he is also at the limit of what can be achieved with his methods and, although he does produce solutions, he is again defining the problems and setting out the challenges for the next generation of researchers. So this chapter is more about how the *Principia* introduces conceptual matters than how it develops exact theories.

24.1. THE SHAPE OF THE EARTH

For many years, both before and after Newton, the question of the geometry of the surface of the Earth was a topic of great interest with importance for surveying and gravitation. (The paper by Chapin, listed in the "Further Reading" section at the end of this chapter, gives a good, short introduction to the history of the topic.) The fundamental questions are: What is the shape of the Earth? and, Why is it that shape? Neither question was definitively answered when Newton wrote the *Principia*, in which he considers them for planets in general, before concentrating on the Earth. He demonstrates his understanding in the first *Proposition* on this topic and in his discussion:

PROPOSITION XVIII.

That the axes of the planets [the lines about which they rotate] *are less than the diameters drawn perpendicular to the axes.*

The equal gravity of the parts on all sides would give a spherical figure to the planets, if it was not for their diurnal revolution in a circle. By that circular motion it comes to pass that the parts receding from the axis endeavour to ascend about the equator; and therefore if the matter is in a fluid state, by its ascent towards the equator it will enlarge the diameters there, and by its descent towards the poles it will shorten the axis. So the diameter of Jupiter (by the concurring observations of astronomers [Flamsteed and Cassini]*) is found shorter betwixt pole and pole than from east to west. And, by the same argument, if our Earth was not higher about the equator than at the poles, the seas would subside about the poles, and, rising towards the equator, would lay all things there under water.*

In short, Newton has explained how simple ideas about dynamics suggest that the planets should flatten at the poles while bulging at the equator. This will be the mechanism for a molten, cooling planet to assume its shape. Newton considers how to quantify these physical effects in *Proposition XIX. Problem III*: "*To Find the Proportion of the Axis of a Planet to the Diameters Perpendicular Thereto.*"

In modern terms, we are looking for the "Shape of a Self-Gravitating Fluid" (a chapter title in Grossman's book), and even assuming simple conditions about density and so on, this is a technically difficult problem. Of course, Newton had none of the mathematical formalism required, but in typical form, he came up with a most ingenious device to tackle the problem. With reference to figure 24.1, Newton introduces a sort of thought experiment:

if APBQ represent the figure of the Earth, now no longer spherical, but generated by the rotation of an ellipse about its lesser axis PQ; and ACQqca a canal full of water, reaching from the pole Qq to the centre Cc, and thence rising to the equator Aa; the weight of water in the leg of the canal ACca will be to the weight of water in the other leg QCcq.

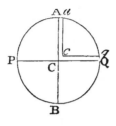

Figure 24.1. Newton's diagram to explain his argument for *Proposition XIX*. (From *The Principia*, trans. Andrew Motte [Amherst, NY: Prometheus Books, 1995].)

Newton gives a long and detailed argument involving centrifugal and gravitational effects in the composite spherical and nonspherical bodies that he discussed in *Sections XII* and *XII* in *Book I* (see chapter 17). He has to make various assumptions, about the density of the Earth, for example, and the interested reader should consult the resource by Chandrasekhar and others listed in the "Further Reading" section at the end of this chapter for details. Newton's conclusion was that:

the diameter of the Earth at the equator $[D_E]$ *is to its diameter from pole to pole* $[D_p]$ *as 230 to 229.*

Thus the polar flattening or oblateness, $(D_E - D_p)/D_E$, is 1/230. The modern value is 1/298, so Newton slightly overestimated the nonspherical effect. He also used that result to give dimensions for the Earth:

And since the mean semi-diameter of the Earth, according to Picart's [Picard's] mensuration, is 19615800 Paris feet, or 3923.16 miles (reckoning 5000 feet to a mile) the Earth will be higher at the equator than at the poles by 85472 feet or 17 1/10 miles.

(Modern values for the mean radius and bulge at the equator are 6,371.02 and 22.4 kilometers, respectively.) It is clear that Newton understood the underlying physics so that he correctly predicted the shape of the Earth and found a reasonably good value for the degree of flattening at the poles or bulging at the equator. He also solved the problem of why centrifugal effects did not lead to bodies flying off the moving Earth (although he did not say as much); he makes the comparison for Paris where "*the total force of gravity in that latitude will be to the centrifugal force of bodies as 289 to 1.*" The overwhelming force of gravity keeps objects on the Earth's surface.

There was considerable debate about the shape of the Earth, and, in fact, the French scientist Jacques Cassini (son of the astronomer) used measurements to claim that the flattening was at the equator and the bulge was at the poles. It required detailed, accurate, large-scale surveying to find the correct result; the fascinating story can be found in the books listed in this chapter's "Further Reading" section. The heroic work in Lapland by Pierre Louis Maupertuis reads like an adventure story. There was also a similarly exciting expedition to Peru. Finally, Newton was found to have correctly solved the puzzle of the shape of the Earth, although the details were studied and debated intensely by a string of mathematicians, including the great Pierre-Simon Laplace.

Newton gave values for the oblateness of Jupiter and the Moon (in *Proposition XXXVIII*), which are very roughly correct.

24.2. THE VARIATION OF GRAVITY OVER THE EARTH

If the Earth is not spherical, the force of gravity experienced on its surface must vary from place to place. Those effects might be calculated (for example, using the theory reviewed in chapter 17), but that is extremely difficult and relies on assumptions about the Earth's constitution. The alternative is experimental measurement, and Newton suggested the effects he predicted "*will appear by the experiments of pendulums related under the following Proposition.*"

The pendulum had long been a vital experimental tool, perhaps most famously linked to Galileo. Around Newton's time, the quest was for a "seconds" pendulum, one that would swing back and forth in exactly one second. Huygens discussed the matter at length in his *Horologium Oscillatorium*, and Riccioli's heroic experiments involved counting over 20,000 swings of a pendulum. Newton himself carried out pendulum experiments early in his career at Cambridge (see the 1986 paper by Westfall listed in the "Further Reading" section at the end of this chapter). This was the background in which Newton came to *Proposition XX*: "*To Find and Compare Together the Weights of Bodies in the Different Regions of Our Earth.*"

He uses his canals argument to evaluate the changes expected over the Earth, so, as before, both gravitational-force effects and centrifugal-force effects are included. He is finding the effective gravitation and gives the results that "*the increase in weight in passing from the equator to the poles is . . . as the square of the right sine of the latitude.*" (See section 24.2.1 for results in modern form.) The theory gives numerical results, which Newton summarized in a table (see figure 24.2). (Note that there are twelve inches in a foot, and a line is one-twelfth of an inch.)

Latitude of the place.	Length of the pendulum		Measure of one degree in the meridian.
Deg.	Feet.	Lines.	Toises.
0	3 .	7,468	56637
5	3 .	7,482	56642
10	3 .	7,526	56659
15	3 .	7,596	56687
20	3 .	7,692	56724
25	3 .	7,812	56769
30	3 .	7,948	56823
35	3 .	8,099	56882
40	3 .	8,261	56945
1	3 .	8,294	56958
2	3 .	8,327	56971
3	3 .	8,361	56984
4	3 .	8,394	56997
45	3 .	8,428	57010
6	3 .	8,461	57022
7	3 .	8,494	57035
8	3 .	8,528	57048
9	3 .	8,561	57061
50	3 .	8,594	57074
55	3 .	8,756	57137
60	3 .	8,907	57196
65	3 .	9,044	57250
70	3 .	9,162	57295
75	3 .	9,258	57332
80	3 .	9,329	57360
85	3 .	9,372	57377
90	3 .	9,387	57382

Figure 24.2. Newton's table showing the length of a "seconds" pendulum and the distance related to one degree at a range of latitudes. (From *The Principia*, trans. Andrew Motte [Amherst, NY: Prometheus Books, 1995].)

Newton reviews the evidence for "seconds" pendulum variations, beginning with the report by Jean Richter of his 1672 visit to Cayenne to carry out astronomical observations. Richter's need to fix times led to speculations about the length of a "seconds" pendulum. Newton then covers the reports by Halley (1677), Varin and des Hayes (1682), Couplet (1697), des Hayes (1699 and 1700), and Feuillé (1704). The evidence supports Newton's conclusions, but the observed pendulum lengths "*are a little greater than by the table.*" He goes on to wonder about the constitution of the Earth and the effects of moving to different climates. Again we find Newton the thorough experimentalist as he analyses the effects of temperature, concluding that "*the total differences of the lengths of isochronal pendulums in different climates cannot be ascribed to the difference of heat.*" He goes on: "*nor indeed to the mistakes of the French astronomers. For although there is not perfect agreement betwixt their observations, yet the errors are so small that they may be neglected; and in this they all agree, that isochronal pendulums are shorter under the equator that at the Royal Observatory of Paris.*" He praises Richter for his careful work and states that his findings do support the idea that "*the Earth is higher under the equator than at the poles, and that by an excess of about 17 miles; as appeared above by the theory.*"

This work reminds us that Newton continually seeks to use data to validate his theories and support them with experimental programs. The *Principia* is not just a mathematical treatise, it is a work of science in all its approaches.

24.2.1. Some Modern Details

The modern calculation combines gravitational attraction and centrifugal effects, as shown in figure 24.3. (See the books by Cook, Kibble, and Tsuboi, listed in the "Further Reading" section at the end of this chapter.) The angle deviation between the attraction and gravity vectors is very small (maximum 0°6′) and the variations with latitude ϕ are given by

$$g = 978.03185 \{1 + 0.0053024 \sin^2(\phi) - 0.0000059 \sin^2(2\phi)\}. \quad (24.1)$$

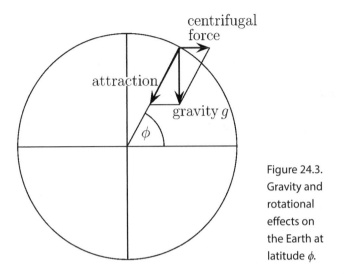

Figure 24.3. Gravity and rotational effects on the Earth at latitude ϕ.

Using equation (24.1) to calculate how Newton's table-pendulum length should increase from equator to pole gives 3 feet 9.797 lines, whereas he gives 3 feet 9.387 lines. Note that the main dependence on latitude ϕ is as Newton predicted.

The shape of the Earth and the resulting nonspherical gravitational field around the Earth gained importance when satellite orbits began to be planned.

24.3. THE TIDES

The tides are a fascinating and complex part of the behavior of Earth's oceans, seas, and rivers. They are vitally important for the natural world and for man's commercial and recreational enterprises. Interest in the tides and their origin dates back a very long time (see the book by Cartwright listed in this chapter's "Further Reading" section). A number of prominent people speculated on the origin of tides. William Gilbert sought to use his magnetic theory to account for interactions of the Moon with the Earth and its seas. Kepler too thought there was some sort of responsible force. Descartes tried to use aspects of his vortex theory to account for the tides.

Galileo took a different tack; he was keen to find evidence for the motion of the Earth, and he claimed that motion generated the tides. He was very critical of Kepler:

> I am more surprised about Kepler than anyone else; . . . he leant his ear and gave his assent to the dominion of the Moon over water, to occult properties, and to similar childish ideas.[1]

That is an interesting quote because it takes us back to the charge that forces are "occult" and, without suitable mechanisms (like Descartes's vortices), they have no part in science. We have seen Newton aware of this point from his *Preface* onward, and it will come up again in his final statements to be reviewed in chapter 26.

In the *Principia*, Newton gives the basic concepts for explaining the tides as we know them today. In *Proposition LXVI* in *Book I*, he used one of his special tricks for going from discrete bodies to fluids, so that *Corollaries 18* to *20* give the theory ready for application in *Book III*. We find:

PROPOSITION XXIV

That the flux and reflux of the sea arise from the actions of the Sun and the Moon.

By Cor. 19 and 20, Prop. LXVI, Book I, it appears that the waters of the sea ought twice to rise and twice to fall every day, as well lunar as solar.

The forces exerted by the Sun and the Moon move the oceans, and by adding their forces as they act on the water and by considering their relative positions, we can understand the variation in the tides.

The two luminaries excite two motions, which will not appear distinctly, but between them will arise one mixed motion compounded out of both. In the conjunction or opposition of the luminaries their forces will be conjoined, and bring on the greatest flood and ebb. In the quadratures the Sun will raise the waters which the Moon depresses, and depress the waters which the Moon raises, and from the difference of their forces the smallest of all tides will follow.

Newton has explained the spring and neap tides, as shown sche-
matically in figure 24.4.

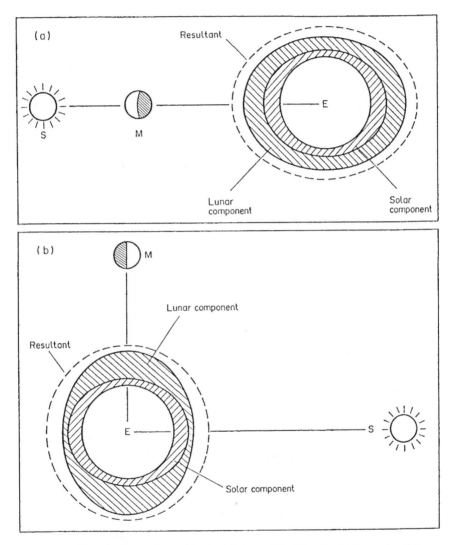

Figure 24.4. How the forces exerted by the Moon and the Sun give (a) the spring tides
and (b) the neap tides. (Reprinted from *Endeavour* 10/4, J. V. Jelley, "The Tides, Their
Origins and Behavior," 184–90, © 1986, with permission of Elsevier.)

Of course, the whole story is made extremely complex when details of the Earth's topography are factored in and the astronomical details are considered. As an example, Newton notes:

> But the effects of the luminaries depend upon their distances from the Earth . . . because the Sun is less distant from the Earth in winter than in the summer, it comes to pass that the greatest and least tides more frequently appear before than after the vernal equinox, and more frequently after than before the autumnal.

There are long discussions about various tides in the *Principia*, but perhaps most important is Newton's evaluation of the relative values of the forces involved in:

PROPOSITION XXXVI

To find the force of the Sun to move the sea,

and

PROPOSITION XXXVII

To find the force of the Moon to move the sea.

Newton calculates that the force of the Sun to move the sea on the Earth's surface is *"to the force of gravity as 1 to 38604600."* According to Cartwright, this is a good result when compared with the modern value of 1 to 39231000. Using tidal data, Newton manages to estimate that the force contributed by the Moon is 4.4815 times that of the Sun. The calculations are not too believable, and the modern value is 2.12. Still, Newton has identified the fact that both the Moon and the Sun must be included when calculating tidal movements. (The Sun is much farther from the Earth than the Moon, but it compensates by being much more massive. However, we must also take into account the results of modern theory, which tell us that it is the variation or gradient of the force over the Earth's surface that is important.)

Newton theories may be wrong in details (hardly surprising, given the great complexity of the tidal phenomena), but he does correctly outline the essential mechanism: the interplay of lunar and solar gravitational forces, along with the rotation of the Earth, explains why tides occur and lets us understand a major part of their daily and seasonal variations. People like Daniel Bernoulli and Laplace gradually refined Newton's theory and developed a theoretical formalism taking us through to present-day computer modeling (see this chapter's "Further Reading" section).

24.4. PRECESSION OF THE EQUINOXES

This is one of the highlights of the *Principia*. To appreciate this topic, we need to combine two facts about the Earth's motion. First, the Earth's orbit lies in a plane, the ecliptic plane, which also contains the Sun. Second, the Earth spins around a south–north axis that makes an angle of 23½° with the normal to the ecliptic plane, so the Earth's equator makes that same angle of 23½° with that plane (see figure 24.5). This tilt of the Earth's axis of rotation controls how parts of the Earth are angled toward or away from the Sun, and that determines how effective the Sun is in those areas. For a given part of the Earth, that effectiveness varies over the year, and, hence, we get the seasons. (The seasons are due to the tilt of the Earth's axis, not the variation in the distance from the Sun.)

At two points on the Earth's orbit around the Sun, the equinoctial points, the tilt is such that the hours of daylight are the same as the hours of nighttime. We get the spring (or vernal) equinox and the autumnal equinox, as shown in figure 24.5. It had been known from ancient times (by the Babylonians, for example—see the Cajori paper in the "Further Reading" section at the end of this chapter) that the equinoxes do not remain fixed on the Earth's orbit, but that they slowly move. Thus the "precession of the equinoxes" was an ancient puzzle. There was seemingly no way to solve the puzzle until Newton came along. The magnitude of the result was aptly summed up by the Astronomer Royal, Sir George Airy (1801–1892):

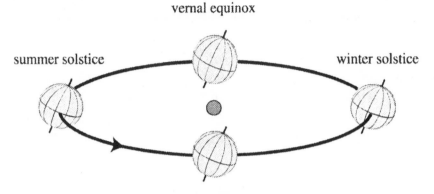

vernal equinox

summer solstice

winter solstice

autumnal equinox

Figure 24.5. The tilt of the Earth's axis as it orbits the Sun. Nomenclature is for the Northern Hemisphere. (From C. M. Linton, *From Eudoxus to Einstein: A History of Mathematical Astronomy* [Cambridge: Cambridge University Press, 2004], reprinted with the permission of Cambridge University Press.)

if at this time we might presume to select the part of the Principia *which probably astonished and delighted and satisfied its readers more than any other, we should fix, without hesitation, on the explanation of the precession of the equinoxes.*[2]

24.4.1. Newton's Solution

Newton knew that the Earth has a bulge at the equator. He reasoned that, as the Sun and the Moon do not lie in the plane containing the Earth's equator, they act differently at different parts of the bulge and so cause it to precess. We may make the analogy with a spinning top where we readily observe that the axis of a "sleeping" top slowly rotates about the vertical (see figure 24.6). (Many books on astronomy give relevant diagrams; for example, see figure 2.24 in the book by Baker and Frederick, or see figure 2.17 in the book by Freedman and Kaufmann, both of which are listed in the "Further Reading" section at the end of this chapter.) This explains why the tilt of the Earth's axis changes direction and, hence, why the equinoxes move. The full rotation takes about twenty-six thousand years, so the movement is about 50″ of arc per year. Projecting the Earth's axis onto the celestial sphere shows that Polaris may be taken today as the North Star, but not into the future (see figures in the cited astronomy books).

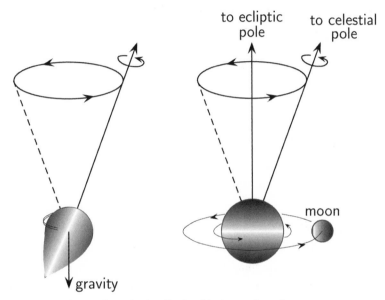

Figure 24.6. Comparing the spinning Earth with the motion of a top.

Newton's method to obtain a quantitative result is ingenious but complex. In *Proposition LXVI* of *Book I* on three-body interactions, he gives *Corollaries 18* through *21*, considering how a particle, a fluid system of particles, and then a solid ring have precessing nodal lines. Then in *Book III*, after three *Lemmas* and *Hypothesis II*, he comes to *Proposition XXXIX:* "*To Find the Precession of the Equinoxes.*"

Thus the bulge around the Earth's equator is modeled as a ring (and the hypothesis suggests that whether it is taken as fluid or rigid is not important), and the effect of the Sun and the Moon on that ring is found using the results from *Book I*. He also discusses the question of what happens when the ring is free or attached to the Earth. The whole business is most complicated (see this chapter's "Further Reading" section for details), but considering both solar and lunar effects, he finds *"the total annual precession from the united forces of both will be 50" 00'" 12^{iv} the quantity of which motion agrees with the phenomena; for the precession of the equinoxes, by astronomical observations, is about 50" yearly."*

Given the many assumptions involved and the accuracy of some of the data he uses, this result is just too good to be true. Newton was clearly fudging the numbers; these rather demeaning actions are discussed in detail by Westfall.

24.4.2. Later Calculations

In modern terms, we say that the problem concerns the rotation of a rigid body, the calculation of angular momentum using moments of inertia, and the action of forces in providing torques to change that angular momentum. There is a whole set of concepts that were unknown to Newton. He had considered the collision of rigid bodies (see references to Herivel, listed in the "Further Reading" section at the end of this chapter), and we saw in chapter 8 how the equal-areas-swept-out-in-equal-times law is equivalent to the conservation of angular momentum, but Newton had no consistent development of these aspects of dynamics. Instead he used his boundless ingenuity to come up with ad hoc methods to estimate what happens in complex rotational situations.

The correct theory for the precession of the equinoxes had to wait for d'Alembert, Euler, and Laplace (see references discussing these men in the "Further Reading" section at the end of this chapter and note the book by Blanco and McCuskey as an introduction to modern calculations). However, it did grow out of Newton's ideas, and I can summarize no better than by quoting that great Newtonian scholar I. Bernard Cohen:

> In retrospect, Newton's explanation of the precession of the equinoxes in relation to the shape of the Earth may be reckoned as one of the greatest intellectual achievements of the *Principia*. At one stroke, Newton gave compelling evidence of the power and validity of the law of universal gravity, he provided supporting evidence for the shape of the Earth as an oblate spheroid, and he proposed a simple reason founded in the prin-

ciples of dynamics for a phenomenon known since the second century BC but never before reduced to its physical cause. Before the *Principia*, there had never been even a suggestion of a physical cause.[3]

24.5. CONCLUSION

Those readers who thought of the *Principia* in terms of Newton's laws of motion, gravity, and a theory of the solar system were in for a great surprise when reading this chapter. Newton shows how the local gravitational force combines with the gravitational forces exerted by the distant Sun and Moon to let us understand terrestrial phenomena. Universal gravity explains the shape of the Earth, the tides, and the precession of the equinoxes. Conceptually, Newton is successful, but technically, his theory is deficient. Nevertheless, he still maintains his rule that theory must be tested against observational data; although rather shamefully some of his outstanding matches seem to rely on some tricky data selection and manipulation.

24.6. FURTHER READING

Chandrasekhar, S. *Newton's* Principia *for the Common Reader*. Oxford: Clarendon Press, 1995. [see ch. 20, 21, and 23]

Cook, A. H. *Gravity and the Earth*. London: Wykeham Publications, 1969.

Gamow, George. *Gravity: Classic and Modern Views*. London: Heinemann, 1962. [Gamow's useful and entertaining "popular" account of the subject]

Grossman, N. *The Sheer Joy of Celestial Mechanics*. Boston: Birkhauser, 1996.

Kibble, T. W. B. *Classical Mechanics*. New York: McGraw-Hill, 1966.

Linton, C. M. *From Eudoxus to Einstein: A History of Mathematical Astronomy*. Cambridge: Cambridge University Press, 2004.

Tsuboi, C. *Gravity*. London: George Allen and Unwin, 1983.

Wilson, C. "The Newtonian Achievement in Astronomy." In *The General History of Astronomy: Volume 2, Planetary Astronomy from the Renaissance to the Rise of Astrophysics—Part A: Tycho Brahe to Newton*. Edited by R. Taton and C. Wilson. Cambridge: Cambridge University Press, 1989.

On the shape of the Earth:

Chapin, S. L. "The Shape of the Earth." In *The General History of Astronomy: Volume 2, Planetary Astronomy from the Renaissance to the Rise of Astrophysics—Part B: The Eighteenth and Nineteenth Centuries*. Edited by R. Taton and C. Wilson Cambridge: Cambridge University Press, 1989.

Greenberg, J. L. "Isaac Newton and the Problem of the Earth's Shape." *Archive for the History of Exact Sciences* 49 (1995): 371–91.

———. *The Problem of the Earth's Shape from Newton to Clairaut*. Cambridge: Cambridge University Press, 1995. [this is a major 780 page study and includes twenty-three pages of biography]

Mignard, F. "The Theory of the Figure of the Earth According to Newton and Huygens." *Vistas in Astronomy* 30 (1987): 291–311.

Terrall, M. *The Man Who Flattened the Earth*. Chicago: University of Chicago Press, 2002.

Pendulum references:

Graney, C. M. "Anatomy of a Fall: Giovanni Battista Riccioli and the Story of g." *Physics Today* 65 (September 2012): 36–40.

Huygens, Christiaan. *The Pendulum Clock; or, Demonstrations concerning the Motion of Pendula as Applied to Clocks*. Translated from the original 1673 *Horologium Oscillatorium* by Richard J. Blackwell. Ames: Iowa State University Press, 1986.

Koyré, Alexandre. "An Experiment in Measurement." *Proceedings of the American Philosophical Society* 97 (1953): 222–37. Reprinted in *Metaphysics and Measurement*. London: Chapman and Hall, 1968.

Matthews, M. R., C. F. Gauld, and A. Stinner, eds. *The Pendulum: Scientific, Historical, Philosophical and Educational Perspectives.* New York: Springer, 2005.

Olmstead, J. W. "The Scientific Expedition of Jean Richter to Cayenne (1672–3)." *Isis* 34 (1942): 11–14.

Olmstead, J. W. "The Voyage of Jean Richter to Acadia in 1670." *Proceedings of the American Philosophical Society* 104 (1960): 612–34.

Westfall, R. S. "Newton and the Acceleration of Gravity." *Archive for the History of Exact Sciences* 35 (1986): 255–72.

For more details on tides:

Aiton, E. J. "The Contribution of Newton, Bernoulli and Euler to the Theory of Tides." *Annals of Science* 11 (1955): 206–23.

Barger, V. D., and M. G. Olsson. *Classical Mechanics: A Modern Perspective.* 2nd ed. New York: McGraw-Hill, 1995. [section 8.2 is a good example of mechanics-textbook presentation]

Burstyn, H. L. "Galileo's Attempt to Prove That the Earth Moves." *Isis* 53 (1962): 161–85.

Cartwright, D. A. *Tides: A Scientific History.* Cambridge: Cambridge University Press, 1999.

Ekman, M. "A Concise History of the Theory of Tides, Precession-Nutation and Polar Motion (from Antiquity to 1950)." *Surveys in Geophysics* 14 (1993): 585–617.

Jelley, J. V. "The Tides, Their Origins and Behavior." *Endeavour* 10 (1986): 184–90.

References for the precession of the equinoxes:

Baker, R. H., and L. W. Fredrick. *Astronomy.* 9th edition. New York: Van Nostrand, 1971.

Blanco, V. M., and S. W. McCuskey. *Basic Physics of the Solar System* Reading, MA: Addison-Wesley, 1961. [see section 5.11 for details of modern calculations]

Cajori, F. "Babylonian Discovery of the Precession of the Equinoxes." *Science* LXV (1927): 184.

Dobson, G. J. "Newton's Problems with Rigid Body Dynamics in the Light of His Treatment of the Precession of the Equinoxes." *Archive for the History of Exact Sciences* 53 (1998): 125–45.

———. "On Lemmas 1 and 2 to Proposition 39 of Book 3 of Newton's *Principia*." *Archive for the History of Exact Sciences* 55 (2001): 345–63.

Freedman, R. A., and W. J. Kaufmann III. *Universe.* 7th edition. New York: W. H. Freeman, 2005.

Goldstein, Herbert, C. Poole, and J. Safko. *Classical Mechanics.* 3rd ed. 1950. Reprint, New York: Addison-Wesley, 2002. [see section 5.8]

Herivel, J. W. "Newton on Rotating Bodies." *Isis* 53 (1962): 212–18.

———. *The Background to Newton's* Principia. Oxford: Clarendon Press, 1965.

Westfall, R. S. "Newton and the Fudge Factor." *Science* 179 (1973): 751–58.

Wilson, C. "D'Alembert versus Euler on the Precession of the Equinoxes and the Mechanics of Rigid Bodies." *Archive for the History of Exact Sciences* 37 (1987): 233–73.

———. "The Precession of the Equinoxes from Newton to D'Alembert and Euler." In *The General History of Astronomy: Volume 2, Planetary Astronomy from the Renaissance to the Rise of Astrophysics—Part B: The Eighteenth and Nineteenth Centuries.* Edited by R. Taton and C. Wilson Cambridge: Cambridge University Press, 1989.

CHALLENGES
THE MOON AND COMETS

This chapter describes Newton's treatment of two remaining problems: the motion of the Moon and the orbits of comets. The significance of these problems may be judged by the fact that together they occupy over 60 percent of *Book III*. These topics have an enduring popular appeal; even today we are fascinated by the Moon, and the appearance of a comet in our skies becomes a major news item. They are also vitally important in the context of establishing the validity and efficacy of the theory of universal gravitation. Can the theory account for the complexity of the motion of the Earth-Moon-Sun system? Will the theory work when it is applied to the vast distances comets travel in the solar system?

25.1. THE MOON

The light reflected from the Moon is a feature of our lives, and we cannot fail to notice the regular changing of the phases of the Moon. However, the Moon does not exhibit the same order of regularity as the Sun; few people would immediately know where to look for it in the night sky, and it will be almost nineteen years before it rises again at the same northernmost point. The movement of the Moon has been recorded for over three thousand years, and eclipse records are very detailed. In the early eighteenth century, it was hoped all this knowledge could lead to a method for determining longitude. There had been naval disasters because sailors failed to judge their position correctly, and the British Longitude

Act of 1714 offered a large reward to anyone who could come up with a suitable, accurate method for finding longitude. Newton, among many others, knew of this challenge and the importance of formulas and tables for giving the position of the Moon.

25.1.1. The Moon's Orbit

The simplest guess would be that the Moon follows an elliptical orbit around the Earth, with the Earth at one focus, and the joint Earth-Moon system follows a far larger elliptical orbit around the Sun. However, the influence of the Sun causes many deviations away from such an ideal, and early astronomers identified a variety of "inequalities" and "variations."

The Moon's orbit around the Earth is roughly elliptical, but the plane of that orbit is not in the plane containing the Sun and the Earth's orbit, the ecliptic plane (see figure 25.1). The angle between the planes is about 5°, but here comes the first variation: that angle oscillates between 4° 58.5′ and 5° 17.5′. The points where the Moon's orbit cuts the ecliptic plane are called the nodes, and the line joining them is called the nodal line (see figure 25.1). That line and its direction compared with the Earth and the Sun is important for determining eclipses. However, the nodes are not constant and, as another inequality, the nodal line rotates with a period of 18.61 years.

At its perigee, the Moon is nearest the Earth; at its apogee, it is farthest away; the line joining those two points on the orbit is called the line of apsides, and it too rotates, taking about 8.85 years to complete one full rotation.

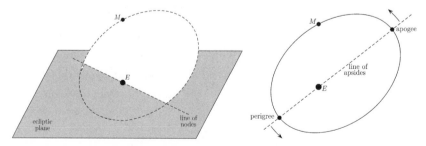

Figure 25.1. The geometry of the Moon's orbit.

Those two rotations are the most famous inequalities, but there are many other—often small but subtle—variations. For example, the speed of the Moon has minor unexpected variations, and Tycho Brahe and Kepler discovered that it speeds up a little in summer and slows down in winter (that is, according to the seasons in the Northern Hemisphere). The fine details of the Moon's orbit were gradually discovered and recorded by astronomers over the centuries leading up to Newton's time.

25.1.2. Implications of Orbital Complexity

The complexity of the Moon's motion meant that even constructing formulas and tables for predicting its position was a difficult business. Good tables were developed, but they were not accurate enough to deal with the longitude problem.

It is clear that the dynamics of this three-body system (Earth, Moon, and Sun) is complicated. It is interesting to compare the two-body case, where Newton had classified the possible motions for inverse-square-law gravitational forces as conic sections; if orbits are bound, they can take the form only of an ellipse (see chapter 11 and section 16.2). The fundamental properties of motion that lead to those beautiful results do not extend far enough to dictate similar simplifications for the gravitational interaction of three bodies. Exact solutions are very rare (see section 16.5), and the predicted motion may be intricate and variable. Thus the complexities of the Earth-Moon-Sun system should come as no surprise. However, it was not a pleasant surprise for Newton. His mathematical battles (not to mention his battles with the first Astronomer Royal, John Flamsteed, over the availability of his accurate lunar observations) caused him great trouble. Allegedly he told mathematician John Machin that "*[my] head never ached but with [my] studies of the Moon*," and he told Halley that the theory of the Moon "*has broke my rest so often I will think of it no more.*"[1] Even that master mathematician Leonard Euler (1707–1783) struggled:

> As often as I have tried these forty years to derive the theory and motion
> of the Moon from the principles of gravitation, there always arose so

many difficulties that I am compelled to break off my work and latest researches.[2]

25.1.3. Newton's Approach

Newton continues the approach he adopted earlier in *Book III*: he has gone "*from the phenomena of motions to investigate the forces of nature*"; now he can use all the mathematical formalism developed in *Book I* to use "*these forces to demonstrate the other phenomena*," in this case, the motion of the Moon. Thus we come to PROPOSITION XXII: "*That All the Motions of the Moon, and All the Inequalities of Those Motions, Follow from the Principles Which We Have Laid Down.*"

This is a bold claim viewed in the light of the many inequalities and complexities involved. Notice that Newton is now moving the science of the Moon's motion in a new direction: he is not asking so much about how to precisely describe the Moon's motion; he is saying that we know **why** the Moon follows such a complicated motion. He is showing an enormous faith in his laws of dynamics and the simple form taken by the theory of universal gravity, even when faced by the vagaries of the Moon. About a hundred years later, Pierre-Simon Laplace explicitly and beautifully expressed this very point:

> Infinitely varied in its effects, nature is simple only in its causes, and its economy consists in producing a great number of phenomena, often very complicated, by means of a small number of general laws.[3]

Newton's discussion of *Proposition XXII* takes the form of a list of the Moon's inequalities and a reference back to the appropriate *Corollary* of *Proposition LXVI, Book I*, which is to be used to account for that inequality. He also suggests that "*there are yet other inequalities not observed by former astronomers*," again giving the form of the inequality and a reference to the responsible *Corollary*. Thus Newton could claim to answer the "why" part of the question about the Moon's motion.

Proposition LXVI, Book I is Newton's perturbation theory for the three-body system. He had previously introduced the idea of perturbing influences when discussing *Proposition XVII; Corollaries 3* and *4* mention the effect on a body moving in a given orbit of an impulse and a continuous force (see section 10.4.1). In *Section IX of Book I*, he considered how changes to an inverse-square-law force could produce rotating orbits, and there he used perturbations around a circular orbit to develop his theory (see chapter 14). Newton was well aware of the unavoidable need for approximations:

> In this demonstration I have supposed the angle BEG, representing the double distance of the nodes from the quadratures, increaseth uniformly; for I cannot descend to every minute circumstance of inequality. (From the discussion of Proposition XXXV.)

An aside for readers not familiar with perturbation theory: when faced with a complex problem, one strategy is to assume that the solution is close to that of a similar but tractable problem. The difference in the two problems is then the perturbation, and methods are developed to assess the effects of that difference. That may take the form of an adjustment of the parameters in the simpler solution or the addition of correction terms to it. Often the theory yields a series of corrections of (hopefully) diminishing importance, and it is expected that the first-order terms will give an accurate enough answer. For example, suppose we want the square root of a number nearly equal to 1, for which we obviously know the answer is again 1. Let the number in question be $1 + x$, so now x is the perturbation on the problem. We can find the formula (the Taylor series)

$$\sqrt{1 + x} = 1 + \frac{x}{2} - \frac{x^2}{8} + \frac{x^3}{16} - \text{higher-order terms in } x^4, x^5, \text{ and so on.}$$

If x is small enough, just keeping the first two terms ($1 + \frac{1}{2}x$) will often give an accurate enough answer. Sometimes more terms must be included. If x gets too large, many terms may be needed to get the required accuracy; and if it becomes larger than 1, no sensible answer will be found at all. Such are the perils of perturbation theory.

Newton has explained why we should expect various inequalities

when we study the Moon's orbit, but of course the validity of his explanations can be assessed only by comparing the numerical predictions with
the actual data. Newton must follow his by-now standard approach.

Before looking at the results, I should comment on the extent of
Newton's work on the Moon. Recall that, for this book, I take the third
edition of the *Principia* as Newton's final word and discuss that edition, only
occasionally mentioning earlier editions or other of his writings. Newton
made extensive researches into the motion of the Moon, and there are
major variations between the first and third editions of the *Principia*. He
also wrote a separate *Theory of the Moon's Motion*, first published in 1702 as
part of a book by David Gregory. To add further to the mix, there are documents in the "Portsmouth Papers" showing that he had produced other
and quite successful calculations. I do not know why Newton chose not
to include some of this in the third edition of the *Principia*—perhaps he
was just too old, too tired, or too depressed by the whole subject. This is
a fascinating part of the Newton story, and I refer you to the references in
the "Further Reading" section at the end of this chapter, especially to the
work of Cohen and Smith, Nauenberg, and Wilson.

25.1.4. Results

To apply his methods, Newton must first determine the perturbing forces,
and that he does in *Proposition XXV*: "*To Find the Forces with Which the Sun
Disturbs the Motion of the Moon.*"

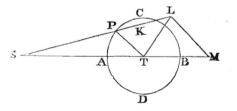

Figure 25.2. Newton's
diagram for *Proposition XXV*.
(From *The* Principia, trans.
Andrew Motte [Amherst, NY:
Prometheus Books, 1995].)

In his diagram, figure 25.2, Newton makes a construction so that "SL
will represent the accelerated force of gravity of the Moon towards the Sun."

He can now break down that force into components using his theory of force resolution to find which parts of it keep the Earth-Moon system in orbit around the Sun and which parts may be taken as perturbing forces for the Moon's orbit around the Earth. The force *SL* of the Sun on the Moon resolves into the components *SM* and *ML*. The force *SM* is broken into the components *ST*, which is the force toward the Earth and which keeps the center of mass motion as it should be, and the force *TM*. Thus he concludes that the perturbing forces are *LM* and *TM*. (Note that these are not orthogonal components; in the following proposition he does introduce orthogonal components.)

Newton goes on to calculate the relative magnitude of the perturbation. He finds that "*the force ML (in its mean quantity) is to the centripetal force by which the Moon may be retained in its orbit revolving about the Earth at rest, at the distance PT, . . . as 1 to 178 29/40.*" The perturbations are of order 2 percent.

These forces are now fed into the methods given in *Proposition LXVI, Book I* to give results contained in *Propositions XXVI* to *XXIX* for various properties of the Moon's orbit. *Propositions XXX* to *XXXIII* deal with the motion of the nodes, and reasonable estimates are produced. Newton also includes supporting work by John Machin. *Propositions XXXIV* and *XXXV* give Newton's calculation of the inclination of the Moon's orbit to the ecliptic plane, and he claims agreement with the observed results. As always, the *Principia* is difficult to follow with its intricate geometric arguments, and the interested reader should consult the usual references for clarifications. Note that the motion of the rotation of the line of apsides is not covered in these *Propositions*. Newton had referred to it in the second *Corollary* to *Proposition XLV, Book I*, concerning rotating orbits (see section 14.3.4). However, he was able to account for only half of the observed rotation, a major failing in his theory of the Moon.

There is no doubt that Newton had made an enormous advance in understanding the Moon's motion. He had shown that the Sun not only carries the Moon along with the Earth in their orbit around the Sun, but it also gives force components that may be taken as those responsible for perturbing the Moon's orbit around the Earth. His evaluations of inequalities were quite respectable in many cases, although his dynamic model certainly could not provide the accuracy required for longitude determination.

25.1.5. After Newton

In the eighteenth century, the dynamical theory of the Moon's motion was attacked by a set of brilliant mathematicians: Alexis Claude Clairaut (1713–1765), Jean-Baptiste le Rond d'Alembert (1717–1783), Leonard Euler (1707–1783), Joseph Luis Lagrange (1736–1813), and Pierre-Simon Laplace (1749–1827). Gradually, algebraic methods took over, and the defining differential equations could be taken as a starting point for study of the Moon's orbit.

Clairaut, d'Alembert, and Euler competed to find a way to account for the rotation of the line of apsides. At one stage, Clairaut and Euler concluded that the inverse-square law of gravity was not the whole story, and a modification depending on $1/r^4$ should be made. However, errors in their working were eventually located, and Newton's mechanism was shown to be correct. It took the inclusion of second-order terms to find Newton's missing half of the rotation's magnitude. Euler was happy with this situation, and no doubt he expressed a generally held view when in 1751 he wrote to Clairaut:

> the more I consider this happy discovery, the more important it seems to me, and in my opinion it is the greatest discovery in the Theory of Astronomy. . . . For it is very certain that it is only since this discovery that one can regard the law of attraction reciprocally proportional to the squares of the distances as solidly established; and on this depends the entire theory of astronomy.[4]

No doubt Newton would have been delighted to read such a sweeping statement by one of the greatest of all mathematicians. (See the references to Bodenmann and Waff, listed in the "Further Reading" section at the end of this chapter, for the full story of the apsides.)

Work on the Moon continues up to the present (see, for example, the books by Cook and Linton in this chapter's "Further Reading" section), and very accurate tables are available. It is interesting to note that a major change in the dynamical theory was made in 1877 by G. W. Hill. Previous work had been based on an initial elliptical orbit, as discussed above. Hill began with an orbit that built in features of the three-body system dynamics, and that innovation proved to be a major improvement.

25.2. COMETS

The year 1665 was a momentous one for Newton—it was the "plague year" when he fled from Cambridge to Lincolnshire to do some of his greatest work. The following year saw the Great Fire of London catastrophe. Newton would have been aware of the comets that accompanied these events and how that illustrated the stories around comets, making them one of the most feared, as well as one of the most spectacular, astronomical phenomena. Here is novelist and journalist Daniel Defoe in his *A Journal of the Plague Year* (1722):

> In the first place, a blazing star or comet appeared for several months before the plague, as there did the year after another, a little before the fire. The old women and the phlegmatic hypochondriac part of the other sex . . . remarked (especially afterward, though not till both those judgments were over) that those two comets passed directly over the city, and that so very near the houses that it was plain they imported something peculiar to the city alone.[5]

Even the appearance of the comets (one "a faint dull, languid colour," the other, "bright and sparkling"[6]) matched the dreadful events they foretold. There is a long history of comets as messengers and merchants of fear (see the books by Brandt and Whipple, for example, listed in this chapter's "Further Reading" section). For scientists like Newton, it would have been an important challenge to establish the true nature and behavior of comets.

25.2.1. The 1680–1681 Comet

In November 1680, a comet was observed heading toward the Sun. It disappeared on December 8. But then on December 10, a comet was seen moving away from the Sun and it was observed until March 1681. There was much debate about whether the two comets were involved, or whether it was a single comet that bent in its obit very close to the Sun. Astronomer Royal Flamsteed favored the latter option but claimed the comet bent in its path before reaching the Sun, so being somehow repelled by the Sun. Eventually it was concluded that one comet was involved and that it turned around the Sun, getting to about one solar radius from it. Figure 25.3 shows Newton's diagram from the *Principia*. The 1680–1681 comet was a wonderful example and challenge for Newton.

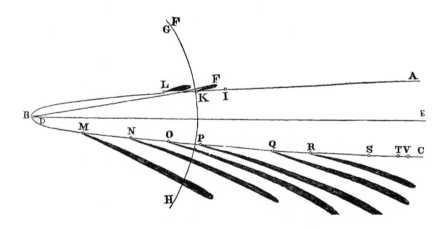

Figure 25.3. Newton's drawing of the 1680–1681 comet. *D* is the Sun, and *GH* is the intersection of the sphere containing the Earth's orbit with the plane of the comet's orbit. (From *The* Principia, trans. Andrew Motte [Amherst, NY: Prometheus Books, 1995].)

There was no satisfactory theory for the orbit of comets or for the mechanism for promoting their motion. Straight-line paths were favored, sometimes as projections of great circles perhaps centered on a star like

Sirius. Some theories invoked magnetic attraction by the Sun. Descartes had his vortex theory (see figure 2.3), but since comets sometimes moved against the vortex (in a retrograde manner), that theory was not reliable. The theory of comets was difficult, and even as late as June 1686, when Newton was trying to complete the *Principia*, he confessed to Halley that he had no final theory and had wasted much time on the matter.

25.2.2. Comet Basics

In the *Principia* Newton begins with *Lemma IV*: "*That the Comets Are Higher Than the Moon, and in the Regions of the Planets.*"

As with everything about comets, Newton accompanies this *Lemma* with a long (almost six pages) discussion presenting a variety of detailed evidence. The statement in the *Lemma* is important because it contradicts the ancient idea of Aristotle that there was an unchanging, perfect heavenly region beyond the Moon. Newton also concludes that:

> COR. 1. *Therefore the comets shine by the Sun's light, which they reflect.*

After comments on the related problem of observing comets near the Sun, we find two major conclusions expressed in:

> COR. 3. *Hence also it is evident that the celestial spaces are void of resistance; for though the comets are carried in oblique paths, and sometimes contrary to the course of the planets, yet they move everyway with the greatest freedom, and preserve their motions for an exceedingly long time, even where contrary to the course of the planets. I am out in my judgment if they are not a sort of planets revolving in orbits returning into themselves with a perpetual motion.*

That second conclusion (that comets are something like planets) naturally leads to *Proposition XL*: "*That the comets move in some of the conic sections, having their foci in the centre of the Sun; and by radii drawn to the Sun describe areas proportional to the times.*"

This naturally follows using the results from *Book I*. It also allows Newton to suggest how comet periods may be estimated and to conclude that they may be very long—240 years, in the example he constructs. For the extremely elongated orbits involved, conic sections (ellipse, parabola, and hyperbola) around the focus are very close to each other (see figure 25.4). This makes the theory difficult but also leads to:

> COR. 2. But their orbits will be so near to parabolas, that parabolas may be used for them without sensible errors.

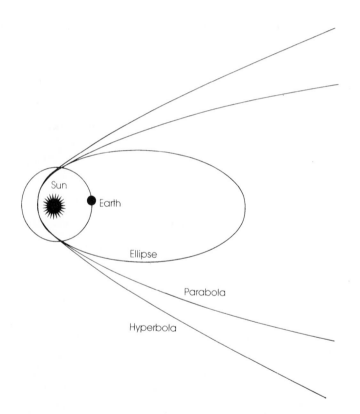

Figure 25.4. An illustration of the similarity of different conic-section orbits near the Sun, which is taken as the focus. (From F. L. Whipple, *The Mystery of Comets* [Cambridge: Cambridge University Press, 1985], reprinted with the permission of Cambridge University Press.)

It is easier to fit parabolas than ellipses, and corrections may be applied later, as discussed below. Newton uses the parabolic section of orbit to discuss comet velocity and orbit properties in two more *Corollaries*.

Thus the great step has been made: the comets are something like planets and the theory of gravity used with *Book I* dynamics tells us about their orbits.

25.2.3. A Mathematical Interlude

Next come seven essentially mathematical *Lemmas* dealing with orbit fitting and properties. Most readers would skip over this part, but that would miss out on a mathematical gem *LEMMA V*: *"To Find a Curve of the Parabolic Kind* [a Polynomial] *Which Shall Pass through Any Given Number of Points."* This curve may then be used to find the value of the curve at intermediate points (see figure 25.5).

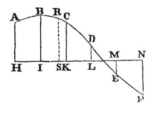

Figure 25.5. The values of the curve at *A*, *B*, *C*, *D*, *E*, and *F* are given, and the value at *R* is to be found. (From *The* Principia, trans. Andrew Motte [Amherst, NY: Prometheus Books, 1995].)

This is the interpolation problem that was so important for table makers; for example, when preparing a table of square roots or logarithms, can certain values be calculated using the appropriate algorithm and then intermediate values be found by the less-time-consuming process of interpolation? Newton brilliantly solved this problem with a method now called the finite-difference method. It is easy to understand why he called it *"one of the prettiest problems I can ever hope to solve."*[7] Newton's method may be found in any textbook on applied mathematics or numerical analysis. The given points need not be evenly spaced. The method is easy to apply and Newton gives examples (see this chapter's "Further Reading" section.)

In a *Corollary*, Newton points out that we can find the area under any given curve simply by fitting a polynomial to it and then integrating that with methods that are "*vulgarly known.*" In about one page, Newton sets out a powerful, simple method for interpolation and then suggests how integrals may be approximated. So, just as an aside, here are foundation principles for numerical analysis!

25.2.4. Fitting Comet Orbits

Newton now gets down to business with *Proposition XLI:* "*From Three Observations Given to Determine the Orbit of a Comet Moving in a Parabola.*" It is instructive to read his opening paragraph:

> *This being a Problem of very great difficulty, I tried many methods of resolving it; and several of those Problems, the composition whereof I have given in the first Book, [Sections IV and V] tended to this purpose. But afterwards I contrived the following solution, which is something more simple.*

There follows some of Newton's usual complex geometrical manipulations, leading to "*Example. Let the comet of the year 1680 be proposed.*" Five pages describe how data collected by Flamsteed and Newton himself were used. It must have been a daunting task, using a combination of methods on a large scale:

> *All of which I determined by scale and compass, and the chords of angles, taken from the table of natural sines, in a pretty large figure, in which, to wit, the radius of the orbis magnus [Earth] (consisting of 10000 parts) was equal to 16 1/3 inches of an English foot.*

He then compared further points on the calculated orbit with other observational data, and the table of comparisons shows a reasonable fit to the data. However, "*afterwards Dr Halley did determine the orbit to a greater accuracy by an arithmetical calculus than could be done by linear* [graphical] *descriptions.*" The comparison table shows that Halley's orbit was a good fit over the whole December–March observational period. There is a long discussion of these results.

Halley also checked the records for comets and found that apparently the 1680–1681 comet had a period of 575 years and had been recorded three times before, including "*in the month of September after Julius Caesar was killed.*" It was using an elliptical orbit based on that idea that Halley had used to fit his orbit data. However, it now appears that Halley was wrong about that periodical behavior, so maybe he was somewhat fortuitous in his working!

Among all the discussion, Newton sets out two major conclusions:

> *And the theory which justly corresponds with a motion so unequable, and through so great a part of the heavens, which observes the same laws with the theory of the planets, and which accurately agrees with astronomical observations, cannot be otherwise than true.*
>
> *Now if one reflects upon the orbits described, and duly considers the other appearances of this comet, he will be easily satisfied that the bodies of comets are solid, compact, fixed, and durable, like the bodies of the planets.*

25.2.5. Orbits and Halley's Comet

Being convinced that "*comets are a sort of planets revolved in very eccentric orbits about the Sun,*" it remains to build on the near-focus parabolic estimation using *Proposition XLII: "To Correct a Comet's Trajectory Found as Above."*

Newton gives a long account of his methods and comparisons with observational data. He concludes that "*the motions of comets are no less accurately represented by our theory than the motions of the planets.*"

We now come to the most famous part of the comet story. Halley computed details of the 1607 and 1682 comets and concluded that there was a single comet—"Halley's comet," as we now call it—with an orbital period of about seventy-five years. (He also referred to a 1531 sighting.) Figure 25.6 shows the orbit of Halley's comet and clearly illustrates the way in which comets travel to the outer reaches of the solar system.

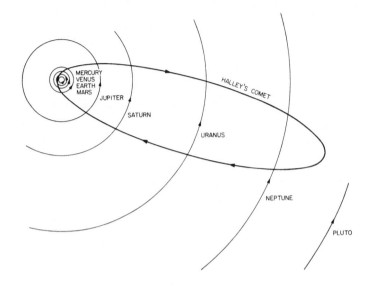

Figure 25.6. The orbit of Halley's comet projected onto the plane of the solar system. Note that it has a retrograde motion—it travels in its orbit in the opposite direction to the planets. (From F. L. Whipple, *The Mystery of Comets* [Cambridge: Cambridge University Press, 1985], reprinted with the permission of Cambridge University Press.)

There are many records of Halley's comet throughout history; most recently, we could observe it in 1986. But surely the most famous "return" was that predicted for the end of 1758 or early in 1759. In fact, it was spotted on Christmas Day 1758. Sadly, both Halley and Newton were dead by then. Undoubtedly they would have enjoyed hearing the lecture about it at the Paris Academy of Sciences on April 25, 1759, which opened with the words:

> the Universe sees this year the most satisfying phenomenon that astronomy has ever offered us; unique event up to this day, it changes our doubts into certainty, and our hypotheses into demonstrations.[8]

It was at that meeting of the academy that Nicolas Lacaille first used the term *Halley's comet*, surely an appropriate recognition for someone who helped to give universal gravitation one of its greatest triumphs.

25.3. CONCLUSIONS

It took many years, but finally the intricate motion of the Moon and the spectacular, vast motion of comets confirmed that Newton had indeed found the theory for universal gravity.

The extent of his *Principia* discussions suggests that Newton was fascinated by comets, and he allowed himself some rare speculations about them. He suggested that they may help to replenish stars and could be responsible for the nova that Tycho Brahe observed. When considering the effect of the Sun's heat on comets, Newton tossed out one of his stunning asides: he suggested a way to calculate the age of the Earth. Considering the cooling of a globe of red-hot iron the same size of the Earth, Newton came up with the figure of fifty thousand years. He discussed cooling mechanisms and commented that "*I should be glad that the true ratio was investigated by experiments.*" About 150 years later, Lord Kelvin was following Newton and calculating an age for the Earth by modeling its cooling processes.

25.4. FURTHER READING

For the Moon:

Cohen, I. Bernard. "A Guide to Newton's *Principia.*" In *Isaac Newton: The* Principia, by I. Bernard Cohen and Anne Whitman. Berkeley: University of California Press, 1999. [see sec. 8.15 by G. E. Smith, "Newton and the Problem of the Moon's Motion"]

Cook, A. *The Motion of the Moon.* Bristol, UK: Adam Hilger, 1988.

Bodenmann, S. "The 18th-Century Battle over Lunar Motion." *Physics Today* (January 2010): 27–32.

The General History of Astronomy: Volume 2, Planetary Astronomy from the Renaissance to the Rise of Astrophysics—Part B: The Eighteenth and Nineteenth Centuries. Edited by R. Taton and C. Wilson Cambridge: Cambridge University Press, 1989.

Gutzwiller, M. C. "Moon-Earth-Sun: The Oldest Three-Body Problem." *Reviews of Modern Physics* 70 (1998): 589–638. [a comprehensive review by a theoretical physicist]

Linton, C. M. *From Eudoxus to Einstein: A History of Mathematical Astronomy.* Cambridge: Cambridge University Press, 2004.

McEvoy, J. P. *Eclipse: The Science and History of Nature's Most Spectacular Phenomenon.* London: Fourth Estate, 1999. ["popular" introduction]

Peterson, I. *Newton's Clock.* New York: W. H. Freeman, 1993. [ch. 6 is a popular account of the study of the Moon's motion]

Whiteside, D. T. "Newton's Lunar Theory: From High Hope to Disenchantment." *Vistas in Astronomy* 19 (1976): 317–28.

Wilson, C. "The Newtonian Achievement in Astronomy." In *The General History of Astronomy: Volume 2, Planetary Astronomy from the Renaissance to the Rise of Astrophysics—Part A: Tycho Brahe to Newton.* Edited by R. Taton and C. Wilson. Cambridge: Cambridge University Press, 1989.

Definitive papers by Forbes, Morando, Waff, and Wilson are contained in:

The General History: Volume 2, Planetary Astonomy from the Renaissance to the Rise of Astrophysics—Part A: Tycho Brahe to Newton. Edited by R. Taton and C. Wilson. Cambridge: Cambridge University Press, 1989.

For detailed analysis of Newton's work on lunar theory, see:

Nauenberg, M. "Newton's Perturbation Methods for the Three-Body Problem and Their Application to Lunar Motion." In *Isaac Newton's Natural Philosophy.* Edited by J. Z. Buchwald and I. Bernard Cohen. Cambridge, MA: MIT Press, 2001.

Wilson, C. "Newton on the Moon's Variation and Apsidal Motion: The Need for a Newer 'New Analysis.'" In *Isaac Newton's Natural Philosophy.* Edited by J. Z. Buchwald and I. Bernard Cohen. Cambridge, MA: MIT Press, 2001.

On comets, see the references above to Linton and Wilson, as well as the following:

Bork, A. "Newton and Comets." *American Journal of Physics* 55 (1987): 1089–95.

Brandt, J. C., and R. D. Chapman. *Introduction to Comets*. Cambridge: Cambridge University Press, 1981.

Hughes, D. W. "The Principia and Comets." *Notes and Records of the Royal Society of London* 42 (1988): 53–74.

Kollerstrom, N. "The Path of Halley's Comet, and Newton's Late Apprehension of the Law of Gravity." *Annals of Science* 56 (1999): 331–56.

Ruffner, J. A. "Newton's Propositions on Comets: Steps in Transition, 1681–84." *Archive for History of the Exact Sciences* 54 (2000): 259–77.

Waff, C. B. "Predicting the Mid-eighteenth-century Return of Halley's Comet." In *The General History of Astronomy: Volume 2, Planetary Astronomy from the Renaissance to the Rise of Astrophysics—Part B: The Eighteenth and Nineteenth Centuries*. Edited by R. Taton and C. Wilson Cambridge: Cambridge University Press, 1989.

Whipple, F. L. *The Mystery of Comets*. Cambridge: Cambridge University Press, 1985.

For more on *Lemma V* and numerical analysis, see:

Goldstine, H. H. *A History of Numerical Analysis*. New York: Springer-Verlag, 1997. [sec. 2.5]

Turnbull, H. W. *The Mathematical Discoveries of Newton*. London: Blackie, 1945. [sec. 9]

Whitehead, D. T. "Patterns of Mathematical Thought in the Later Seventeenth Century." *Archive for the History of the Exact Sciences* 1 (1961): 179–388.

THE CONCLUDING *GENERAL SCHOLIUM*

We expect to find a summing up and some general conclusions at the end of such an extensive book as the *Principia*. There was nothing in the first edition, but then Newton had produced the whole book in a comparatively short period, so maybe he had no time or energy left for a final conclusion. He wrote the *General Scholium* for the second edition, and it remained, after a few small changes, in the final, third edition. It takes up only about four pages. We know that Newton did consider other and longer concluding statements (see the reference to Cohen, listed in the "Further Reading" section at the end of this chapter, for a discussion), but what we get is the *General Scholium*. I take the third edition of the *Principia* to be Newton's own treatise on dynamics and its applications, and, as such, it contains his work and opinions as he wished them to be known generally and publicly. He expressed other views in private conversations, in letters, or in other documents, but presumably *Principia* is what he wished us to have as a public record of his work.

The *General Scholium*, or excerpts from it, is possibly the best known part of the *Principia* and may be the only part that many people actually read! I recommend that you too read the *General Scholium* before going on. Some statements in it have led to extensive debates and analysis, and I will say more on some points in the chapters in part 7.

26.1. REVIEW OF DESCARTES'S VORTICES

We recall from chapter 2 that the work of Galileo and others was essentially kinematics: a description of motion, but with little or no causal mechanism. The only major theory for motion of the planets and other entities was Descartes's system of vortices in the heavens to carry bodies around in them. The Sun's vortex sweeps the planets around it. Newton had critically discussed Descartes's mechanism in *Book II*, and so he opens the *General Scholium* with the following words: "*The hypothesis of vortices is pressed with many difficulties.*"

He makes three points. The vortex theory does not give the correct behavior (as summarized in Kepler's laws) for the planets. There would need to be smaller vortices "*undisturbed in the greater vortex of the sun,*" but moving around with it, to account for the moons of Saturn and Jupiter. Lastly, the "*comets are carried with very eccentric motions through all parts of the heavens indifferently, with a freedom that is incompatible with the notion of a vortex.*" (Surprisingly Newton does not mention the retrograde comets that move oppositely to the supposed vortex around the Sun.) Descartes never gave a quantitative theory of vortices, and Newton has shown that even qualitatively it does not make sense.

26.2. THE SOLAR SYSTEM

Newton first reminds us about the medium in which motion takes place. The motions of bodies locally ("*in our air*") suffer air resistance. However the motion in "*Mr. Boyle's vacuum*" reveals no resistance forces, and "*a bit of down and a piece of solid gold descend with the same velocity.*" Now we can consider "*celestial places*" as free of air, and there will be no resistance forces hindering the motion of planets and comets. Thus:

> the planets and comets will constantly pursue their revolutions in orbits given in kind and position, according to the laws above explained.

Newton is claiming that his theory as applied in *Book III* successfully replaces Descartes's vortex theory and explains the motion of planets and comets. However, there is one problem.

26.2.1. Initial Conditions

Recall that in section 13.5 I explained how the solution of a dynamical problem involves two steps: first, use the laws of motion with the appropriate forces to find the possible orbits; second, pick out the required orbit by using the initial conditions, that is, the positions and velocities of the bodies at some given time. It is to the question of initial conditions that Newton turns when he writes,

> but though these bodies may, indeed, persevere in their orbits by the mere laws of gravity, yet they could by no means have at first derived the regular position of the orbits from those laws.

The laws of motion do not contain the initial conditions. He then goes on to remind us about the intricate, beautiful nature of the solar system. There are six planets orbiting the Sun in (almost) circular orbits, all going around in the same direction and in (almost) the same plane. Similarly and fitting in are the moons of Jupiter and Saturn. "*But it is not to be conceived,*" Newton says, "*that mere mechanical causes could give birth to so many regular motions.*" There are also comets traveling vast distances and going slowly at their aphelions. Newton suggests his famous resolution:

> This most beautiful system of the sun, planets, and comets, could only proceed from the counsel and dominion of an intelligent and powerful Being.

Newton believes that God is the "intelligent designer" (as some would now put it) of the solar system. The suggestion is that only God could make such a system and set it in motion in such a beautiful way. It is interesting that Pierre-Simon Laplace (1749–1827), the man known as the "French Newton," quoted Newton on this matter but showed how things had advanced by his time with his viewpoint:

> But could not this arrangement of the planets itself be a result of the laws of motion, and could not the Supreme Intelligence, which Newton makes to interfere, have made them all depend on more general phenomena?[1]

Laplace was putting forward his nebular hypothesis. The origin of the solar system is still not fully understood today (see the "Further Reading" reference to Freedman and Kaufmann at the end of this chapter).

26.2.2. The Fixed Stars

I commented earlier (section 23.6) that there is little in the *Principia* about the stars, but Newton does now make three points about them. First, there is the possibility that they could act like our Sun:

> And if the fixed stars are centres of other like systems, these, being formed by the like wise counsel, must be subject to the dominion of One.

This is backed up by a use of *Rule of Reasoning II* (see section 22.2):

> Especially since the light of the fixed stars is of the same nature with the light of the sun, and from every system light passes into all other systems.

The third point concerns the stability of the universe, which we know Newton worried about but restricted himself to this public statement in the *Principia*:

> and lest the systems of the fixed stars should, by their gravity, fall on each other mutually, he hath placed those systems at immense distance from one another.

God has taken care of things. I return to this matter in chapter 28.

26.3. GOD

Newton was an intensely religious man, as discussed in chapter 1, but in the *Principia* he writes about God only in the *General Scholium*. Now his views are clear and forcefully expressed, as these few examples reveal:

This Being governs all things, not as the soul of the world, but as Lord over all. The Supreme God is a being eternal, infinite, absolutely perfect.

He is eternal and infinite, omnipotent and omniscient; that is, his duration reaches from eternity to eternity; his presence from infinity to infinity; he governs all things, and knows all things that are or can be done.

After a lengthy discussion of such matters, Newton turns to the question of how we can know God:

We have ideas of his attributes, but what the real substance of any thing is we know not. In bodies, we see only their figures and colours, we hear only the sounds, we touch only their outward surfaces, we smell only the smells, and taste the savours; but their inward substances are not to be known either by our senses, or by any reflex act of our minds; much less, then, have we any idea of the substance of God.

We can have limited knowledge of God:

We know him only by his most wise and excellent contrivances of things, and final causes.

And we come back to that belief that God must be responsible for the state of the world:

Blind metaphysical necessity, which is certainly the same always and everywhere, could produce no variety of things. All the diversity of natural things which we find suited to different times and places could arise from nothing but the ideas and will of a Being necessarily existing.

Having explained his ideas about God, Newton now states that he supports the concept of knowing God through God's works:

And thus much concerning God; to discourse of whom from the appearances of things, does certainly belong to Natural Philosophy.

Some three hundred years later, we still debate this view of God, his relevance for science, and whether a scientist may also be a religious person. Much has been written on this topic, and to go deeper you might start with the recent long discussion by Snobelen and its guide to other relevant literature. I return to the subject in the next chapter.

26.4. GRAVITY

If the *General Scholium* is the most read part of the *Principia*, we now come to the most famous of all paragraphs. Newton faces a problem that has been bubbling away in the background for the whole book. The paragraph begins:

> *Hitherto we have explained the phenomena of the heavens and of our sea by the power of gravity, but have not yet assigned the cause of this power.*

He goes on to what we can be certain about: gravity has all those properties first established in *Book III* (see chapter 22); it is an inverse-square-law effect that extends from the Sun to immense distances, as we may deduce from the motion of comets. Newton has used his theory of dynamics and an inverse-square-law force to successfully investigate the effects of gravity, "*but hitherto I have not been able to discover the cause of those properties of gravity from phenomena.*" Added to that is his most famous statement:

> *and I frame no hypotheses.*

I believe too much has been read into this statement. Newton is saying that specifically he will set out no hypothesis for the mechanism behind the gravitational force. He is not saying anything about hypotheses in general—we have seen how he set out several hypotheses in the *Principia*.

Confusion may arise because he goes on to comment about hypotheses and their use:

for whatever is not deduced from the phenomena is to be called an hypothesis; and hypotheses, whether metaphysical or physical, whether of occult qualities or mechanical, have no place in experimental philosophy. In this philosophy particular propositions are inferred from the phenomena, and afterwards rendered general by induction.

This is not an easy point for Newton to explain, but he is saying that the theory-phenomena link is crucial, and if it cannot be made, then there is no value in the form of the theory or any hypothesis involved in it. As an example, he claims:

thus it was the impenetrability, the mobility, and the impulsive force of bodies, and the laws of motion and gravitation, were discovered.

However, he has not been able to use the same approach to discover a mechanism for gravity. He can give no underlying theory involving the exchange of particles between interacting bodies, or waves, or deviations in some aether-like medium in which the bodies move. He had already been clear in the *Scholium* at the end of *Section XI* (see section 16.6) that he was dealing with mathematical forces, not various interaction mechanisms. He now says that within his methodological framework, he cannot go beyond the forces to those mechanisms. Here was an admission that would leave him open to all sorts of challenges, as he foresaw in his *Preface*. His way out of this difficulty is set out in what may well be the most profound statement in the whole *Principia*:

And to us it is enough that gravity really does exist, and act according to the laws which we have explained, and abundantly serves to account for all the motions of celestial bodies, and of our sea.

Newton has given us a scheme for describing and predicting properties of the motions of bodies over a breathtaking range of instances. In effect, he is saying we should glory in that success and accept the elements of the theory, even though we might wish to have reasons why those elements take the form that they do. "*Satis est—it is enough.*" It is a philosophy that has become a part of modern physics. For example, the theory of electromagnetic fields is enormously successful, but in his section 1–5, "What Are the Fields?" even the great Richard Feynman finally states:

> The best way is to use the abstract field idea. That it is abstract is unfortunate, but necessary. The attempts to try to represent the electric field as the motion of some kind of gear wheels, or in terms of lines, or as stresses in some kind of material have used up more effort of physicists than it would have taken simply to get the right answers about electrodynamics.[2]

That is exactly the position Newton advocated, and it is used over and over again in our modern physics.

26.4.1. Gravitational Worries

That was Newton's public position. We know that privately he did try to invent mechanisms that would result in the gravitational force. The fact is, he could not do it, and even today we have no accepted mechanism for gravitational effects. We have a meson exchange theory for telling us how nuclear forces operate, and we have a photon theory for explaining electric forces, but we have no equivalent for gravity. Newton was worried by the concept of gravity, as he expressed in his famous statement in a letter to Richard Bentley (see section 27.5):

> *That gravity should be innate, inherent, and essential to matter, so that one body may act upon another at a distance through a vacuum without the mediation of anything else, by and through which their action and force may be conveyed from one to another, is to me so great an absurdity, that I believe no man who has in philosophical matters a competent faculty of thinking can ever fall into it. Gravity must be caused by an agent acting constantly according to certain laws; but whether this agent be material or immaterial, I have left to the consideration of my readers.[3]*

Christiaan Huygens, a man greatly respected by Newton, certainly picked up on that "*absurdity*" in a way that may not have surprised Newton but would surely have dismayed him:

> Concerning the cause of the tides given by M. Newton, I am by no means satisfied, nor by all the other theories he builds upon his principle of Attraction, which to me seems absurd. . . . And I have often wondered how he could have given himself all the trouble of making such a number of investigations and difficult calculations that have no other foundation than this very principle.[4]

Absurd or not, Newton's theory still holds up today, and men have used it to travel to the moon and guide spacecraft to the outer solar system.

26.5. FINAL MUSINGS

There is one final paragraph in the *General Scholium*, and (to me at least) it seems surprising that Newton included it. It is really about electrical effects and "*a certain most subtle Spirit which pervades and lies hid in all gross bodies.*" Newton muses about this spirit and, for example, "*all sensation is excited, and members of animal bodies move at the command of the will, namely, by vibrations of this Spirit, mutually propagated along the solid filaments of the nerves, from the outward organs of sense to the brain, and from the brain to the muscles.*" Perhaps Newton is wishing to say that this new field of investigation should come within his philosophy of science. Maybe he is thinking of some sort of research program. Certainly it is something for the future as, he concludes, "*these are things that cannot be explained in a few words, nor are we furnished with the sufficiency of experiments which is required to an accurate determination and demonstration of the laws by which this electric and elastic Spirit operates.*"

26.7. SUPPORT BY COTES

Newton's position was explained, supported, and defended by Roger Cotes, who, in addition to editing the second edition of the *Principia*, wrote a long preface to it. As I am concentrating on what Newton himself wrote, I have not referred to that preface. However, a reader will find lots of interesting history and science there, and it is well worth reading alongside Newton's *General Scholium*.

26.8. FURTHER READING

Cohen, I. Bernard. "A Guide to Newton's *Principia*." In *Isaac Newton: The Principia*, by I. Bernard Cohen and Anne Whitman. Berkeley: University of California Press, 1999.

Freedman, R. A., and W. J. Kaufmann III. *Universe*. 7th edition. New York: W. H. Freeman, 2005. [see ch. 8 for a discussion on the origin of the solar system]

Snobelen, S. D. "'God of Gods, and Lord of Lords:' The Theology of Isaac Newton's General Scholium to the *Principia*." *Osiris* 16 (2001): 169–208.

PART 7

RECEPTION AND INFLUENCE OF *PRINCIPIA*

The publication of the *Principia* was one of mankind's greatest steps, and this part outlines some of the reactions to it and how it changed the way people thought and acted in different areas of intellectual life. The *Principia* laid the foundations for the science of mechanics, and chapter 28 (the second chapter in this section) reviews how subsequent developments were built on it. These are enormous topics, so obviously these chapters can provide only a brief introduction and suggest suitable references for further study. The book ends with some parting words from the author!

Reading plans: If you have reached this point, you should have an idea about the style and contents of the Principia. *If you wonder what came next, these chapters give some answers plus some final thoughts from the author.*

RECEPTION AND INFLUENCE

Even today it is impossible not to be impressed—perhaps overawed—by the sweep and power of Newton's achievements in the *Principia*. Add to that his *Opticks* and other writings, and we should have no doubts about the widespread impact of Newton and his work. Already in 1687 David Gregory had set the scene when he thanked Newton "for having been at pains to teach the world that which I never expected any man should have knowne."[1]

Just how so many new things had been discovered was sure to bring both admiration and curiosity. We might also expect other responses, such as jealousy, belittling, and disbelief. People of all kinds, not just the academically minded, would soon begin to look for a suitable introduction to this new wonder.

The publication of the *Principia* came at the start of the period known as the Enlightenment. In fact Bacon, Newton, and Locke are sometimes known as the "patron saints" of the Enlightenment. A variety of scientific, religious, and philosophical matters became of importance, as I discuss in sections 5 through 7 below.

This is already a large book, and to deal fully with the "What came after the *Principia*?" question would require another equally large book. This chapter has very limited aims: to pick out a few themes for brief comment and to list examples from the expansive literature on this topic.

27.1. NEWTON'S WORK: SOURCES OF WONDER AND DISPUTES

Kepler and Galileo gave us kinematics, mathematical descriptions of some important examples of bodies in motion. Newton made the enormous step to dynamics, explaining how motion "works." In Einstein's words (see section 3.5), Newton produced a "self-contained system of physical causality."[2] This was based on the mechanical philosophy, supplemented by the notion of force. As this book has revealed, Newton's success is breathtaking.

Newton set out his approach in the *Preface* to the *Principia*: the use of mathematics to develop and explore theories, plus the essential interplay between theory and experiment. He began with definitions and axioms (or laws), and gradually built up a whole structure for dynamics and its applications. He followed the same approach in his *Opticks*, which opens with the statement:

> My Design in this Book is not to explain the Properties of Light by Hypotheses, but to propose and prove them by Reason and Experiments: In order to which I shall premise the following Definitions and Axioms.

It is both interesting and important to note that it is only at the end of his books (in the *General Scholium* in the *Principia*, and in *Query 31* in *Opticks*), that Newton fully discusses some more philosophical and religious matters.

This respect for mathematics and the experimental method was supported and dramatically illustrated by the Enlightenment philosopher David Hume (1711–1776) in the final paragraph of his *An Enquiry concerning Human Understanding*:

> When we run over libraries, persuaded of these principles, what havoc must we make? If we take in our hand any volume; of divinity or school metaphysics, for instance; let us ask "Does it contain any abstract reasoning concerning quantity or number?" No. "Does it contain any experimental reasoning concerning matter of fact and existence?" No. Commit then to the flames; for it can contain nothing but sophistry and illusion.

Newton's basic methodology has survived and remains in place today. Here is Newton's approach, set out again in the twentieth century by Einstein:

> Science is the attempt to make the chaotic diversity of our sense experience correspond to a logically uniform system of thought. In this system single experiences must be correlated with the theoretic structure in such a way that the resulting coordination is unique and convincing.
>
> The sense-experiences are the given subject matter. But the theory that shall interpret them is man-made. It is the result of an extremely laborious process of adaptation: hypothetical, never completely final, always subject to question and doubt.
>
> The justification (truth content) of the system rests in the proof of usefulness of the resulting theorems on the basis of sense experiences, where the relation of the latter to the former can only be comprehended intuitively.[3]

The data were clear, and Newton's theory fits magnificently. However, there is a problem in the phrase "the theory that shall interpret them is man-made." Newton built his theory on the concept of force and attractive forces (gravity) that act over large distances. These aspects of his theory were not explained in terms of mechanical models or causes. For some people, the fact that the theory works so well was never enough; not for them, "*satis est—it is enough.*" Equally, Newton set out "*to explain the Properties of Light,*" rather than definitively establish the absolute nature of light.

Thus, the scene is set. There is room for both wonder and disputes.

27.2. RECEPTION AND REVIEWS

The 1687 first edition of the *Principia* attracted four reviews in 1687 and 1688. The first was by Edmond Halley in the *Philosophical Transactions*, and, hardly surprisingly, it was somewhat extravagant in its praise. The second, in the *Bibliothéque Universelle*, published anonymously but attributed to John Locke, was largely a summary of the contents and not particularly penetrating. The anonymous review in *Acta Eruditorum* was, according to I. Bernard Cohen, written by Christoph Pfautz. It was a serious and detailed review, and it gave many their first ideas about the *Principia* and its contents. The fourth review appeared in the *Journal des sçavans*, and it too was anonymous, although it is clear that the author strongly favored the Cartesian philosophy. The actual technical achievements are of less interest than the underlying approach. This is where we find the famous comments about Newton having written a perfect work in mechanics, which now needed an equally impressive work in physics. If we seek the first indication of the battle lines between the rival approaches of Newton and Descartes, this review could well be the answer.

There now followed a period during which experts such as Christiaan Huygens, Johann Bernoulli, and Gottfried Leibnitz subjected both the mathematics and the science in the *Principia* to a searching analysis. Further details are given in Hall's paper "Correcting the *Principia*." Obviously care had to be taken, and fortunately Newton had Roger Cotes to help him prepare the 1713 second edition. This time there were three reviews. The *Journal des sçavans* review was less critical and more complimentary, but the 1718 one in *Mémoires de Trévoux* did make the good mathematics—but poor physics judgment. The *Acta Eruditorum* noted the changes from the first edition, especially reviewing the *General Scholium*. The *Acta Eruditorum* similarly reviewed the 1726 third edition.

The printing figures also tell a story about interest and demand. Around 300 copies of the first edition were printed, 750 for the second, and 1,200 for the third. At that time, apparently (see the Speiser reference in the "Further Reading" section at the end of this chapter), 250 was a large printing, so the *Principia* must have been wanted by many people. (Of course, how many of them actually read, or could fully understand, the book was a different matter!)

It seems safe to say that the *Principia* was received enthusiastically in Britain but raised opposition and many doubts in continental Europe. The gradual development of dynamics is covered in the next chapter, and some of the philosophical matters are referred to below.

27.3. STYLE AND INFLUENCE

Chapter 28 deals in more detail with the development and application of mechanics following the *Principia*. In Britain, Newton's use of forces and geometrical formalism continued to be taught for many years, with the modifications by Euler and Lagrange and the mathematical notation and ideas of Leibnitz and others gradually finding acceptance. (See the paper by Harman and other references in this chapter's "Further Reading" section.) As an example, you may look at the readily available *Principles of Mechanics and Dynamics* by Sir William Thomson (later known as Lord Kelvin) and Peter Guthrie Tait. This leading textbook was published in 1879 as *Treatise on Natural Philosophy*, and it is unashamedly Newtonian; although the introduction of new dynamical formalism, the concentration on the conservation of energy, the use of algebra and analysis, and the range of applications make it mostly a world away from the *Principia*. It is fascinating to read their introduction to *Dynamical Laws and Principles*, written almost two hundred years after the *Principia* first appeared:

> We cannot do better, at all events in commencing, than follow Newton somewhat closely. Indeed the introduction in the *Principia* contains in a most lucid form the general foundations of Dynamics. The Definitiones and Axiomata sive Leges Motûs, there laid down, require only few amplifications and additional illustrations, suggested by subsequent developments, to suit them to the present state of science, and to make a much better introduction to dynamics that we find in even some of the best modern treatises.[4]

The tradition lives on today. There are textbooks (by Landau and Lifshitz, for example) that begin with the principle of least action and Lagrange's equations (see the next chapter for details), but many are still very much in sympathy with Thomson and Tait.

Newtonian mechanics, along with its various extensions, covers a vast amount of classical physics (that is, pre-quantum and pre-relativity physics). The big gap is electric and magnetic phenomena, the study of which was well under way in Newton's time. The great step was made in the nineteenth century with the work of Faraday, Kelvin, and Maxwell, culminating in the theory of the electromagnetic field. This also provides the wave theory of light. The story of this discovery has parallels with the work of Newton on gravitation. The mechanical philosophy was famously espoused by Kelvin in his Baltimore lectures: "I never satisfy myself until I can make a mechanical model of a thing. If I can make a mechanical model I can understand it. As long as I cannot make a mechanical model all the way through I cannot understand; and that is why I cannot get the electromagnetic theory."[5] Maxwell used equations from fluid dynamics to develop theory for the electromagnetic field, but stressing that this was a technique using analogy to find appropriate mathematics rather than some underlying physical reality (see my 2003 paper in the "Further Reading" section at the end of this chapter). To this day, there is no mechanical model for the electromagnetic field, and we are still left with Heinrich Hertz's concise summary: Maxwell's theory is Maxwell's system of equations.[6] As with his theory of gravity, Newton might well add "*satis est—it is enough.*"

The success of Newton's approach as outlined above also influenced those working in other disciplines. Just as Newton clearly followed Euclid, there were those who sought to emulate Newton in other fields. David Hume wished to become the Newton of the social sciences. In his *An Enquiry concerning Human Understanding,* he referred to Newton's method and success in astronomy and continued: "The like has been performed with regard to other parts of nature. And there is no reason to despair of equal success in our enquiries concerning the mental powers and economy if prosecuted with equal capacity and caution."[7] Raphael has discussed the influence of Newton on Adam Smith as he wrote *An Inquiry into the Nature and Causes of the Wealth of Nations.* There were even attempts to deal with medicine in a Newtonian style (see Brown's paper in the "Further Reading" section at the end of this chapter).

27.4. BOOKS AND READERS

First, a reminder of a few dates. The three editions of the *Principia* appeared in 1687, 1713, and 1726. They were all in Latin, and Motte's English translation did not appear until 1729. The *Opticks* was published (in English) in 1704, although papers on Newton's theories had appeared much earlier.

Obviously, news of Newton's achievements had spread far and wide, and the demand for details would have been enormous. A number of people soon saw the opportunity for courses of lectures aimed at undergraduate students and the general public. Those lectures were also converted into books. (See the paper by Snobelen and the catalogue of publications given by Gjertsen, both listed in the "Further Reading" section at the end of this chapter.) Thus there were popular and highly successful books in English by William Whiston (1696, 1715, and 1716), Willem 's Gravesande (1723), John Theophilus Desaguliers (1728 and 1734), Benjamin Martin (1743) and James Ferguson (1756). The title of Ferguson's book is revealing: *Astronomy Explained upon Sir Isaac Newton's Principles, and Made Easy to Those Who Have Not Studied Mathematics*. It is sometimes said that the mathematical barrier between the professional and public appreciation of scientific advances began in Newton's time—and it is as strong as ever today.

The wonderful *The Newtonian System of Philosophy: Explained by Familiar Objects in an Entertaining Manner for the Use of Young Persons* appeared in 1761. The author is "Tom Telescope," usually reckoned to be Oliver Goldsmith. (A facsimile of a later edition is available today through Nabu Public Domain Reprints.) There are six lectures and an *Introduction*, which begins:

> A party of young people of both sexes being invited to spend the holidays with a friend in the country, were sometimes at a loss for amusement, especially when rainy weather confined them to the house.

Enter Tom Telescope to save the day with his exciting lectures. Try this book for fun and to get an idea of the way things were presented at that time.

There were also more serious textbooks and commentaries, beginning with John Keill's 1701 *Introductio ad veram physicam*; David Gregory's 1702 *Astronomiae physicae et geometricae elementa*; John Maxwell's *Course concerning God*, which was the first book containing English translations of parts of the *Principia*, including the *General Scholium*; Henry Pemberton's 1728 *A View of Sir Isaac Newton's Philosophy*; and Colin Maclaurin's 1748 *An Account of Sir Isaac Newton's Philosophy*. John Desaguliers was in charge of experiments at the Royal Society meetings when Newton was president, and his influential *A Course of Experimental Philosophy* appeared in 1733.

In his introduction to *Book III*, Newton writes that he had composed a book "*in a popular method that it might be read by many*." However, to avoid controversies (see section 7.5), he wrote the more mathematical version that forms the last book of the *Principia*. His book "*in a popular method*" did not appear until 1728 (in both Latin and English). Other great scientists followed suit, with Leonard Euler's wonderful *Letters to a German Princess* appearing in 1772 and Laplace's *System of the World* in 1796. Euler's book was translated into eight languages and published in over a hundred editions. (Those books by Newton, Euler, and Laplace are readily available today and make wonderful reading.)

Thus Newton's work became available in a large variety of presentations suitable for a wide audience. It is noticeable that women were involved and catered for—see the chapter *Newtonian Women* in Feingold's book and chapters 6 and 10 in Alic's book (both in the "Further Reading" section at the end of this chapter). Franceso Algarotti's 1737 *Newtonianism for Ladies* was extremely popular.

Of course, not everyone was thrilled by Newton's work and fame. In the literary world, the reaction was mixed. (For an outline, see the section "Poetic Tradition" in Gjertsen's book.) At one extreme, we have Alexander Pope's famous couplet:

> *Nature and Nature's laws, lay hid in night:*
> *God said, Let Newton be! and all was light.*

At the other extreme, we might put Keats, who almost equally famously wrote in the poem "Lamia":

Do not all charms fly
At the mere touch of cold philosophy?
There was an awful rainbow once in heaven:
We know her woof, her texture; she is given
In the dull catalogue of common things.
Philosophy will clip an Angel's wings,
Conquer all mysteries by rule and line,
Empty the haunted air, and gnomed mine—
Unweave a rainbow.

The power of reason and mathematics did not appeal to everyone!

27.5. RELIGIOUS MATTERS

The debate about religion and science is very much alive today with a stream of books and articles regularly appearing. (See the Oxford VSI *Science and Religion*, listed in this chapter's "Further Reading" section, for a short modern introduction to the topic.) However, in Newton's time, the work of scientists like Boyle and Newton was quite intimately linked to views about God and his powers. Unlike Descartes, who began his *Principia Philosophiae* with considerations of God, Newton left his explicit comments until the end of the *Principia* and *Opticks*. Nevertheless, his views on and respect for God are clear (see the previous chapter).

Newton saw his work as revealing the power and glory of God, but for some, it also raised questions.

If God made everything and set it going, why does he need to exist from then on? If he does exist, and perhaps needs to reset things (as Newton hinted at in the *Opticks*), can we see him as a less-than-perfect creator of the universe? If observation, reason, and analysis reveal the workings of nature, why not see God in that and dispense with the God of faith and revelation as in the Bible? The philosopher David Hume presented a powerful argument against the belief in miracles (see the next section).

The idea that reason might challenge faith (instead of the reverse, as we saw in section 2.2) can lead to ideas of a new type of God or religion,

and maybe even to atheism. One person with such worries was Robert
Boyle. In his will, he left funds "To settle an annual salary for some divine
or preaching minister to preach eight sermons in the year for proving
the Christian religion against notorious infidels, viz. Atheists, Deists,
Pagans, [and so on]."[8] "A Confutation of Atheism" was the title of the
first such lecture, given in 1692 by Richard Bentley, who became master
of Newton's old college, Trinity at Cambridge University. Bentley sought
Newton's help when preparing his lectures, and the four letters Newton
sent to Bentley may be read in the book edited by Janiak (listed in this
chapter's "Further Reading" section). Many important points are discussed, including the comments on the absurdity of action-at-a-distance
forces, as quoted in section 26.4.1. The first words in the first letter make
Newton's feelings immediately known:

> When I wrote my treatise about our system, I had an eye upon such principles
> as might work with considering men, for the belief of a deity, and nothing can
> rejoice me more than to find it useful for that purpose.

In contrast to that, some hundred years later, when Napoleon
questioned Pierre-Simon Laplace about the absence of God in his great
treatise on celestial mechanics, Laplace is reputed to have replied "Sire, I
have no need of that hypothesis."[9]

The debate about God and his role in science is closely allied to the
philosophical issues to which I next turn.

27.6. SOME PHILOSOPHICAL POINTS

The period under discussion saw the appearance of some of the great
works of philosophy:

Locke, John. *An Essay concerning Human Understanding* (1690).
Berkeley, Bishop George. *A Treatise concerning the Principles of Human
    Knowledge* (1710).

Hume, David. *An Enquiry Concerning Human Understanding* (1748).
Kant, Immanuel. *The Critique of Pure Reason* (1781).

In some ways, the titles tell the story. It was time for deliberations on some of the great questions: How do we know things? What can we know? How do we reason about things? What part do faith and reason play? Where and how does God fit into the world picture? And so on.

Locke and Newton were friends, and it has been suggested by Rogers (see this chapter's "Further Reading" section) that "between them they did supply such a long-lived, and influential, not to say intellectually respectable, foundation for much of the eighteenth century, and later, thought."[10]

Newton had shown how reasoning (using a mathematical formalism) can be combined with observational and experimental data to create a science or "*system of the world.*" Obviously the merits and limits of that approach called out for philosophical analysis. Few could dispute the success of Newton's methods, but there were aspects of it that caused unease. They were largely in the fact that "the sense-experiences are the given subject matter. But the theory that shall interpret them is man-made," as Einstein put it.[11] Philosophical problems arose when considering the following: the nature of space, the existence of absolute space and time, whether inertia is a primary or secondary quality, what the nature of force is, whether matter is active or passive, if force can act only through direct contact, and so on.

Some of these matters placed Newton in direct conflict with Descartes and Leibnitz. Not surprisingly, there were continental philosophers ready and willing to try to discredit Newton's work. The most famous clash was between Newton and Leibnitz. There are letters between Newton and Leibnitz (see the book edited by Janiak listed in the "Further Reading" section at the end of this chapter), but as the disputes became more and more bitter, it was Samuel Clarke who wrote on Newton's behalf. (On a personal note: I have always wondered how much time and energy Newton wished to dedicate to such things, even if he did find the challenges extremely irritating.)

This leads us directly to the next section.

27.7. NEWTON AND THE ENLIGHTENMENT

A good, short introduction to the period known as the Enlightenment is given by Dorinda Outram (see this chapter's "Further Reading" section). It is clear that Newton would be a major influence when we read her characterization of the period:

> Enlightenment was a desire for human affairs to be guided by rationality rather than by faith, superstition, or revelation; a belief in the power of human reason to change society and liberate the individual from the constraints of custom or arbitrary authority; all backed up by a world view increasingly validated by science rather than religion or tradition.

It was in France that the Enlightenment was so important, leading up, as it did, to the French Revolution. It also spread to Germany, Italy, and America.

In order for that basic approach set out by Newton and Locke to conquer Europe, it was necessary to overthrow the entrenched Cartesian philosophy. That was a slow process. (See the recent major study by Shank, listed in the "Further Reading" section at the end of this chapter.) The first senior scientist who became a confessed Newtonian was Pierre Louis Maupertuis, whom we met in chapter 24 when discussing the dispute over the shape of the Earth. It was not until 1732 that Maupertuis used Newtonian ideas in his book *Discours sur la figure des asters*.

The leading champion of Newtonianism was the French writer universally known as Voltaire. He had escaped to England in 1728–1729 to avoid personal problems, and there he became acquainted with Newton's work. He enthusiastically presented his observations of Newton and the scientific and philosophical scene in England in his *Les Lettres Philosophiques*, also known as *Letters from England*. For many Frenchmen, these gave their first contact with Newton and his achievements. Voltaire followed up in 1738 with *Elémens de la philosophie de Newton*. Voltaire took advice from Maupertuis, but his greatest assistance came from his mistress Émilie du Châtelet. In contrast to Voltaire, Émilie really did understand mathematics and physics and, in 1749, with the help of Alexis Claude Clairaut, she published a French translation of the *Principia*. In fact, it remains the only French translation. (For the wonderful story of Émilie and Voltaire, see the recent books by Bodanis and Arianrhod, both of which are listed in this chapter's "Further Reading" section.) Gradually Newton was becoming triumphant over Descartes and Leibnitz.

Perhaps the greatest of all French writing in the Enlightenment is the *Encyclopédie*. It was edited largely by Denis Diderot, and its seventeen folio volumes of text and eleven volumes of plates were published in 1751–1772 (see the book by Blom in the "Further Reading" section at the end of this chapter). At first, Diderot was assisted by Jean-Baptiste le Rond d'Alembert, whom we met in chapters 24 and 25. D'Alembert wrote the great introduction or "Preliminary Discourse," which is a brilliant survey of knowledge, philosophy, and scientific methods, setting the scene for the whole work. It is available in translation today and is well worth reading. Here at last is a clear statement that Newton's approach to the gravitational force is acceptable; when writing about mechanics, d'Alembert states:

> We extend our investigations even to the movement of bodies animated by unknown driving forces or causes, provided the law whereby these causes act is known or supposed to be known.[12]

He puts astronomy at the head of "*the physic-mathematical sciences.*" After commenting on its successes, he writes:

> Thus it may justly be regarded as the most sublime and the most reliable application of Geometry and Mechanics in combination, and its progress may be considered the most incontestable monument to the success to which the human mind can rise by its efforts.[13]

The triumphs of Newton and those who followed him are finally given their rightful recognition.

The publication of the *Encyclopédie* was an event of enormous importance. Historian Philipp Blom summarizes it as follows:

What makes it the most significant event in the entire intellectual history of the Enlightenment is the particular constellation of politics, economics, stubbornness, heroism, and revolutionary ideas that prevailed, for the first time ever, against the accumulated determination of Church and Crown, of all the established forces in France taken together, to become a triumph of free thought, secular principle, and private enterprise. The victory of the *Encyclopédie* presaged not only the Revolution, but the values of the two centuries to come.[14]

27.8. CONCLUSION

The work begun by Newton and Locke forms the basis for much of science as we know it today. Newton's contribution may not be known in complete detail, but his name remains synonymous with much in our scientific and intellectual world. There are still cartoons about Newton and falling apples, and newly discovered aspects of his stranger alchemist and religious studies still lead to articles in our daily newspapers. (See the article by McNeil in the "Further Reading" section at the end of this chapter for more.) In my (and Newton's) old town of Grantham, there is, naturally, a Sir Isaac Newton pub, and even the Isaac Newton Shopping Center. I doubt that Newton would approve of either!

Even today, only Einstein can rival Newton as the great icon of science. In 2005, the Royal Society conducted a poll to assess opinions on who had made the bigger contribution to science, given the state of knowledge during his time, and who made a bigger contribution to humankind. More than 1,300 members of the general public voted for Newton by 60.9 percent to 39.1 percent. It was much closer for the 345 members of the Royal Society: Newton, 50.1 percent; Einstein, 49.9 percent. There have been other polls, but always Newton and Einstein are in a class of their own. I wonder how you would cast your vote?

27.9. FURTHER READING

A selection of general references. (If I had to nominate just one book for the general reader, it would be the recent, large-format, beautifully illustrated book by Mordechai Feingold.)

Cohen, I. Bernard, and G. E. Smith. *The Cambridge Companion to Newton*. Cambridge: Cambridge University Press, 2002.

Cohen, I. Bernard, and R. S. Westfall, eds. *Newton: Texts Background Commentaries*. New York: Norton, 1995.

Dobbs, B. J. T., and M. C. Jacob. *Newton and the Culture of Newtonianism*. New York: Prometheus Books, 1995.

Fauvel, J., R. Flood, M. Shortland, and R. Wilson, eds. *Let Newton Be!* Oxford: Oxford University Press, 1988.

Feingold, M. *The Newtonian Moment: Isaac Newton and the Making of Modern Culture*. Oxford: Oxford University Press, 2004.

Hankins, T. L. *Science and the Enlightenment*. Cambridge: Cambridge University Press, 1985.

Janiak, A., ed. *Newton: Philosophical Writing*. Cambridge: Cambridge University Press, 2004.

Koyré, A. "The Significance of the Newtonian Synthesis." In *Newton: Texts Background Commentaries*. Edited by I. Bernard Cohen and R. S. Westfall. New York: Norton, 1995.

McNeil, M. "Newton as National Hero." In *Let Newton Be!* Edited by J. Fauvel, R. Flood, M. Shortland, and R. Wilson. Oxford: Oxford University Press, 1988.

Schaffer, S. "Newtonianism." In *Companion to the History of Modern Science*. Edited by R. C. Olby, G. N. Cantor, J. R. R. Christie, and M. J. S. Hodge. London: Routledge, 1990.

Westfall, R. S. *Never at Rest: A Biography of Isaac Newton*. Cambridge: Cambridge University Press, 1980.

For section 27.2:

Cohen, I. Bernard. "The Review of the First Edition of Newton's *Principia* in the *Acta Eruditorium*, with Notes on the Other Reviews." In *The Investigation of Difficult Things*. Edited by P. M. Harman, Alan E. Shapiro, and D. T. Whiteside. Cambridge: Cambridge University Press, 1992.

Gjertsen, D. *The Newton Handbook*. London: Routledge and Kegan Paul, 1986.

Hall, A. R. "Correcting the *Principia*." *Osiris* 13 (1958): 291–326.

Maglo, Koffi. "The Reception of Newton's Gravitational Theory by Huygens, Varignon, and Maupertuis: How Normal Science May Be Revolutionary." *Perspectives on Science* 11 (2003): 135–69.

Speiser, D. "Newton's *Principia*." In *Discovering the Principles of Mechanics 1600–1800*. Edited by K. Williams and S. Caparrini. Basel, Switz.: Birkhauser, 2008.

For section 27.3, see the book by Hankins and:

Brown, T. M. "Medicine in the Shadow of the *Principia*." *Journal of the History of Ideas* 48 (1987): 629–48.

Harman, P. M. "Newton to Maxwell: The *Principia* and British Physics." *Notes and Records of the Royal Society of London* 42 (1988): 75–96.

Pask, C. "Mathematics and the Science of Analogies." *American Journal of Physics* 71 (2003): 526–34.

Raphael, D. D. "Newton and Adam Smith." In *Newton's Dream*. Edited by M. S. Stayer. Toronto: McGill-Queens University Press, 1988.

Thomson, Sir W., and P. G. Tait. *Principles of Mechanics and Dynamics*. New York: Dover, 1962.

For section 27.4, see the general references (especially Feingold and Gjersten) and:

Alic, M. *Hypatia's Heritage: A History of Women in Science from Antiquity to the Late Nineteenth Century.* Reading, Berkshire, UK: Cox and Wyman, 2001.

Nicolson, M. H. *Newton Demands the Muse.* Princeton: Princeton University Press, 1966.

Rogers, P., ed. *The Context of English Literature: The Eighteenth Century.* London: Methuen 1978. [see particularly the chapters "Religion and Ideas" by W. A. Speck and "Science" by G. S. Rousseau]

Snobelen, S. D. "On Reading Isaac Newton's *Principia* in the 18th Century." *Endeavour* 22 (1998): 159–63.

For section 27.5, see articles in the Cohen and Smith *Companion*, as well as the Cohen and Westfall and Janiak books. Also see:

Dixon, T. *Science and Religion: A Very Short Introduction.* Oxford: Oxford University Press, 2008.

Jacob, M. C. *The Newtonians and the English Revolution 1689–1720.* Hassocks, Sussex, UK: Harvester Press, 1976. [see especially chs. 4 and 5]

Strong, E. W. "Newton and God." *Journal of the History of Ideas* 13 (1952): 147–67.

Westfall, R. S. "Newton and Christianity." In *Newton: Texts Background Commentaries.* Edited by I. Bernard Cohen and R. S. Westfall. New York: Norton, 1995.

For section 27.6, this is a topic covered in many of the general references. See the various articles in the Cohen and Smith, Cohen and Westfall, and Janiak books.

Forbes, E. G. "Newton's Science and the Newtonian Philosophy." *Vistas in Astronomy* 22 (1978): 413–18.

Rogers, G. A. J. "Locke's *Essay* and Newton's *Principia*." *Journal for the History of Ideas* 39 (1978): 217–32.

———. "Locke, Newton and the Enlightenment." *Vistas in Astronomy* 22 (1979): 471–76.

For section 27.7, see Feingold (especially the chapter "The Voltaire Effect") and:

Arianrhod, R. *Seduced by Logic: Émilie du Châtelet, Mary Somerville and the Newtonian Revolution*. Brisbane, Aus.: University of Queensland Press, 2011.

Blom, P. *Enlightening the World: Encyclopédie, the Book That Changed the Course of History*. New York: Palgrave Macmillan, 2004.

Bodanis, D. *Passionate Minds: The Great Enlightenment Love Affair*. London: Little, Brown, 2006.

Bristow, W. "Enlightenment." In *Stanford Encyclopedia of Philosophy*. Stanford University Metaphysics Research Lab. Article first published in 2010. http://plato.stanford.edu/entries/enlightenment (accessed May 22, 2013).

d'Alembert, Jean-Baptiste le Rond. *Preliminary Discourse to the Encyclopedia of Diderot*. Translated by R. N. Schwab. Chicago: University of Chicago Press, 1995.

Jacob, M. C. "Newtonian Science and the Radical Enlightenment." *Vistas in Astronomy* 22 (1979): 545–55.

Johnson, W., and S. Chandrasekhar. "Voltaire's Contribution to the Spread of Newtonianism." *International Journal of Mechanical Science* 35 (1990): 423–53.

Outram, Dorinda. *The Enlightenment*. 2nd ed. Cambridge: Cambridge University Press, 2005.

Shank, J. B. *The Newton Wars and the Beginning of the French Enlightenment*. Chicago: University of Chicago Press, 2008.

MECHANICS AFTER NEWTON

The work of Kepler, Galileo, Huygens, and Newton provided the basis for our modern theory of dynamics and indeed for our approach to science as a whole. In this chapter, I will briefly survey some of the key developments that followed on from Newton's work.

28.1. NEWTON'S LEGACY

Newton made the decisive step from the kinematics of Kepler and Galileo (among others) to a theory of dynamics. He began with inertial frames of reference and how transformations are made between such frames. He then introduced the concept of force, which accounted for changes away from uniform inertial motion. The effect of a force on a body depends on its mass, so it was necessary to define the mass of a body. Forces are vector quantities, and the parallelogram law tells us how to combine forces or resolve a force into components. The third law allows us to use those ideas for interacting bodies by assuming that they exert equal and opposite forces on each other. In particular, Newton discovered the universal force of gravity.

Those concepts are used mathematically when the laws of motion are applied. Newton demonstrated how this leads to general results, like the conservation of linear momentum and the uniform motion of the center of mass, and to specific results, like the orbits of planets, moons, and comets.

While describing Newton's work, I have also pointed out various places where he had come to barriers limiting his further progress in particular topics. These deficiencies fall under three main headings:

mathematical and technical problems;
the need to extend the theory in order to cover more applications; and
fundamental and conceptual problems in the theory.

Addressing these issues lies at the heart of the history of dynamics.

28.1.1 Immediate Mathematical Problems

As we are now painfully aware, following the details in the *Principia* is
made difficult by Newton's use of limiting procedures set in a geometric
approach that is alien to the modern reader. In this book, I have given
introductions to a modern version of Newton's theory, which involves
calculus and differential equations. Although it is obvious that Newton
was using part of that modern formalism, it is almost never stated explicitly
and, in places, apparently quite deliberately hidden away. Thus a first
significant step was the conversion of Newton's dynamics into the form
presented in chapter 11.

Pierre Varignon (1654–1722) was a French professor of mathematics
who corresponded with Newton and received from him a second-edition
copy of the *Principia* and even a portrait of Newton! He used calculus
methods to derive some of Newton's results for centripetal and centrifugal
forces. He was also involved in the Newton–Johann Bernoulli disputes
about the correctness of parts of the *Principia*. The first systematic presentation of Newton's mechanics in a new mathematical form was probably
that given by Jacob Herrmann in his book *Phoromonia, sive de viribus corporum solidorum et fluidorum*, published in 1716. It is usually agreed that
the formulation of Newton's mechanics that we would recognize today
was first set out by Leonard Euler (1707–1783) in his 1736 *Mechanica sive
motus scientia analytice exposita*. From then on, scientists like Jean-Baptiste
le Rond d'Alembert and Alexis Claude Clairaut could use differential
equations and the algebraic or analytical theory to investigate motion.
This new formalism greatly helped with perturbation theory for tackling
problems such as those mentioned in chapters 23–25.

28.2 NEW APPROACHES: POTENTIALS

Although Newton was concerned about the nature and origin of forces, his major achievement was to use the force concept to successfully describe the dynamics of a dazzling array of phenomena. In one sense, he put aside the philosophical difficulties with his "*satis est—it is enough*" statement in reference to those successes. I now come to a different approach to forces, one that makes the applications easier, paves the way for a more generalized approach, and ultimately (see section 28.8) tackles the worrying conceptual problems.

28.2.1. Defining the Potential

Newton introduced the vector force $\mathbf{F} = (F_x, F_y, F_z)$, and the components are used when the force is resolved in different directions to study the motion of a body acted upon by that force in that direction. We now introduce a scalar potential function $V(x, y, z)$, which is related to the force by differentiation:

$$F_x = -\frac{\partial V}{\partial x}, \quad F_y = -\frac{\partial V}{\partial y}, \quad F_z = -\frac{\partial V}{\partial z}, \quad \text{or generally} \quad \mathbf{F} = -\nabla V(x,y,z). \quad (28.1)$$

Thus, mathematically, giving $V(x, y, z)$ is equivalent to specifying $\mathbf{F} = (F_x, F_y, F_z)$. But what has been added to the physical picture?

28.2.2. Work Done by Forces

The work done by a force acting on a body is the magnitude of the force multiplied by the distance moved, which will require an integral in the general case. Suppose the force acts on the body moving from x_0 to x_1. For the one-dimensional case, the work done is

$$\int_{x_0}^{x_1} F \, dx = -\int_{x_0}^{x_1} \frac{dV}{dx} \, dx = V(x_0) - V(x_1).$$

A similar result holds in the general three-dimensional case: the work done is given by the change in the potential function.

28.2.3. The Equations of Motion

Simply using equation (28.1) in Newton's second law gives an equation of motion for a particle in terms of the potential,

$$m\frac{d^2x}{dt^2} = -\frac{dV(x)}{dx} \quad \text{or generally } m\frac{d^2\mathbf{r}}{dt^2} = -\nabla V(\mathbf{r}). \quad (28.2)$$

We now speak of a body moving in a force field specified by its potential function V.

If we consider the body moving from x_0 to x_1 and integrate the equation of motion, we find

$$\tfrac{1}{2}m\left(\frac{dx_1}{dt}\right)^2 - \tfrac{1}{2}m\left(\frac{dx_0}{dt}\right)^2 = V(x_0) - V(x_1),$$

$$\text{or} \quad \tfrac{1}{2}m\left(\frac{dx_1}{dt}\right)^2 + V(x_1) = \tfrac{1}{2}m\left(\frac{dx_0}{dt}\right)^2 + V(x_0), \quad (28.3)$$

$$T_1 + V(x_1) = T_0 + V(x_0).$$

Two important observations can now be made. First, the equation of motion begins as a second-order differential equation in (28.2), but is only a first-order differential equation in (28.3). That takes us a long way toward finding the particle trajectory, as explained in chapter 11.

Second, the second form in equation (28.3) is just the conservation-of-energy law: if a body moves under the action of a force that may be described in terms of a potential function V, then the sum of the kinetic energy T and the potential energy V is a constant of motion. The potential has now been interpreted as an energy term. Note that forces that cannot be written in terms of a potential, as in equation (28.1), are dissipative, like the resistance forces studied in *Book II*, and they cause a loss of energy.

28.3. GENERALIZING THE EQUATIONS OF MOTION

Introducing the potential has given a way to eliminate forces; a body now moves in a force field characterized by a potential function. Physically, we can relate the potential to work done and create a definition of energy for a moving particle. Old questions in the *vis viva* dispute (see chapter 6) are also settled by this change of emphasis.

The next step was made by Joseph Louis Lagrange (1736–1813), who introduced "generalized coordinates" to specify the state of a dynamical system. A single coordinate q will specify the situation if we can write the kinetic energy T and potential energy V in terms of it to obtain $T(\dot{q}, q)$, where $\dot{q} = dq/dt$, and $V(q)$. (Newton's fluxions may have been long gone, but his dot notation for a derivative is still a useful abbreviation.) We now form what has come to be known as the Lagrangian L:

$$L(\dot{q}, q) = T(\dot{q}, q) - V(q). \qquad (28.4)$$

The Lagrangian is simply the difference between the kinetic energy and the potential energy. Then the Lagrange equation of motion is

$$\frac{d}{dt}\left(\frac{\partial L}{\partial \dot{q}}\right) - \left(\frac{\partial L}{\partial q}\right) = 0. \qquad (28.5)$$

If q is taken to be the usual spatial coordinate x, Lagrange's equation becomes identical with Newton's equation of motion. Most often it will take several coordinates $q_1, q_2, q_3, \ldots q_n$ to fully specify the system of interest if it has n "degrees of freedom." In that case, L will depend on all those coordinates, and there will be an equation like (28.5) for each q_i.

We are now moving into a technical area of the theory, and it is not appropriate to go into details here; refer to the dynamics texts recommended in section 6.9.1 for further information. However, we can note the key advances: the theory now uses any appropriate and convenient coordinates, and the dynamical behavior is determined by giving expressions for the kinetic and potential energies. The vectors have gone and there is no messy resolving of forces. The new formulation is sometimes called analytical mechanics. Here is the way Lagrange expressed it in his *Le Méchanique Analitique* (1788):

> There are several treatises on mechanics, but the plan of this one is entirely new. I have set myself the task of reducing the theory of this science, and the art of solving problems concerned with it, to general formulae, the simple development of which gives all the equations necessary to the solution of each problem.
>
> No drawings are to be found in this work. The methods which I present require neither constructions nor geometrical or mechanical arguments, but only algebraic operations, subject to a regular and uniform process.

So no geometry and not a single diagram—surely a progression as far as possible away from the *Principia*!

This formal development was extended into a highly sophisticated approach by William Rowan Hamilton (1805–1865) and Carl Gustav Jacobi (1804–1851). Mathematical techniques, such as the perturbation theory initiated by Newton, were developed by many people, but particularly by Pierre-Simon Laplace (1749–1827) in his five-volume *Traité de mécanique céleste*. Here is how he summarized things in the preface to volume 1:

> Towards the end of the seventeenth century Newton published his discovery of universal gravitation. Mathematicians have, since that epoch, succeeded in reducing to this great law of nature all the known phenomena of the system of the world and have thus given to the theories of the heavenly bodies, and astronomical tables, an unexpected degree of precision.

For more on this topic, see Bruno Morando's article "The Golden Age of Celestial Mechanics." This area has continued to be of importance as satellites were launched and space exploration began. The advent of computers also extended the range of problems that can be tackled.

28.4. A DIFFERENT PERSPECTIVE

Newton's approach is to consider at each point in a particle's motion the combination of an inertial component and a force-created component. Expressing the second law of motion in analytical terms gives a differential equation that contains the essence of Newton's approach, as illustrated in equation (11.2) and discussed in section 11.1. The differential equation may be integrated to give whole possible trajectories; initial conditions are used to pick out actual paths. This provides the link to whole-path results like Kepler's first law.

There is another approach to mechanics that seems quite different from Newton's. It is still based on the Lagrangian, but now the fundamentals of mechanics are expressed in:

Hamilton's Principle: The motion of the system from time t_1 to time t_2 is such that the line integral, or action integral,

$$I = \int_{t_1}^{t_2} L \, dt,$$

has a stationary value for the actual path of the motion.

A stationary value means either a maximum or minimum, but it is most often a minimum. This way of looking at mechanics says that for a given time interval in the motion, you can.try out all possible paths or ways in which the whole system can move, but if you evaluate the integral of the Lagrangian over that selected interval, then you will find that only the actual, physically realizable path gives a minimum for that integral. Hamilton's principle picks out the actual paths for a dynamical system. This is a "whole-path" formulation.

What a strange and beautiful result. It turns out that most of physics can be formulated in terms of similar "variational principles." See the "Further Reading" section at the end of this chapter for references—I particularly recommend Richard Feynman's wonderful lecture titled "The Principle of Least Action." Some people see a mystic or religious side to these kinds of principles, with thoughts about a God creating the best possible world or one understandable in terms of meaningful principles.

This approach to mechanics is based on the mathematics known as the calculus of variations. Newton was a pioneer in this area, as we saw in section 23.3.1. Problems involving the functions required to maximize or minimize an integral can be shown to be equivalent to the solution of a certain set of differential equations, named after their discoverers as the Euler-Lagrange equations. In this way, Hamilton's principle leads to the Lagrange equations, equation (28.5), and, hence, back to Newton's laws of motion. Thus, for many situations, all approaches to mechanics are equivalent (as they must be), but they may have different values in terms of computation or interpretation. For example, dissipative systems may be described using Newton's laws (as we saw in chapter 19), but they are not covered by Hamilton's principle.

28.5. CONSTANTS OF MOTION

Quantities that remain constant throughout the motion of a physical system, or conserved quantities as they are often called, play a vital role in mechanics and, in fact, in science as a whole (see the reference to Feynman). A typical example is the energy $T + V$, as in equation (28.3). In this book, we have seen how they play a crucial part in solving the equations of motion. Newton began the theory of constants of motion when he showed quite generally that linear momentum is conserved (see sections 6.3 and 6.6.1). I mention three points on this topic (the details may be found in the recommended texts).

The form of the Lagrangian can immediately indicate that there are conserved quantities, and then manipulation of Lagrange's equations (28.5) will lead to an expression for those quantities. For example, if L does not depend explicitly on the time t, so that $\partial L/\partial t = 0$, then energy will be conserved.

We know that the search for more constants of motion is futile in certain cases. Most importantly, for a system of discrete bodies with motions described by classical mechanics, Henri Poincaré (1854–1912) showed that there are only ten such constants. We saw earlier that this makes the solution of the three-body problem very difficult.

The origin of conserved quantities was given a physical meaning by Emmy Noether (1882–1935). Her main result, now known as Noether's theorem, states that a symmetry in the dynamical system will be associated with a conservation law. For example, if the Lagrangian is unchanged by linear shifts in time or space specification, then we will get conservation of energy and linear momentum, respectively (see the resource by Goldstein for the technical details). Einstein built in much of the physics directly into his mathematical formulation, as given in "The Foundation of the General Theory of Relativity," so it is no surprise to find that he naturally comes, in section 17, to "The Laws of Conservation in the General Case."

28.6. EXTENSIONS

Newton began his theory of mechanics, as we still do today, by considering the motion of a particle or a body that is so small that it may be described by giving point coordinates. He deals with the two- and three-body problems and comments on the general case (see chapter 16). He does offer some examples, moving beyond that formulation with his work on large, composite bodies (see chapter 17) and several discussions of nonspherical bodies, gases, and liquids in *Books II* and *III*. As we have repeatedly seen, he reaches the limits of his methods for these cases, and over the following century or so, it was a major part of mathematical science to address those deficiencies. If there are many particles, can we identify physical attributes of the collective system that allows a theory to be developed?

28.6.1. Gases

In chapter 21, we saw Newton beginning a theory of gases that involved "*particles fleeing from each other*" and the construction of scaling arguments leading to Boyle's law. However, it is usually accepted that the first contribution to what is now called the kinetic theory of gases was made by Daniel Bernoulli in his 1738 book *Hydrodynamica*. (Readers should consult the book by Brush for original papers and commentary, as well as Truesdell's book, both of which are listed in the "Further Reading" section at the end of this chapter.) The idea that many particles may be treated statistically, and that their collisions produce a distribution of velocities, came over one hundred years later with Maxwell's *Dynamical Theory of Gases*. Here was a satisfactory derivation of Boyle's law and the properties of gases from mechanical principles. The theory was then extensively developed by people like physicists Rudolf Clausius and Ludwig Boltzmann, and eventually it evolved into the discipline known as statistical mechanics.

28.6.2. Rigid Bodies

Rigid bodies are at the other extreme; the constituent particles are taken to be fixed in place so that they form a single, unchanging body. Newton was obviously aware of this case; right from the start when discussing *Law I*, he writes of "*a top, whose parts by their cohesion are perpetually drawn aside from rectilinear motions.*" Newton also introduced the idea of summing or intergrating over all the constituent parts in order to produce an equivalent yet simpler theoretical entity. In chapter 17, we saw that the interacting particles in large spherical bodies could be replaced, for calculational purposes, by a single force acting between the centers of those bodies.

Newton did consider nonspherical bodies and their attractions, but he produced no general theory for their motion that involved their orientation in space. We saw in chapter 24 that he used ad hoc methods when discussing the shape of the Earth and its tides. The rotation of bodies is understood using the notions of angular momentum and moments of forces, or torques. The idea of the lever is ancient; Huygens had begun by considering the composite pendulum and Jacob Bernoulli wrote about the center of oscillation using ideas based around the lever. However, it was Leonard Euler who gave the theory to parallel Newton's theory of linear momentum and the center of mass.

For a nonspherical body, it is necessary to specify its orientation in space and to describe in some overall way the distribution of mass around any given axis. This is what Euler did in *Opera omnia* with *Theoria motus corporum solidorum seu rigidorum*, appearing in 1765. Look in a modern textbook and you will find "Euler angles," "moments of inertia," and the "inertia tensor." There are now equations of motion for the rotational mechanics of rigid bodies, with famous and beautiful applications to spinning tops. The theory for the precession of the equinoxes was then successfully presented by d'Alembert and Euler.

28.6.3. Deformable Bodies and Elasticity

Newton had used deformable bodies in the collision experiments discussed in the *Scholium* following the introduction of the laws of motion. However, he gave no theory for extended, deformable systems, which became an intense area of research in the eighteenth century (see this chapter's "Further Reading" references to Speiser and Truesdell).

Perhaps the simplest (and certainly the most important and famous) problem is the vibration of a stretched string. If a length of string (or wire, or whatever) is stretched along the x axis, the deviation from a straight line may be quantified in terms of $y(x,t)$. Because the deflection depends on both position along the string x and the time t, it is necessary to introduce a partial differential equation for the dynamics. Thus a whole new area of mathematics must be developed. The equation of motion for the string involves an elasticity constant determining how the string resists stretching. The history of this problem is covered by Cannon and Dostrovsky. Gradually it was shown that there are specific modes of vibration with individual frequencies. The validity of the superposition principle, in which those modes are linearly combined to give the total motion of the string, is one of the great problems in early applied mathematics. These ideas were extended to a variety of distortion and vibration problems in two and three dimensions, and they assisted with the development of the theory of waves.

28.6.4. Fluids

This is another part of mechanics that Newton struggled with in *Book II* without producing any real, general, dynamical theory. The continuous system is now deformable and flows according to complex dynamics. The fluid must be characterized by its local density and motion, and its properties are linked to compressibility and viscosity. Obviously, partial differential equations are again required. Euler was able to consider the motion of an element in the fluid to derive the Euler equation relating fluid motion to pressure and any external forces. This area has become of major importance in applied mathematics and several branches of engineering.

28.7. THE NATURE OF MOTION

Newton solved problems about motion along conic sections and other relatively simple curves. He knew that the analysis of the motion of systems comprising three or more bodies was difficult, and he appreciated that even small perturbing interactions might produce complex results, as for the motion of the Sun-Jupiter-Saturn system, for example (see section 23.5). He also knew that the Earth was nonspherical and its influence was not entirely described by one simple, central force. Gradually these various complications led to questions about the nature of the trajectories for bodies predicted by classical mechanics and realized in nature.

A new approach is often associated with Henri Poincaré. One aspect of this is to look not at the shape of curves traced out in space, but at the way the canonical variables, position and momentum, change as the motion takes place. As a simple example, figure 28.1 shows the curves traced out in the $\dot{\theta}, \theta$ plane for the pendulum with various amplitudes (see figure 20.2), and in the $\dot{r}, r$ plane for a planet (see equation 11.26). These phase diagrams allow us to study the nature of the motion and how positions and speeds vary during the motion. We can go on to plot whole sets of curves on such diagrams showing how the motion properties change as initial conditions or force parameters change. In this way, the emphasis is placed on the nature of the motion for whole sets of orbits rather than the specific time-evolution details for motion along a single trajectory.

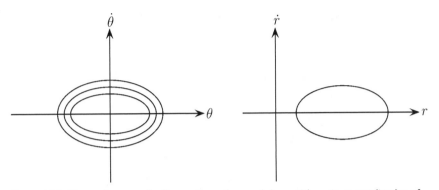

Figure 28.1. Phase diagrams for the motion of a pendulum with various amplitudes of oscillation and a planet.

It may be possible to study the nature of motion considered in this way even when the complete details of any one example are extremely difficult to generate. It has been discovered that the straightforward, regular sort of motions familiar to Newton may not be present when problems deviate from their simplest forms. For example, what happens when the forces are no longer central or centripetal? Or what happens when three or four particles interact?

It is now known that even simple systems may produce very complicated motions. Furthermore, it is sometimes the case that even very small changes in the initial conditions may produce wildly different trajectories. This is now often referred to as chaotic motion. (The books by Stewart and Peterson, listed in the "Further Reading" section at the end of this chapter, provide simple and relatively nontechnical introductions.)

One of the driving applications for this area of research is the study of the stability of the solar system. When all the various planets and the occasional disturbing comets are taken together, as one large interacting system, will we get the stable "clockwork heavens" that appears to be the case? Of course, it depends on the timescales involved, and computations projecting millions of years into the future are now made. Laplace believed he had proven the stability of the solar system, but today there are doubts (see the article by Laskar in this chapter's "Further Reading" section). Whatever the final outcome, we do now know that the regular, "deterministic" motion discovered by Newton is not the whole story.

28.8. EINSTEIN AND RELATIVITY THEORY

Over two hundred years passed from the first publication of the *Principia* before there were serious conceptual challenges to Newton's mechanics requiring a change in the fundamental formulation. Early in the twentieth century, Albert Einstein produced his theory of relativity.

The special theory followed Einstein's study of electromagnetic theory, and brought that too into the framework in which physical laws were to be valid in any inertial frame of reference. However, Einstein also had to build in the fact that the speed of light is the same in any such frame. The result is that the simple transition between inertial frames, as illustrated in Newton's sailor-on-a-ship example (see section 5.5.3), must be replaced with the Lorentz transformation. It also becomes necessary to modify thinking about mass and energy.

Einstein's 1916 general theory of relativity deals with major conceptual matters. They take us back to matters raised by Newton. First, there is the question of why the inertial and gravitational masses are equal (see section 22.6.1). Einstein used that fact in his famous deliberations about the comparison of observations made in free fall or in a uniform gravitational field. (For his lovely discussion, see chapter 20 in his *Relativity: The Special and General Theory*.) Thus it was that he related these matters to geometry and the properties of space-time. Newton had pondered the nature of space and time (see section 5.5), but the conceptual framework he proposed was attacked by Leibnitz and has been challenged ever since. He was also troubled right to the end (see section 26.4.1) by the nature and origin of the gravitational force. In Einstein's theory, a body travels in a space with metric properties controlled by gravitation; its path is determined by those properties rather than by a force. In the memorable words of theoretical physicist John Wheeler, "Space-time tells matter how to move; matter tells space-time how to curve."[1]

So it was that Einstein tied up Newton's worries in one major step—mass, the nature of space, and the operation of gravitation are inextricably linked. Here is Einstein's summary:

> There can be no space nor any part of space without gravitational potentials; for these confer upon space its metrical qualities, without which it cannot be imagined at all. The existence of the gravitational field is inseparably bound up with the existence of space.[2]

This is a difficult matter and, obviously in response to that, in 1952 Einstein added an appendix on "Relativity and the Problem of Space" to his 1916 book. It is well worth reading.

Before leaving this topic, I would like to make three points. First, it was one of Einstein's requirements that his theory should reduce to Newton's in certain circumstances. That is indeed the case, and it is the gravitational potential that appears as a common quantity in the two theories. (See section 21, "Newton's Theory as a First Approximation" in Einstein's 1916 foundational paper.)

Second, the logic and physical requirements, so beautifully used by Einstein to produce his theory, lead naturally to an inverse-square-law gravitational force in the Newtonian approximation. Einstein has explained why gravity follows that particular force law, with all its wonderful ramifications, as discussed in chapter 18.

Third, I must reiterate the point made in the preface and in section 5.7: it is only necessary to move from Newton's mechanics to Einstein's in extreme circumstances; in virtually all day-to-day situations, Newton's theory is completely adequate. The famous tiny defect in the celestial-mechanics theory for the motion of Mercury can be corrected by using the relativistic modification to Newton's theory, as discussed in section 14.4.2.

28.9. CONCLUSION

The theories that Newton gave us in the *Principia* are often incomplete, formulated in obscure ways, and tied up in models and analogies that only such a genius could invent. History has shown us that the basic ideas are correct and can be developed into a powerful and widely applicable part of science. The *Principia* remains a great and stimulating document; it is with that book that the science of dynamics truly began.

28.10. FURTHER READING

The mechanics texts recommended in section 6.9.1 cover the material in this chapter. The book by Moulton has brief historical sections. For a more historical treatment see:

Feynman, R. *The Character of Physical Law*. Cambridge, MA: MIT Press, 1987. [see particularly the chapter titled "*The Great Conservation Principles*"]

Laskar, J. "The Stability of the Solar System from Laplace to the Present." In *The General History of Astronomy: Volume 2, Planetary Astronomy from the Renaissance to the Rise of Astrophysics—Part A: Tycho Brahe to Newton*. Edited by R. Taton and C. Wilson. Cambridge: Cambridge University Press, 1989.

Lindsay, R. B., and H. Margenau. *Foundations of Physics*. New York: Dover, 1957.

Linton, C. M. *From Eudoxus to Einstein. A History of Mathematical Astronomy*. Cambridge: Cambridge University Press, 2004.

Maglo, Koffi. "The Reception of Newton's Gravitational Theory by Huygens, Varignon, and Maupertuis: How Normal Science May Be Revolutionary." *Perspectives on Science* 11 (2003): 135–69.

Morando, B. *The Golden Age of Celestial Mechanics*." In *The General History of Astronomy: Volume 2, Planetary Astronomy from the Renaissance to the Rise of Astrophysics—Part A: Tycho Brahe to Newton*. Edited by R. Taton and C. Wilson. Cambridge: Cambridge University Press, 1989.

Nauenberg, M. "The Early Application of the Calculus to the Inverse Square Force Problem." *Archive for the History of Exact Sciences* 64 (2010): 269–300.

Peterson, I. *Newton's Clock: Chaos in the Solar System.* New York: W. H. Freeman, 1993.

Speiser, D. *Discovering the Principles of Mechanics 1600–1800.* Edited by K. Williams and S. Caparrini. Basel, Switz.: Birkhauser, 2008.

———. "The Kepler Problem from Newton to Johann Bernoulli." *Archive for the History of Exact Sciences* 50 (1996): 103–16.

Steele, B. D. "Muskets and Pendulums: Benjamin Robbins, Leonard Euler, and the Ballistics Revolution." *Technology and Culture* 35 (19995): 348–82.

Stewart, I. *Does God Play Dice? The New Mathematics of Chaos.* London: Penguin Books, 1990.

Taton, R., and C. Wilson, eds. *The General History of Astronomy: Volume 2, Planetary Astronomy from the Renaissance to the Rise of Astrophysics— Part A: Tycho Brahe to Newton.* Edited by R. Taton and C. Wilson. Cambridge: Cambridge University Press, 1989. [see parts 6 and 8]

Truesdell, C. "History of Classical Mechanics Part I." *Die Naturwissenschaften* 63 (1976): 53–62.

———. "History of Classical Mechanics Part II." *Die Naturwissenschaften* 63 (1976): 119–30.

———. *Essays in the History of Mechanics.* Berlin: Springer-Verlag, 1968.

Whittaker, E. *From Euclid to Eddington: A Study of Conceptions of the External World.* Cambridge: Cambridge University Press, 1949.

For action principles and the calculus of variations as used in mechanics:

Feynman, R. "The Principle of Least Action." In *The Feynman Lectures on Physics.* Edited by R. P. Feynman, R. B. Leighton, and M. Sands. Vol. 2. Reading, MA: Addison-Wesley, 1963.

Lanczos, C. *The Variational Principles of Mechanics.* New York: Dover, 1970.

Yourgrau, W., and S. Mandelstam. *Variational Principles in Dynamics and Quantum Theory.* New York: Dover, 1968.

Many particle systems and continuum mechanics:

Brush, S. G. *Kinetic Theory Volume 1: The Nature of Gases and Heat.* Oxford: Pergamon Press, 1965.

Cannon, J. T., and S. Dostrovsky. *The Evolution of Dynamics: Vibration Theory from 1687 to 1742.* New York: Springer-Verlag, 1981.

Darrigol, O. *Worlds of Flow: A History of Hydrodynamics from the Bernoullis to Prandtl.* Oxford: Oxford University Press, 2005.

Darrigol, O., and U. Frisch. "From Newton's Mechanics to Euler's Equations." In *Physica D—Nonlinear Phenomena* 237 (2008): 1855–69.

Neményi, P. F. "The Main Concepts and Ideas of Fluid Mechanics in Their Historical Development." *Archive for the History of Exact Sciences* 2 (1962): 52–86.

Rouse, H., and S. Ince. *History of Hydraulics.* New York: Dover, 1957.

Trefil, J. S. *Introduction to the Physics of Fluids and Solids.* New York: Pergamon Press, 1975.

Wilson, C. "The Precession of the Equinoxes from Newton to d'Alembert and Euler." *The General History of Astronomy: Volume 2, Planetary Astronomy from the Renaissance to the Rise of Astrophysics—Part A: Tycho Brahe to Newton.* Edited by R. Taton and C. Wilson. Cambridge: Cambridge University Press, 1989.

On relativity see:

Eddington, Sir Arthur. *Space, Time and Gravitation: An Outline of the General Relativity Theory.* Cambridge: Cambridge University Press, 1920. [later reprinted by other publishers]

Einstein, Albert. *Relativity: The Special and General Theory.* 15th ed. New York: Crown Publishers, 1952.

———. "The Foundation of the General Theory of Relativity." 1916. Reprinted in *The Principle of Relativity.* New York: Dover, 1923.

Lambourne, R. J. A. *Relativity, Gravitation and Cosmology.* Cambridge: Cambridge University. Press, 2010.

Weinberg, Steven. *Gravitation and Cosmology.* New York: John Wiley, 1972.

Will, Clifford. *Was Einstein Right?* 2nd ed. New York: Basic Books, 1993.

EPILOGUE

"**N**ewton was the greatest genius who ever lived, and the most fortunate, for we cannot find more than once a system of the world."[1]

So wrote the great French mathematical scientist Joseph Louis Lagrange. The *Principia* reflects Newton's genius and does give us the "system of the world" much as we know it today.

Lagrange's statement implies that if Newton had not found that "system," then someone else would have. That has inspired comparison with William Shakespeare, with the comparative thought that he alone could have written the plays in the wonderful *First Folio*. The poet Coleridge suggested that "the Souls of 500 Sir Isaac Newtons would go to the making up of a Shakespeare or a Milton."[2] Be that as it may, Newton's *Principia* has survived to this very day, alongside Shakespeare's plays, as one of the glories of our civilization. Certainly it is barely known compared with those plays, but maybe this book has convinced you that the *Principia* too has its glories and inspirations.

Shakespeare gave us new words and a sublime use of language; Newton gave us a conceptual framework and methods for science that equally demand our respect and admiration. The amazing thing is that the *Principia* contains such a coherent and comprehensive story, beginning with Newton's ideas about science and concluding with an explanation of the mystery of comets. Today, such developments would be spread over hundreds of research papers with thousands of different authors.

Many people may have a feeling for the great sweep of Newton's achievement, but little knowledge of the originality and brilliance of the working and steps involved. It was my intention to show you some of those achievements. In retrospect, they are so many and so diverse that you may have lost track of them. Perhaps a final review is in order. What has Newton given us? Here are some of my favorite examples:

Conceptual: the concept of mass; importance of space, time, frames of reference, and the first law of motion; the concept of force and the second law of motion taking us from the kinematics of Kepler and Galileo to dynamics; how to combine or resolve forces; the same physics applies on earth and in the heavens; the moons of Jupiter and Saturn provide examples of little solar systems; comets follow the same laws as do planets.

Methodology: going from phenomena to forces, and then from those forces to other phenomena; exploring a formalism mathematically and producing general results; setting out *Rules of Reasoning*; insisting on testing theory against physical data.

Theoretical results: the third law of motion, the resulting conservation of linear momentum and determination of center-of-mass motion; the complete solution for the two-body problem; identification of forces giving closed and rotating orbits; integrations to deal with composite bodies; theory of resisted motion.

Applications: the Moon Test and universal gravitation; explanation of Kepler's laws and a theory of the solar system; finding the mass of planets; the shape of the Earth, its gravity, and its tides; precession of the equinoxes; trajectories of comets; lunar theory and perturbation theory; calculating the speed of sound.

Experiments: testing the equality of inertial and gravitational masses; the use of pendulums and falling bodies to measure resistance forces; testing for an aether.

Mathematics: recognizing the importance of if-and-only-if results; limiting processes to derive general dynamical laws; the non-integrability of algebraic ovals; an application of the calculus of variations; scaling laws and similarity theory.

Numerical analysis: iteration methods for solving equations; interpolation methods and approximation methods for integrals.

Einstein called Newton's grand applications in *Book III* "a deductive achievement of unique magnificence."[3] Newton's friend, the philosopher John Locke, wrote of "the incomparable Mr. Newton."[4] For me, their assessments remain equally valid today. I hope you have enjoyed finding out about the *Principia* and what makes Newton worthy of those accolades.

NOTES

Preface: Why You Should Read This Book

1. Steven Hawking, "Newton's *Principia*," in *Three Hundred Years of Gravitation*, ed. S. W. Hawking and W. Israel (Cambridge: Cambridge University Press, 1987).

2. Albert Einstein, "The Mechanics of Newton and Their Influence on the Development of Theoretical Physics," *Ideas and Opinions* (New York: Crown Publishers, 1959).

3. Story related by Martin Folkes (1690–1754) who was vice president of the Royal Society when Newton was president.

4. I. Bernard Cohen, "A Guide to Newton's *Principia*," introduction to Isaac Newton, *The* Principia: *Mathematical Principles of Natural Philosophy*, trans. I. Bernard Cohen and Anne Whitman (Berkeley: University of California Press, 1999).

5. Steven Weinberg, "Newtonianism and Today's Physics," in Hawking and Israel, *Three Hundred Years of Gravitation*.

6. Einstein, "Mechanics of Newton."

7. Albert Einstein, "Time, Space, and Gravitation," *Out of My Later Years* (New York: Philosophical Library, 1950).

8. Questions to John Arbuthnot, circa 1698. See Derek Gjertsen, *The Newton Handbook* (London: Routledge and Kegan Paul, 1986).

PART 1. INTRODUCTORY MATERIAL

Chapter 1. Introducing Our Hero

1. Albert Einstein, "The Mechanics of Newton and Their Influence on the Development of Theoretical Physics," *Ideas and Opinions* (New York: Crown Publishers, 1959).

2. John Maynard Keynes, "Newton, the Man," available (in part) in *The World of Mathematics*, vol. 1, ed. James R. Newman (London: Allen and Unwin, 1956).

3. Ibid.

4. I. Newton, *Twelve Articles on Religion*, Newton Project, http://www.newtonproject.sussex.ac.uk/view/texts/normalized/THEM00008 (accessed April 22, 2013).

5. I. Newton, *Seven Statements on Religion*, Newton Project, http://www.newton project.sussex.ac.uk/view/texts/normalized/THEM00006 (accessed April 22, 2013).

6. Quoted in A. Janiak, *Isaac Newton: Philosophical Writings* (Cambridge: Cambridge University Press, 2004), p. 64. Also available online at I. Newton, *Letter to Richard Bentley Dated 10 December, 1692*, Newton Project, http://www.newtonproject.sussex.ac.uk/view/texts/normalized/THEM00254 (accessed April 22, 2013).

7. Humphrey Newton, *Two Letters from Humphrey Newton to John Conduit*, Newton Project, http://www.newtonproject.sussex.ac.uk/view/texts/normalized/THEM00033 (accessed April 22, 2013).

8. John Flamsteed, *Suppressed Preface to the Historia Coelestis Britannica*, Newton Project, http://www.newtonproject.sussex.ac.uk/view/texts/normalized/OTHE00033 (accessed April 22, 2013).

9. Quoted by Westfall in *Never at Rest: A Biography of Isaac Newton* (Cambridge: Cambridge University Press, 1980), p. 650 (taken from William Whiston in *Authentic Records*).

10. Quoted from William Whiston's 1749 *Memoirs*, in *The Newton Handbook*, by D. Gjertsen (London: Routledge and Kegan Paul, 1986), p. 610.

11. George Bernard Shaw, comments made at a 1930 banquet in honor of Einstein, recorded by Shaw's secretary, B. Patch, in *Thirty Years with G. B. S.* by B. Patch (London: Gollancz, 1951).

Chapter 2. Setting the Scene

1. Quoted in Andrew Gregory, *Eureka! The Birth of Science* (London: Icon Books, 2001), p. 30.

2. Aristotle, *Physics*, in *The Works of Aristotle*, ed. W. D. Ross (Oxford: Clarendon Press, 1930).

3. Ibid.

4. Ibid.

5. Ibid.

6. Ibid.

7. Ibid.

8. Ibid.

9. Ibid.

10. Ibid.

11. Ibid.

12. Ibid.

13. Quoted in *The Beginnings of Western Science*, 2nd ed., by D. C. Lindberg (Chicago: University of Chicago Press, 2007), p. 310.

14. Letter from Saint Augustine, in *What Augustine Says*, by N. L. Geisler (Grand Rapids, MI: Baker Book House, 1982).

15. N. Copernicus, preface and dedication to Pope Paul III in *On the Revolution of the Heavenly Spheres* (n.p.: 1543), reprinted in *On the Shoulders of Giants*, ed. Stephen Hawking (London: Penguin Books, 2002).

16. M. Luther, quoted in O. Gingerich, *The Book Nobody Read* (New York: Penguin Books, 2004).

17. A. Osiander, "To the Reader: Concerning the Hypotheses of This Work," introduction to Copernicus, *On the Revolutions*.

18. J. Kepler, letter to Michael Maestlin, in *Oxford Dictionary of Scientific Quotations*, ed. W. F. Bynum and Roy Porter (Oxford: Oxford University Press, 2005).

19. J. Kepler, *Epitome of Copernican Astronomy* (n.p.: 1618).

20. J. Kepler, letter to J. G. Herwart von Hohenburg, in Bynum and Porter, *Oxford Dictionary of Scientific Quotations*.

21. Galileo Galilei, *The Assayer* (n.p.: 1623). See M. A. Finocchiaro, ed., *The Essential Galileo* (Indianapolis, IN: Hackett Publishing, 2008).

22. F. Bacon, *A Proposition Touching the Compiling and Amendment of the Laws of England* (n.p.: 1616). Also available in Bynum and Porter, *Oxford Dictionary of Scientific Quotations*.

23. Albert Einstein, letter to J. S. Switzer, April 1953. Quoted in *Albert Einstein: Creator and Rebel*, by Banesh Hoffman (Saint Albans, UK: Paladin, 1977).

24. Robert Boyle, *About the Excellency and Grounds of the Mechanical Hypothesis* (n.p.: 1674), in *Selected Philosophical Papers of Robert Boyle*, ed. M. A. Stewart (Manchester, UK: Manchester University Press, 1979).

25. R. Descartes, *Principles of Philosophy* (n.p.: 1644). Reprinted with translation by V. R. Miller and R. P. Miller (Dordrecht, Neth.: Reidel Publishing, 1983), part IV, sec. 203.

26. Galileo Galilei, *Letter to the Grand Duchess Christina* (n.p.: 1615). Also available in M. A. Finocchiaro, ed., *The Essential Galileo*.

27. Galileo Galilei, *Two New Sciences* (n.p.: 1638). Reprinted with translation by Drake Stillman (Madison: University of Wisconsin Press, 1974).

28. Ibid.

29. Ibid.

30. R. Descartes, letter to M. Mersenne, quoted in *Augustine to Galileo*, by A. C. Crombie (London: Penguin Books, 1959).

31. Descartes, *Principles of Philosophy*, part III, sec. 30.

Chapter 3. A First Look at the *Principia*

1. A. de Moivre, quoted in Westfall, *Never at Rest: A Biography of Isaac Newton* (Cambridge: Cambridge University Press, 1980) p. 403.

2. E. Halley, quoted in *The Newton Handbook*, by Derek Gjertsen (London: Routledge and Kegan Paul, 1986), p. 460.

3. W. Derham, quoted in Westfall, *Never at Rest*, p. 459.

4. D. T. Whiteside, *The Mathematical Principles underlying Newton's* Principia Mathematica, 9th lecture in the Gibson Lectureship in the History of Mathematics (Glasgow, UK: University of Glasgow, 1970).

5. Albert Einstein, "The Mechanics of Newton and Their Influence on the Development of Theoretical Physics," *Ideas and Opinions* (New York: Crown Publishers, 1959).

6. I. Bernard Cohen, "A Guide to Newton's *Principia*," introduction to Isaac Newton, *The* Principia: *Mathematical Principles of Natural Philosophy*, trans. I. Bernard Cohen and Anne Whitman (Berkeley: University of California Press, 1999), ch. 11.

7. Albert Einstein, "The Theory of Relativity," *Essays in Physics* (New York: Philosophical Library, 1950).

8. Ernst Mach, *The Science of Mechanics: A Critical and Historical Account of Its Development* (1893; repr., La Salle, IL: Open Court Publishing, 1960), p. 226.

9. A. Sommerfeld, introduction to *Mechanics: Lectures on Theoretical Physics*, vol. 1 (1942; repr., New York: Academic Press, 1964).

PART 2. HOW THE *PRINCIPIA* BEGINS

Chapter 4. Newton's *Preface*

1. Richard Feynman, ch. 1 in *The Feynman Lectures on Physics*, Richard Feynman, R. B. Leighton, and M. Sands, vol. 1 (Reading, MA: Addison Wesley, 1972), pp. 1–2.

2. Various sources, including certain translations of the *Principia*, present Halley's first name as *Edmund* rather than *Edmond*. Unless in quoted material, his name will be spelled *Edmond*.

Chapter 5. Fundamentals

1. J. Thewlis, *Concise Dictionary of Physics and Related Subjects* (Oxford: Pergamon Press, 1979), s.v. "mass."

2. F. Wilczek, *The Lightness of Being: Mass, Ether, and the Unification of Forces* (New York: Basic Books, 2008), s.v. "mass."

3. Charles Lamb, letter to Thomas Manning, 1810. Quoted in *Oxford Dictionary of Scientific Quotations*, ed. W. F. Bynum and Roy Porter (Oxford: Oxford University Press, 2005).

4. Saint Augustine, *Confessions*, based on a translation by J. G. Pilkington, ed. Justin Lovill (London: Folio Society, 1993), book XI, ch. 14. See also W. J. Oats, ed., *Basic Writings of Saint Augustine* (New York: Random House, 1948).

5. Murray Gell-Mann, *The Quark and the Jaguar* (New York: W. H. Freeman, 1994).

6. Sir William Thomson and P. G. Tait, *Principles of Mechanics and Dynamics* (1888; repr., New York: Dover Press, 1962).

7. Herbert Goldstein, C. Poole, and J. Safko, *Classical Mechanics*, 3rd ed. (1950; repr., New York: Addison-Wesley, 2002).

Chapter 6. Newton's Laws of Motion and Their Immediate Consequences

1. R. Descartes, *Principles of Philosophy* (n.p.: 1644). Reprinted with translation by V. R. Miller and R. P. Miller (Dordrecht, Neth.: Reidel Publishing, 1983), sec. 203.

2. Rudolf Peierls, *The Laws of Nature* (London: George Allen and Unwin, 1955).

3. E. F. Taylor and J. A. Wheeler, *Spacetime Physics* (San Francisco: W. H. Freeman, 1963).

4. R. Feynman, R. B. Leighton, and M. Sands. *The Feynman Lectures on Physics*, vol. 1 (Reading, MA: Addison-Wesley, 1963), ch. 12.

5. Albert Einstein, "The Mechanics of Newton and Their Influence on the Development of Theoretical Physics," *Ideas and Opinions* (New York: Crown Publishers, 1959).

Chapter 7. Mathematical Methods

1. Albert Einstein, "The Mechanics of Newton and Their Influence on the Development of Theoretical Physics," *Ideas and Opinions* (New York: Crown Publishers, 1959).

2. I. Newton, quoted in "'Gigantic Implements of War': Images of Newton as a Mathematician," by N. Guicciardini, in *The Oxford Handbook of the History of Mathematics*, ed. Eleanor Robson and Jacqueline Stedall (Oxford: Oxford University Press, 2009).

3. Newton to Richard Bentley, quoted in John Roche, *Let Newton Be!* (Oxford: Oxford University Press, 1988).

4. H. Pemberton, quoted in *Reading the* Principia: *The Debate on Newton's Mathematical Methods for Natural Philosophy from 1687 to 1736*, N. Guicciardini (Cambridge: Cambridge University Press, 1999) p. 30.

5. Ibid., p. 2.

6. C. Truesdell, *Essays in the History of Mechanics* (Berlin: Springer-Verlag, 1968).

7. Guicciardini, *Reading the* Principia, p. 4.

8. R. Descartes, letter to M. Mersenne, in *The Geometry of René Descartes* (New York: Dover, 1980).

9. William Whewell, *History of the Inductive Sciences*, vol. 2 (Cambridge: Parker and Deighton, 1837).

PART 3. DEVELOPING THE BASICS OF DYNAMICS

Chapter 10. The Inverse-Square Law

1. Johann Bernoulli, quoted in *Reading the* Principia, by N. Guicciardini (Cambridge: Cambridge University Press, 1999), sec. 8.6.2.7.

2. Ibid.

3. V. I. Arnol'd, *Huygens and Barrow, Newton and Hooke* (Basel: Birkhauser Verlag, 1990), p. 33.

4. Newton, quoted in Guicciardini, *Reading the* Principia, sec. 8.6.2.7.

Chapter 12. Time and a Mathematical Gem

1. V. I. Arnol'd, *Huygens and Barrow, Newton and Hooke* (Basel: Birkhauser Verlag, 1990), p. 83.

2. Ibid.

Chapter 13. Completing the Single-Body Formalism

1. Christiaan Huygens, *The Pendulum Clock; or, Demonstrations concerning the Motion of Pendula as Applied to Clocks*, translated from the original 1673 *Horologium Oscillatorium* by Richard J. Blackwell (Ames: Iowa State University Press, 1986), pp. 45, 46.

2. Eugene Wigner, *Symmetries and Reflections: Scientific Essays of Eugene P. Wigner* (Bloomington: Indiana University Press, 1967).

3. Ibid.

4. William Whewell, *History of the Inductive Sciences*, vol. 2 (Cambridge: Parker and Deighton, 1837).

5. Isaac Newton, quoted in N. Guicciardini, *Isaac Newton on Mathematical Certainty and Method* (Cambridge MA: MIT Press, 2011), p. 255.

PART 4. ON TO MORE COMPLEX SITUATIONS

Chapter 16. Many Bodies

1. E. T. Whittaker, *A Treatise on the Analytical Dynamics of Particles and Rigid Bodies*, 4th ed. (Cambridge: Cambridge University Press, 1904; New York: Dover, 1944), p. 339.

Chapter 17. Big Bodies and Superb Theorems

1. Albert Einstein, "Physics and Reality," *Essays in Physics* (New York: Philosophical Library, 1950).

2. See W. W. Rouse Ball, *An Essay on Newton's* Principia (1893; repr., New York: Johnson Reprint Corporation, 1972), pp. 157–58.

3. Ibid., p. 159.

4. J. E. Littlewood, "Newton and the Attraction of a Sphere," *Mathematical Gazette* 32 (1948). Reprinted in *Littlewood's Miscellany*, ed. B. Bollobas (Cambridge: Cambridge University Press, 1982), p. 180.

5. V. D. Barger and M. G. Olsson, *Classical Mechanics: A Modern Perspective*, 2nd. ed. (New York: McGraw-Hill, 1995), p. 287.

6. Rouse Ball, *Essay on Newton's* Principia, p. 61.

PART 5. ABOUT *PRINCIPIA BOOK II*

Chapter 19. Starting on *Book II*

1. Clifford Truesdell in "History of Classical Mechanics Part I." *Die Naturwissenschaften* 63 (1976): 53–62.

2. Galileo Galilei, Day IV in *Two New Sciences*. Madison: University of Wisconsin Press, 1974.

3. R. Descartes, *Principles of Philosophy* (n.p.: 1644). Reprinted with translation by V. R. Miller and R. P. Miller (Dordrecht, Neth.: Reidel Publishing, 1983), part II, sec. 38.

Chapter 20. Newton the Experimentalist

1. William Stukeley, *Memoirs of Sir Isaac Newton's Life*, 1752. Also available online at William Stukeley, "Revised Memoir of Newton," Newton Project, http://www.newton project.sussex.ac.uk/view/texts/normalized/OTHE00001 (accessed May 21, 2013).

2. Einstein, quoted in *Physics and Beyond: Encounters and Conversations*, Werner Heisenberg (New York: Harper and Row, 1971), p. 77; see *Oxford Dictionary of Scientific Quotations*, ed. W. F. Bynum and Roy Porter (Oxford: Oxford University Press, 2005), s.v. "Albert Einstein."

Chapter 21. What Lies Beneath

1. E. T. Whittaker, *A Treatise on the Analytical Dynamics of Particles and Rigid Bodies*, 4th ed. (Cambridge: Cambridge University Press, 1904; New York: Dover, 1944), p. 47.

2. Herman H. Goldstine, *A History of the Calculus of Variations* (New York: Springer-Verlag, 1980), sec. 1.2.

3. S. Chandrasekhar, *Newton's* Principia *for the Common Reader* (Oxford: Clarendon Press, 1995), pp. 567–68, sec. 145.

4. Clifford Truesdell, "History of Classical Mechanics: Part 1, to 1800," *Die Naturwissenschaften* 63 (1976): 53–62.

PART 6. THE MAJESTIC *PRINCIPIA* BOOK III

Chapter 22. *Book III* and Gravity

1. Widely attributed to Richard Feynman; see http://en.wikipedia.org/wiki/Philosophy_of_Science (accessed May 21, 2013).

2. Albert Einstein, "On the Method of Theoretical Physics," Herbert Spencer lecture, June 1933, available in *Ideas and Opinions*, Albert Einstein (New York: Crown Publishers, 1959).

3. William Stukeley, *Memoirs of Sir Isaac Newton's Life*, 1752. Also available online at William Stukeley, "Revised Memoir of Newton," Newton Project, http://www.newton project.sussex.ac.uk/view/texts/normalized/OTHE00001 (accessed May 21, 2013).

4. Albert Einstein, "On the Theory of Relativity," lecture, 1921, available in Einstein, *Ideas and Opinions*.

5. Albert Einstein, "The Equality of Inertial and Gravitational Mass as an Argument for the General Postulate of Relativity," in *Relativity: The Special and the General Theory*, 15th ed. (New York: Crown Publishers, 1952), p. 69.

6. William Whewell, *History of the Inductive Sciences*, vol. 2 (Cambridge: Parker and Deighton, 1837).

7. Steven Weinberg, "Newtonianism and Today's Physics," in *Three Hundred Years of Gravitation*, ed. S. W. Hawking and W. Israel (Cambridge: Cambridge University Press, 1987).

8. R. Feynman, R. B. Leighton, and M. Sands. *The Feynman Lectures on Physics*, vol. 1 (Reading, MA: Addison-Wesley, 1963), ch. 7.

9. Einstein, "On the Method of Theoretical Physics."

10. Albert Einstein, quoted in *Subtle Is the Lord*, Abraham Pais (Oxford: Oxford University Press, 1982).

Chapter 23. Theory of the Solar System

1. Newton, *De Motu*, in "The Newtonian Achievement in Astronomy," *The General History of Astronomy: Volume 2, Planetary Astronomy from the Renaissance to the Rise of Astrophysics—Part A: Tycho Brahe to Newton*, ed. R. Taton and C. Wilson (Cambridge: Cambridge University Press, 1989).

2. Pierre-Simon Laplace, *Exposition du Système du Monde* (Paris: 1796).

Chapter 24. Earthly Phenomena

1. Galileo Galilei, "Day 4," in *Dialogue on the Two Chief World Systems* (n.p.: 1632). See M. A. Finocchiaro, ed., *The Essential Galileo* (Indianapolis, IN: Hackett, 2008), p. 268.

2. Sir George Airy, quoted in *From Eudoxus to Einstein: A History of Mathematical Astronomy*, C. M. Linton (Cambridge: Cambridge University Press, 2004), p. 281.

3. I. Bernard Cohen, "A Guide to Newton's *Principia*," introduction to Isaac Newton, *The* Principia: *Mathematical Principles of Natural Philosophy*, trans. I. Bernard Cohen and Anne Whitman (Berkeley: University of California Press, 1999).

Chapter 25. Challenges

1. I. Newton, quoted in D. T. Whiteside, "Newton's Lunar Theory: From High Hope to Disenchantment," *Vistas in Astronomy* 19 (1976): 317–28.

2. Quoted by C. M. Linton, *From Eudoxus to Einstein: A History of Mathematical Astronomy* (Cambridge: Cambridge University Press, 2004), p. 295.

3. Pierre-Simon Laplace, *A Philosophical Essay on Probabilities* (1825; repr., New York: Dover, 1951).

4. Leonard Euler, quoted in Linton, *From Eudoxus to Einstein*, p. 215.

5. Daniel Defoe, *A Visitation of the Plague* (London: Penguin Books, 1986), pp. 24–25. Also in Daniel Defoe, *A Journal of the Plague Year* (London, 1722).

6. Ibid.

7. Newton, quoted in H. W. Turnbull, *The Mathematical Discoveries of Newton* (London: Blackie, 1945).

8. Joseph-Jérôme Lefrançais de Lalande, lecture to the Paris Academy of Sciences, April 25, 1759.

Chapter 26. The Concluding *General Scholium*

1. Pierre-Simon Laplace, "The Evolution of the Solar System," in *Physical Thought: An Anthology*, edited by S. Sambursky (New York: Pica Press, 1975).

2. R. Feynman, R. B. Leighton, and M. Sands. *The Feynman Lectures on Physics*, vol. 1 (Reading, MA: Addison-Wesley, 1963), ch. 1.

3. Newton, letter to Richard Bentley, February 11, 1692, available in *Isaac Newton: Philosophical Writings*, ed. Andrew Janiak (Cambridge: Cambridge University Press, 2004).

4. C. Huygens, letter to Gottfried Leibnitz, November 18, 1690, quoted in I. Bernard Cohen, *The Newtonian Revolution* (Cambridge: Cambridge University Press, 1980).

PART 7. RECEPTION AND INFLUENCE OF *PRINCIPIA*

Chapter 27. Reception and Influence

1. David Gregory, letter to Newton, referenced in Westfall, *Never at Rest: A Biography of Isaac Newton* (Cambridge: Cambridge University Press, 1980), p. 470.

2. Albert Einstein, "The Mechanics of Newton and Their Influence on the Development of Theoretical Physics," *Ideas and Opinions* (New York: Crown Publishers, 1959).

3. Albert Einstein, *Out of My Later Years* (New York: Philosophical Library, 1950).

4. Sir W. Thomson and P. G. Tait, *Principles of Mechanics and Dynamics* (New York: Dover, 1962), sec. 206.

5. Lord Kelvin, *Baltimore Lectures on Molecular Dynamics and the Wave Theory of Light* (Cambridge, MA: C. J. Clay and Sons, 1904).

6. Heinrich Hertz, *Electric Waves* (London: Macmillan, 1893).

7. David Hume, "An Enquiry concerning Human Understanding" (n.p.: 1751), in *Great Books of the Western World* (Chicago: Encyclopedia Britannica, 1952).

8. Robert Boyle, quoted in *Boyle: Between God and Science*, by Michael Hunter (New Haven, CT: Yale University Press, 2009), p. 241.

9. W. F. Bynum and R. Porter, eds., *Oxford Dictionary of Scientific Quotations* (Oxford: Oxford University Press, 2005), p. 366.

10. G. A. J. Rogers, "Locke, Newton and the Enlightenment." *Vistas in Astronomy* 22 (1979): 471–76.

11. Albert Einstein, "The Fundamentals of Theoretical Physics," in *Essays in Physics* (New York: Philosophical Library, 1950).

12. Jean-Baptiste le Rond d'Alembert, *Preliminary Discourse to the Encyclopedia of Diderot*, trans. R. N. Schwab (Chicago: University of Chicago Press, 1995), p. 21.

13. Ibid.

14. P. Blom, *Enlightening the World: Encyclopédie, the Book That Changed the Course of History* (New York: Palgrave Macmillan, 2004), p. xiii.

Chapter 28. Mechanics after Newton

1. J. A. Wheeler, *Geons, Black Holes and Quantum Foam* (New York: Norton, 2000).

2. Albert Einstein, "Ether and the Theory of Relativity," *Sidelights on Relativity* (New York: Dover, 1923).

Epilogue

1. J. B. Shank, *The Newtonian Wars* (Chicago: University of Chicago Press, 2008), p. 4.

2. W. F. Bynum and R. Porter, eds., *Oxford Dictionary of Scientific Quotations* (Oxford: Oxford University Press, 2005), p. 127.

3. Albert Einstein, "The Mechanics of Newton and Their Influence on the Development of Theoretical Physics," *Ideas and Opinions* (New York: Crown publishers, 1959).

4. John Locke, "An Essay concerning Human Understanding" (n.p.: 1690), in *Great Books of the Western World* (Chicago: Encyclopedia Britannica, 1952).

INDEX

9/13